Porewater Toxicity Testing:
Biological, Chemical, and Ecological Considerations

Porewater Toxicity Testing:
Biological, Chemical, and Ecological Considerations

Edited by

R Scott Carr
U.S. Geological Survey
Marine Ecotoxicology Research Station
Texas A & M University-Corpus Christi
Corpus Christi, Texas, USA

Marion Nipper
Center for Coastal Studies
Texas A & M University-Corpus Christi
Corpus Christi, Texas, USA

Proceedings from the Workshop on Sediment Porewater Toxicity Testing:
Biological, Chemical, and Ecological Considerations
18–22 March 2000
Pensacola, Florida, USA

Coordinating Editor of SETAC Books
Andrew Green
International Lead Zinc Research Organization
Department of Environment and Health
Durham, North Carolina, USA

Publication sponsored by the Society of Environmental
Toxicology and Chemistry (SETAC)

Cover by Michael Kenney Graphic Design and Advertising
Copyediting and typesetting by Wordsmiths Unlimited
Indexing by IRIS

Library of Congress Cataloging-in-Publication Data

Porewater toxicity testing : biological, chemical, and ecological considerations / edited by R. Scott Carr, Marion G. Nipper
    p. cm.
Includes bibliographical references and index.
    ISBN 1-880611-65-1 (alk. paper)
       1. Water quality bioassay. 2. Pore water--Toxicology--Testing.   I. Carr, R. Scott (Robert Scott), 1951- II.
Nipper, Marion G., 1955-

QH91.57.B5P67 2003
628.1'61--dc21

                                  2003041641

© 2003 Society of Environmental Toxicology and Chemistry (SETAC)
This publication is printed on recycled paper using soy ink.
SETAC Press is an imprint of the Society of Environmental Toxicology and Chemistry.
No claim is made to original U.S. Government works.

International Standard Book Number 1-880611-65-1
Printed in the United States of America
09 08 07 06 05 04 03   10 9 8 7 6 5 4 3 2 1

∞ The paper used in this publication meets the minimum requirements of the American National Standard for
   Information Sciences—Permanence of Paper for Printed Library Materials, ANSI Z39.48-1984.

Reference Listing: Carr RS, Nipper M. 2003. Porewater toxicity testing: Biological, chemical, and ecological considerations. Pensacola FL, USA: Society of Environmental Toxicology and Chemistry (SETAC). 346 p.

SETAC Publications

The publication of books by the Society of Environmental Toxicology and Chemistry (SETAC) provides in-depth reviews and critical appraisals on scientific subjects relevant to understanding the impacts of chemicals and technology on the environment. The books explore topics reviewed and recommended by the Publications Advisory Council and approved by the SETAC North America Board of Directors and the SETAC World Council for their importance, timeliness, and contribution to multidisciplinary approaches to solving environmental problems. The diversity and breadth of subjects covered in the series reflect the wide range of disciplines encompassed by environmental toxicology, environmental chemistry, hazard and risk assessment, and life-cycle assessment. SETAC books attempt to present the reader with authoritative coverage of the literature, as well as paradigms, methodologies, and controversies; research needs; and new developments specific to the featured topics. The books are generally peer reviewed for SETAC by acknowledged experts.

SETAC Publications, which include Technical Issue Papers (TIPs), workshop summaries, newsletter (*SETAC Globe*), and journal (*Environmental Toxicology and Chemistry*), are useful to environmental scientists in research, research management, chemical manufacturing and regulation, risk assessment, life-cycle assessment, and education, as well as to students considering or preparing for careers in these areas. The publications provide information for keeping abreast of recent developments in familiar subject areas and for rapid introduction to principles and approaches in new subject areas.

SETAC would like to recognize the past SETAC Special Publication Series editors:

C.G. Ingersoll, Midwest Science Center
U.S. Geological Survey, Columbia, Missouri, USA

T.W. La Point, Institute of Applied Sciences
University of North Texas, Denton, Texas, USA

B.T. Walton, U.S. Environmental Protection Agency
Research Triangle Park, North Carolina, USA

C.H. Ward, Department of Environmental Sciences and Engineering
Rice University, Houston, Texas, USA

Contents

Chapter 4 Sediment and Porewater Chemistry 63
Bruce Williamson, Robert M Burgess

Chapter 5 Porewater Chemistry: Effects of Sampling, Storage, Handling, and Toxicity Testing 95
William J Adams (Workgroup Leader), Robert M Burgess, Gerardo Gold-Bouchot, Lawrence LeBlanc, Karsten Liber, Bruce Williamson

Chapter 8 Porewater Toxicity Tests: Value as a Component of Sediment Quality Triad Assessments 163
Edward R Long, R Scott Carr, Paul A Montagna

Chapter 9 Uses of Porewater Toxicity Tests in Sediment Quality Triad Studies 201
R Scott Carr (Workgroup Leader), Edward R Long, Julie A Mondon, Paul A Montagna, Pasquale F Roscigno

Chapter 10 Experiences with Porewater Toxicity Tests from a Regulatory Perspective 229
Linda Porebski

Chapter 11 Comparison of Sediment Quality Guideline Values Derived using Sea Urchin Porewater Toxicity Test Data with Existing Guidelines 249
R Scott Carr, James M Biedenbach, Donald D MacDonald

List of Figures

List of Tables

Acknowledgments

We would first like to thank all the workshop participants without whose expertise and scientific knowledge this workshop would not have been possible. In addition, we would like to especially thank all of the chapter authors and workgroup leaders who worked diligently under tight time constraints to accomplish their assignments. The SETAC North America staff did an excellent job of providing equipment, facilities, and logistical support. Finally, we would like to thank Drs. Jim Shine, Jeff Hyland, Ted DeWitt, and Gary Ankley for their thorough review of the book and valuable comments. The Workshop on Porewater Toxicity Testing was made possible through the financial support of the following organizations:

> Electric Power Research Institute
>
> E.I. DuPont deNemours
>
> MacDonald Environmental Sciences Ltd.
>
> Kennecott Utah Copper
>
> Environment Canada
>
> Exxon Company
>
> Minerals Management Service
>
> U.S. Environmental Protection Agency
>
> U.S. Geological Survey

——*R Scott Carr and Marion Nipper*

About the Editors

R Scott Carr is a marine ecotoxicologist with the U.S. Geological Survey's (USGS) Columbia Environmental Research Center and is Station Leader at the Marine Ecotoxicology Research Station in Corpus Christi, Texas, USA. His research interests have focused on the sublethal responses of marine animals to pollutants, particularly those involving growth and reproductive processes. His recent research has focused on the development and application of the porewater toxicity test approach for assessing the quality of marine and estuarine sediments. Dr. Carr has authored or co-authored more than 125 publications and technical reports.

Marion Nipper is a senior research scientist with Texas A & M University, Corpus Christi. Her main research interest has always been marine ecotoxicology, although she has had some past experience with fresh water as well. At her previous jobs in Brazil and New Zealand, she had a chance to work with a variety of toxicity test systems, organisms. and endpoints, concentrating on comparisons of whole sediment and porewater toxicity assessments. In recent years, she has focused on the toxicity of energetic materials in marine waters, sediments, and pore waters. Dr. Nipper has authored or co-authored more than 70 publications and technical reports and 2 books.

Workshop Participants*

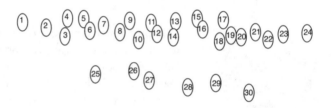

Standing: 1) Scott Carr, 2) Marion Nipper, 3) Kristen Milligan,
4) Brian Anderson, 5) Don Morrisey, 6) Julie Mondon,
7) Paul Montagna, 8) Francesca Bona, 9) Joe Germano,
10) Linda Porebski, 11) Mick Hamer, 12) Rick Scroggins, 13) Ken Doe,
14) Allen Burton, 15) Parley Winger, 16) Duane Chapman,
17) Bob Hoke, 18) Steve Bay, 19) Barbara Albrecht, 20) Karsten Liber,
21) Pat Roscigno, 22) Bill Adams, 23) Bruce Williamson, 24) Ed Long

Seated: 25) Linda Longsworth, 26) Greg Schiefer, 27) Rob Burgess,
28) Kay Ho, 29) Larry LeBlanc, 30) Walter Berry

Not pictured: Gerardo Gold-Bouchot, Don MacDonald,
and Gladys Stephenson

William J Adams
  Kennecott Utah Copper Corporation
  Magna UT, USA

Barbara Albrecht
  EnSafe, Inc.
  Pensacola FL, USA

Brian S Anderson
  University of California-Davis
  Environmental Toxicology
  Monterey CA, USA

Steven M Bay
  Southern California Coastal Water
    Research Project
  Toxicology Department
  Westminster CA, USA

Walter J Berry
  U.S. Environmental Protection Agency
  Narragansett RI, USA

Francesca Bona
  Università di Torino
  Dip. Biologia Animale
  Torino, Italy

Robert M Burgess
  U.S. Environmental Protection Agency
  NHEERL / Atlantic Ecology Division
  Narragansett RI, USA

G Allen Burton Jr
  Wright State University
  Institute for Environmental Quality
  Dayton OH, USA

R Scott Carr
  U.S. Geological Survey
  Marine Ecotoxicology Research Station
  Texas A & M University-Corpus Christi
  Corpus Christi TX, USA

Duane C Chapman
  U.S. Geological Survey
  Biological Resources Division
  Columbia Environmental Research
    Center
  Columbia MO, USA

Kenneth G Doe
  Environment Canada
  Moncton NB, Canada

Joe Germano
  Germano & Associates, Inc.
  Bellevue WA, USA

Gerardo Gold-Bouchot
  CINVESTAV'IPN Unidad Merida
  Marine Resources Department
  Merida, Yucatan, Mexico

Mick Hamer
  Zeneca AgroChemicals
  Ecological Risk Assessment Section
  Bracknell, Berks, UK

Kay T Ho
  U.S. Environmental Protection Agency
  Atlantic Ecology Division
  Narragansett RI, USA

Robert A Hoke
  E.I. DuPont deNemours Company
  Haskell Laboratory
  Newark DE, USA

Lawrence LeBlanc
  U.S. Geological Survey, WRD
  Sacramento CA, USA

Karsten Liber
  University of Saskatchewan
  Toxicology Centre
  Saskatoon SK, Canada

Edward R Long
  ERL Environmental
  Salem OR, USA

Donald D MacDonald
  MacDonald Environmental Sciences Ltd.
  Nanaimo BC, Canada

Kristen Milligan
  Clean Ocean Action
  Highlands NJ, USA

Julie A Mondon
  University of Tasmania
  Department of Aquaculture
  Newnham, Launceston, Australia

Paul A Montagna
  University of Texas-Austin
  Marine Science Institute
  Port Aransas TX, USA

Donald J Morrisey
  National Institute for Water and
    Atmospheric Research
  Nelson, New Zealand

Marion Nipper
  Center for Coastal Studies
  Texas A & M University-Corpus Christi
  Corpus Christi TX, USA

Linda Porebski
  Environment Canada
  Marine Environment Division
  Ottawa ON, Canada

Pasquale F Roscigno
  DDI - Minerals Management Service
  Gulf of Mexico OCS Region
  New Orleans LA, USA

Richard Scroggins
  Environment Canada
  Environmental Technology Centre
  Glouster ON, Canada

Gladys L Stephenson
  ESG International
  Department of Environmental Biology
  University of Guelph
  Guelph, ON, Canada

Bruce Williamson
  Diffuse Sources Ltd.
  Auckland, New Zealand

Parley V Winger
  U.S. Geological Survey
  Patuxent Wildlife Research Center
  University of Georgia
  Warnell School of Forest Resources
  Athens GA, USA

---

*Affiliations were current at the time of the workshop.

Workshop on Porewater Toxicity Testing: Biological, Chemical, and Ecological Considerations

R Scott Carr, Marion Nipper

Scope and Objectives

This book is the result of, and is based on, discussions that took place at the Workshop on Porewater Toxicity Testing: Biological, Chemical, and Ecological Considerations, with a review of methods and applications and recommendations for future areas of research. The workshop, sponsored by the Society of Environmental Toxicology and Chemistry (SETAC), was held 18–22 March 2000 in Pensacola, Florida, USA. Thirty participants came from the U.S., Canada, Mexico, the United Kingdom, Italy, Australia, and New Zealand to examine the current uses of porewater toxicity tests in sediment quality assessments and testing. The workshop followed the format of previous SETAC Technical Workshops in that a wide array of specialists in different aspects of porewater testing research and applications were invited to participate. These participants, selected on the basis of their expertise and affiliation, included chemists, toxicologists, benthic ecologists, and regulators; participants represented universities, government research facilities, regulatory agencies, mining and chemical industries, and consulting agencies, as well as a nongovernmental organization (NGO).

The workshop began with a series of plenary talks that provided reviews of several porewater-related issues and identified key questions to be discussed by the participants during the workshop. Following the opening talks, the participants were assigned to 1 of 5 workgroups:

1) Comparison of Porewater and Solid-Phase Sediment Toxicity Tests
2) Porewater Chemistry: Effects of Sampling, Storage, Handling, and Toxicity Testing
3) Porewater Toxicity Testing: Methodological Uncertainties, Confounding Factors, and Toxicity Identification Evaluation (TIE) Procedures
4) Uses of Porewater Toxicity Tests in Sediment Quality Triad (SQT) Studies
5) Regulatory Applications of Porewater Toxicity Testing.

The chapters in this book include written versions of the plenary talks followed by a synopsis of the discussions from each workgroup for the respective topic. It is our hope that this volume provides a useful summary of the current state of knowledge regarding studies with pore water and will serve as a basis for studies on this topic into the foreseeable future.

Introduction

The porewater approach for assessing the quality of sediments is a recent development in the field of aquatic ecotoxicology. Pore water has been studied, by geochemists in particular, for considerably longer than it has been used in toxicity testing. Many of the recent ecotoxicological studies with pore water have been conducted without the benefit of collaborations among geological and environmental chemists and ecologists. The purpose of this summary is to provide a brief overview of the research that has been conducted with sediment pore water and to describe the objectives for the workshop that has resulted in this volume.

The first porewater extractors were mechanical squeezing devices made of stainless steel, whose use resulted in trace metal contamination problems. A variety of different methods has been employed, including centrifugation, pneumatic extraction, suction, and equilibration with a dialysis membrane. All of the methods have their pros and cons, which are discussed throughout the book.

The first studies to assess the applicability of using pore water to evaluate the quality of sediments were conducted in the late 1980s. Numerous studies over the past decade have demonstrated that a wide variety of test species and endpoints are amenable to testing with pore water. TIE procedures, which originally were developed for use with effluents disposed into freshwater bodies, have more recently been modified for use with pore water. Porewater TIEs have been shown to be powerful tools for identifying the primary toxic constituents in a complex matrix of sediment-associated contaminants.

Comparison of Porewater and Solid-Phase Sediment Toxicity Tests

Chapter 2 contains the plenary presentation for comparing porewater and solid-phase sediment toxicity tests, while Chapter 3 contains a summary of the workgroup discussion.

Sediments provide habitats that support populations of benthic and infaunal organisms, the status of which reflects the quality and health of the ecosystem. The assessment of sediment quality generally involves an evaluation of solid-phase sediments, but pore water is also an important matrix because it represents a major route of exposure to sediment-dwelling organisms.

Investigation of concordance between porewater and solid-phase toxicity tests provides a tool to evaluate the efficacy and ecological relevance of porewater tests. Complete agreement between solid-phase and porewater toxicity tests is neither required nor necessarily desirable. Situations in which porewater and solid-phase toxicity tests would be expected to give similar results include studies of sediments from highly contaminated sites, reference sites, situations in which pore water is the primary route of exposure, and studies in which the same species is tested in both porewater and solid-phase matrices.

One recommendation generated by this workgroup was the simultaneous use of both solid-phase sediment and porewater tests whenever possible, using species with different sensitivities. Incorporating these conditions in the assessment enhances the ability to discriminate sediment quality and identify exposure route and contributes to the weight-of-evidence approach. Contaminants in pore water and in whole sediment should be measured as often as possible, in order to provide information on exposure concentration and routes of exposure, aid in the interpretation of test results, and identify sources of toxicity. Potential confounding factors in pore water should also be identified and measured in order to help interpret test results, understand the contribution of these factors to concordance or discordance between methods, and contribute to the interpretation of TIE data.

Because a lack of concordance between toxicity tests often occurs, some areas of research were suggested. Naturally occurring porewater characteristics that influence toxicity test results, as well as acceptable ranges for these variables, should be established. The role of porewater toxicity tests should be defined for situations in which the aqueous phase is not the primary route of exposure, and comparative sensitivities to contaminants of porewater and solid-phase test species should be evaluated. Porewater toxicity test methods should be improved, validated, and standardized, and the analytical methods for chemical analysis of pore water should be improved.

Sediment and Porewater Chemistry

Chapter 4 contains the plenary presentation for this topic, while Chapter 5 presents a summary of the workgroup discussion.

It is nearly impossible to remove a porewater sample from sediment and expose organisms to it in a toxicity-testing vessel without altering the chemistry of its natural and anthropogenic organic and inorganic constituents. Chapter 4 provides guidance on the advantages and disadvantages of the various procedures used in sediment porewater sampling, processing, and storage from the viewpoint of how they affect porewater chemistry.

It is particularly difficult to devise generic guidance for the storage and handling of sediment samples for porewater extraction because porewater extraction and testing often introduce artifacts that are unavoidable. The most fundamental questions regarding chemical aspects of porewater testing revolve around the limitations of porewater toxicity tests that are imposed by the chemical changes that inevitably occur during testing procedures. These include oxidation of dissolved Fe^{2+}, Mn^{2+}, HS^-, and dissolved organic carbon (DOC); precipitation of iron and manganese oxides and sulfur; adsorption of metals and hydrophobic organic compounds (HOCs); and degassing and volatilization of CO_2, H_2S, and volatile hydrocarbons.

The advancement of the science of porewater chemistry in the special context of the toxicity testing of pore waters would require research in some specific areas. In terms of extraction and storage, experiments were recommended to assess the validity of extraction methods and provide information necessary for method standardization. Experiments were proposed using simple systems (i.e., spiked solutions and sediments) to evaluate volatilization and sorption of HOCs, metals, and NH_3 to the apparatus. Simple aqueous-phase experiments were also suggested to examine the extent of degassing and volatilization of CO_2, H_2S, and NH_3 with sampling, extraction, and storage procedures. The effects of sample holding time on the fundamental constituents in both pore water and whole sediments (e.g., dissolved oxygen [DO], redox, CO_2, DOC, H_2S, NH_3, Fe, and Mn) should also be assessed.

In terms of toxicity test systems, a quantitative understanding of the interactions of colloids with surfaces and with dissolved contaminants is needed, for example, regarding the concentrations and distributions of prominent colloid species in porewater tests, including DOC and iron and manganese oxides. Pore water is frequently anoxic, and oxidation and the resulting precipitation reactions are commonly observed. These changes are inevitable in many porewater toxicity tests, so it may be more effective to let these reactions occur in a prescribed manner during the test procedure. It is necessary to establish a standard procedure allowing for the oxidation and precipitation of iron and manganese oxides and sulfides, along with alterations to DOC, prior to the initiation of toxicity testing. There is also a need to investigate the impact of iron and manganese oxide precipitation on toxicity test results for both metals and organic substances.

Porewater Toxicity Testing

Chapter 6 documents the plenary presentation for this topic, while Chapter 7 contains a summary of the workgroup discussion.

The chemical changes associated with porewater extraction and sediment handling procedures described in the previous chapters represent major concerns for porewater toxicity testing and its use for TIE procedures. These changes can affect the equilibrium of contaminants contained in the sediment and the pore water, which affects the toxicity in the pore water and solid-phase sediment alike. Porewater storage time and conditions can also introduce a series of artifacts that can considerably affect its toxicity; this is clearly an area that needs more research.

Several of the identified research needs would improve the understanding of the effects of methodological variables, confounding factors, and TIE manipulations on the results of porewater toxicity tests. In terms of porewater sampling, improved (larger-volume) in situ porewater collection devices are needed. Research on acceptable sediment storage time and conditions prior to porewater extraction is needed, as is further evaluation of the effect of freezing and other storage methods on chemistry and toxicity of contaminants contained in pore water. The effect of porewater extraction and storage methods on microbial processes such as degradation of contaminants and ammonia production also needs to be better understood.

Once porewater samples are collected, they may need to be aerated before they are used in toxicity testing. Different ways of introducing oxygen into samples while minimizing changes to the pore water and contaminants contained in it need to be assessed, and the effect of the oxidation of metals and organics on porewater toxicity needs to be evaluated.

Regarding toxicity test systems, there are the needs to identify the effect of the sorption of contaminants to test and storage vessels and to establish the relation of sorption to container type and surface area to water volume. The tolerance of test organisms to a variety of confounding factors—for example, ammonia, sulfides, and pH—needs to be determined.

Uses of Porewater Toxicity Tests in Sediment Quality Triad Studies

Chapter 8 contains the plenary presentation for this topic, while Chapter 9 presents a summary of the workgroup discussion.

The SQT approach relies upon the analyses of data from chemical analyses, toxicity tests, and infaunal benthic community assessments, which can be used by sediment assessors to form a weight-of-evidence basis for comparing and ranking the relative quality of sediment samples and regions of a study area. Possible applications of porewater toxicity tests in triad studies include but are not restricted to the following:

1) Identifying spatial trends in sediment quality within a specified study area (i.e., an assessment of status and trends)

2) Determining the magnitude (e.g., degree and severity) of degraded sediment quality

3) Identifying and justifying designation of hotspots with unacceptable sediment quality

4) Determining sediment quality adjacent to a designated upland waste site (e.g., off-site migration of contaminants)

5) Determining the biological significance of known contamination

6) Determining temporal trends in sediment quality within a specified area

7) Developing sediment quality guidelines (SQGs) or field-validating their predictive ability.

The improvement of the use of porewater toxicity tests in SQT studies depends on research in several different fields. More synoptic porewater chemistry and toxicity data are needed, as is a better understanding of the relationships between concentrations of chemical mixtures in sediments and pore waters and porewater toxicity. The development of chemical-specific porewater toxicity tests would be desirable, and research on the toxicological mechanism at the cellular level for porewater test organisms is needed.

The use of generalized multiparameter patterns in ecological structure and function, as opposed to species differences in different locations, relative to porewater toxicity, needs to be determined. The need for an index to summarize benthic data that is easy to understand and not easily misused was also identified. And last, validation studies with both whole sediment and porewater toxicity tests should be performed, to demonstrate that these are indeed valid, accurate metrics.

Regulatory Applications of Porewater Toxicity Testing

Chapters 10 and 11 include the plenary presentations for this topic, while Chapter 12 contains a summary of the workgroup discussion.

In the regulatory arena, decisions often are based as much on legal and programmatic considerations as on scientific considerations. The objective of this discussion group was to evaluate the use of porewater toxicity tests in regulatory applications, including their potential use in the development of SQG values.

The principal reasons for using porewater toxicity testing in a regulatory framework are that these tests provide information not currently provided by solid-phase, elutriate, or sediment-extract tests and that there are more aqueous-phase tests and test species available for sublethal or chronic effects measurement in pore water than there are epifaunal or infaunal test species for equivalent solid-phase tests. Another advantage of porewater tests is that they are generally cheaper and more rapid to conduct than solid-phase tests, which is particularly important in a regulatory context because cost and ease of test performance are important considerations for both testing laboratories and regulated entities. Potential regulatory uses for porewater toxicity tests include screening, compliance testing, application of TIE procedures, and environmental monitoring and assessment programs. It was suggested that it is worthwhile to pursue the development and standardization of appropriate porewater toxicity tests for use in regulatory programs.

Advantages for the use of pore water as a test phase in a regulatory context were identified, including these:
- Direct contact with a sediment fraction (versus elutriate)
- Ability to use or adapt small water-based tests with developed methods

- Accessibility to a wider range of sublethal endpoints
- Fast, low-cost, simple commercial use and screening
- Easier use of TIE methods
- Ability to perform dilution series experiments
- Less manipulative than solvent extracts (no residual solvent concerns)
- Ability to assess dissolved phase of sediment (i.e., route of exposure information).

However, some constraints for the use of pore water as a test phase in a regulatory context include small volume; artifacts from sampling, extraction, and storage (oxidation changes, etc.); short shelf life; the fact that most porewater tests are short term; the higher frequency of reported sensitivity to contaminants such as ammonia; and fewer developed and standardized methods.

The incorporation of porewater toxicity test results into SQGs might generate guidelines that are more predictive of porewater effects and better explain the results of porewater tests. They also might be more protective if, for example, the porewater toxicity tests were to pick up a biological response to a chemical or class of chemicals that was missed by the solid-phase testing. SQGs derived from the equilibrium-partitioning (EqP) approach might benefit from further evaluation provided by additional synoptic porewater chemistry and toxicity data.

The Way Forward

Chapter 13 presents the status of knowledge and applications of porewater toxicity tests and summarizes the main recommendations for research related to biological, chemical, and ecological aspects of sediment pore water. Consensus was achieved on several points.

Concordance between the results of solid-phase and porewater toxicity tests should not always be expected, and discordance is indicative of different routes of exposure and/or species sensitivity, rather than inaccuracy in the results of one type of test. It is important to conduct both porewater and solid-phase tests whenever possible. This not only enhances the ability to discriminate sediment quality but also forms a weight of evidence and helps determine concordance among the triad components.

Sampling, extraction, and storage techniques are critically important for achieving the most field-representative samples of pore water, and method selection should be based on the objective of the study. Because it is nearly impossible to avoid artifacts and chemical changes when removing pore water from sediment and using it in a toxicity test, the determination of chemical concentrations in the pore water is recommended, in addition to the regular contaminant measurements conducted in the whole sediment. This provides information on routes and level of exposure, aiding in the interpretation of test results and identifying sources of toxicity. The measurement of several porewater features, a number of which can act as confounding factors (e.g., salinity, alkalinity, pH, conductivity, DO, NH_3, H_2S, Eh), should be recorded shortly after porewater collection and after storage. This also would help in interpreting test results, understanding the contribution of these factors to concordance or discordance between solid-phase and porewater test methods, and contribute to TIE procedures.

The use of a variety of test species was recommended, in order to enrich the database and help account for different modes of action and species sensitivity. The use of indigenous species, however, is not recommended or suggested as important for the understanding of potential biological impacts identified from the results of porewater toxicity tests. The use of water column organisms for porewater toxicity tests was considered scientifically appropriate.

Regulatory aspects of the use of porewater toxicity tests included the need for prior determination of the purpose for the testing in a specific regulatory program and the need to determine what questions are being asked, ensuring that the questions are appropriate for the specific regulatory application. Porewater testing was considered suitable for several types of frameworks but unsuitable for others, for example, as a stand-alone pass/fail method or as a substitute for a solid-phase test. This corroborates the findings that the 2 tests represent different routes of exposure and that the feeding mode of a test species can be of critical importance for the exposure to certain chemicals. The incorporation of porewater toxicity test results into empirically derived SQGs might generate guidelines that are more predictive of porewater effects, better explain the results of porewater tests, and be more protective if the porewater toxicity tests were to pick up a biological response to a chemical or class of chemicals that was missed by the solid-phase testing.

The different workgroups identified several information gaps and research needs to improve the accuracy and enhance the understanding of porewater toxicity test results. Some of the fundamental needs included the improvement of the understanding of the toxicological effects of artifacts and changes introduced when pore water is extracted; the effects of porewater storage methods and time; the effects of naturally occurring porewater characteristics that influence toxicity test results (confounding factors); the validation and standardization of porewater toxicity test methods as well as the need for more short-term chronic test methods; the need for more synoptic porewater chemistry and toxicity data for better understanding of the relationships between concentrations of chemical mixtures in sediments and pore water and porewater toxicity; and the identification of the most appropriate benthic indicator species and the best toxicity endpoints for predicting impacts on the benthos.

CHAPTER 1

Historical Overview of Porewater Toxicity Testing

R Scott Carr, Marion Nipper

The porewater approach for assessing the quality of sediments is a recent development in aquatic ecotoxicology (Figure 1-1). Pore water (also known as "interstitial water") has been studied, by geochemists particularly, for considerably longer than it has been used in toxicity testing. Many recent ecotoxicological studies with pore water have been conducted without the benefit of collaborations among geological and environmental chemists and ecologists. Because porewater toxicity testing methods are relatively new, they have not come into common use in the regulatory arena. Information in the scientific literature concerning porewater studies tends to be scattered and difficult to locate. For these and other reasons, a Society of Environmental Toxicology and Chemistry (SETAC) Technical Workshop was convened to assemble and assess the current state of the science regarding porewater studies. This introductory chapter provides a brief overview of the research that has been conducted with sediment pore water and describes the objectives for the workshop that resulted in this volume.

Porewater Extraction Methods

The first reported studies with pore water were performed by geochemists whose aim was to characterize chemical constituents in the interstitial water of marine sediments. The first porewater extractors were mechanical squeezing devices made of stainless steel (Reeburgh 1967), which resulted in trace metal contamination problems (Presley et al. 1967). The materials used for the construction of these mechanical squeezers then evolved from Teflon-coated stainless steel (Presley et al. 1967), to Lucite (Bender et al. 1987), to all Teflon or polyvinyl chloride (PVC) squeezers that use pneumatic rather than mechanical pressure (Carr and Chapman 1995). Recent studies indicate that if materials are carefully selected to eliminate leaching of contaminants from the extractor and to minimize adsorption of contaminants to the interior surfaces of the extractor, the toxicity of the sample is not noticeably affected. For example, no difference was found in the toxicity of pore water obtained with Teflon and PVC pneumatic extractors (Carr and Chapman

Porewater Toxicity Testing: Biological, Chemical, and Ecological Considerations. R. Scott Carr and Marion Nipper, editors.
© 2003 Society of Environmental Toxicology and Chemistry (SETAC). ISBN 1-880611-65-1

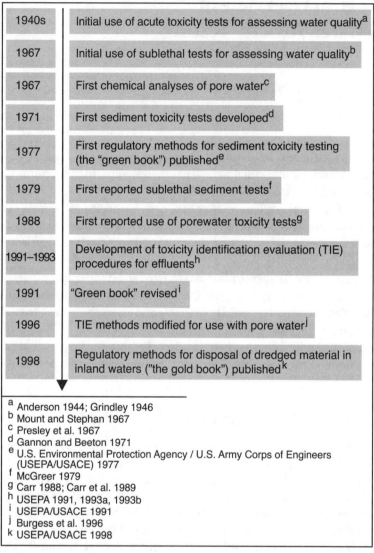

1940s	Initial use of acute toxicity tests for assessing water quality[a]
1967	Initial use of sublethal tests for assessing water quality[b]
1967	First chemical analyses of pore water[c]
1971	First sediment toxicity tests developed[d]
1977	First regulatory methods for sediment toxicity testing (the "green book") published[e]
1979	First reported sublethal sediment tests[f]
1988	First reported use of porewater toxicity tests[g]
1991–1993	Development of toxicity identification evaluation (TIE) procedures for effluents[h]
1991	"Green book" revised[i]
1996	TIE methods modified for use with pore water[j]
1998	Regulatory methods for disposal of dredged material in inland waters ("the gold book") published[k]

[a] Anderson 1944; Grindley 1946
[b] Mount and Stephan 1967
[c] Presley et al. 1967
[d] Gannon and Beeton 1971
[e] U.S. Environmental Protection Agency / U.S. Army Corps of Engineers (USEPA/USACE) 1977
[f] McGreer 1979
[g] Carr 1988; Carr et al. 1989
[h] USEPA 1991, 1993a, 1993b
[i] USEPA/USACE 1991
[j] Burgess et al. 1996
[k] USEPA/USACE 1998

Figure 1-1 Brief chronology of significant events in the development of aquatic toxicology relevant to porewater toxicity testing

1995). Significant differences have been observed with different types of filters that are used with all the squeeze extraction devices. Glass fiber filters adsorb the majority of hydrophobic contaminants, for example, DDT (Word et al. 1987), whereas 5 μm polyester filters appear to adsorb fewer contaminants than most of the commercially available filters that have been tested (Carr and Chapman 1995). Pneumatic extractors can be used with sediments of all textures and are among the most efficient methods for processing large numbers of samples.

The other most commonly employed porewater extraction technique is centrifugation (Edmunds and Bath 1976; Landrum et al. 1987; Word et al. 1987; Giesy et al. 1988, 1990; Schults et al. 1992; Burgess et al. 1993). In order to remove colloids from suspension, high speeds ($\geq$10,000 g) are required (Chin and Gschwend 1991), but colloid removal may not be necessary or desirable in toxicity testing. Centrifugation speeds of 2000 to 5000 g for 10 to 60 minutes are more commonly employed (Word et al. 1987; Giesy et al. 1988). Regardless of the speed used for the initial extraction, it may not be possible to remove all the fine particulate material because of electrostatic repulsion unless the supernatant is decanted and recentrifuged (Carr and Chapman 1995). Centrifugation at different speeds does not appear to affect the chemistry of pore water (Adams 1991). One drawback to this approach is that coarse-grained sediments are relatively incompressible, and therefore, centrifugation yields little if any pore water.

Another extraction technique that has been periodically employed both in the laboratory and in situ is the vacuum extraction method (Sayles et al. 1976; Knezovich and Harrison 1987; Winger and Lasier 1991). This technique generally involves the use of a syringe to supply the vacuum to a needle or filtering port, which extracts the pore water from different depths, depending on the design of the device. The device developed by Winger and Lasier (1991) is the most useful for toxicity testing purposes because it allows larger volumes to be collected than do the other vacuum methods. In order to obtain sufficient pore water for toxicity testing within a reasonable length of time, numerous suction devices are usually employed simultaneously. One of the biggest advantages of the Winger and Lasier apparatus (which is composed of a plastic syringe, aquarium tubing, and a crushed glass air stone) is its low cost.

Collection devices containing a dialysis membrane or filter, sometimes called "peepers," have been used to obtain pore water by passive diffusion (Hoepner 1981; Carignan 1984; Carignan et al. 1985; Di Toro et al. 1990). One of the primary disadvantages of this technique is that it may take a week or more to achieve equilibrium. Another drawback, with regard to toxicity testing, is that only small volumes can be obtained with the passive samplers used in the studies cited above, and a large ratio of sediment to pore water collected is required. More recent studies have shown that larger-volume samplers with shorter equilibration times can be used in freshwater stream environments (Sarda and Burton 1995).

Toxicity Testing with Pore Water

Toxicity tests with aquatic organisms have been used for several decades as a tool for the analysis of adverse effects of chemicals and complex mixtures discharged into water bodies around the world (Figure 1-1). It was recognized in the 1970s that sediments act as a sink for such contaminants, and sediment toxicity tests started to be developed (Gannon and Beeton 1971; Swartz et al. 1979). Historically, the standard approach for assessing the quality or potential toxicity of marine or estuarine sediments has been to expose macrobenthic organisms directly to the whole sediments for a specified time, after which the survival of the test species is determined. This has been the standard approach for assessing the suitability of dredged material for different disposal options in the regulatory arena in the U.S. since the mid 1970s (U.S. Environmental Protection Agency/U.S. Army Corps of Engineers [USEPA/USACE] 1977) and continues in an updated form (USEPA/USACE 1991) to be the most commonly employed sediment toxicity assessment method.

The equilibrium-partitioning theory predicts that pore water is the controlling exposure medium in the toxicity of sediments to infaunal organisms (Adams et al. 1985; Di Toro et al. 1991; USEPA 2000). Except for the sediment-associated contaminants that may be released in the gut of sediment-ingesting animals, only the contaminants that are solubilized in the interstitial water are bioavailable. Exposing organisms to pore water provides a direct measure of contaminant exposure that incorporates all the physical and chemical parameters (e.g., pH, salinity, grain size, total organic carbon [TOC], acid volatile sulfide [AVS]) that affect bioavailability.

The first studies to assess the applicability of using pore water to evaluate the quality of sediments were conducted in the late 1980s (Carr et al. 1989; Long et al. 1990). The first test used with pore water was the polychaete life-cycle test with *Dinophilus gyrociliatus* (Carr 1988; Carr et al. 1989). A recommendation from these initial studies was that additional test organisms be evaluated for use with this medium. Subsequent studies have demonstrated the applicability of the porewater approach for testing with sea urchin gametes and embryos (Nipper et al. 1990; Carr and Chapman 1992), benthic amphipods (Winger and Lasier 1991), fish embryos (Carr and Chapman 1992), copepod nauplii (Carr, Chapman, Presley et al. 1996), algal zoospores (Hooten and Carr 1998), and other microscale test species (Carr 1998). It is apparent that a wide variety of test species and endpoints are amenable to testing with pore water.

Porewater toxicity tests have been used in numerous comprehensive sediment quality assessment studies in coastal areas (Long, Sloane et al. 1995; Long, Wolf et al. 1995; Carr, Chapman, Howard, Biedenbach 1996; Carr, Long et al. 1996; Long et al. 1997; Long, Hameedi et al. 1999; Long, Sloane et al. 1999; Carr et al. 2000; Long et al. 2001) and on the continental shelf of North America (Carr, Chapman, Presley et al. 1996; Carr and Nipper 1998; U.S. Geological Survey [USGS] 1999, 2000; Nipper and Carr 2000, 2001). Many of the studies have afforded the opportunity for comparisons of the sensitivity and concordance with sediment chemistry and benthic data among different toxicity test methods (Long et al. 1990; Carr and Chapman 1992; Long, Sloane et al. 1995; Long, Wolfe et al. 1995; Sarda and Burton 1995; Carr, Chapman, Howard, Biedenbach 1996; Carr, Chapman, Presley et al. 1996; Long et al. 1997; Long, Hameedi et al. 1999; Long, Sloane et al. 1999; Carr et al. 2000; Long et al. 2001). The results of these comparisons will be discussed in Chapters 2 and 3 in this book.

Toxicity Identification Evaluation (TIE) procedures were originally developed for use with effluents disposed into freshwater bodies (USEPA 1991, 1993a, 1993b). More recently, these procedures have been modified for use with effluents disposed into marine environments (Burgess et al. 1996). The methods used for effluents can be applied to pore waters to aid in determining cause-and-effect relationships for toxic sediments. Porewater TIEs have been shown to be powerful tools for identifying the primary toxic constituents in a complex matrix of sediment-associated contaminants (Carr, Biedenbach, Hooten 2001; Carr, Nipper et al. 2001). Sediment porewater tests have also been used extensively in Sediment Quality Triad (SQT) studies (Kennicutt 1995; Carr, Chapman, Howard, Biedenbach 1996; Carr, Chapman, Presley et al. 1996; Carr, Long et al. 1996; Long, Sloane et al. 1999; Carr et al. 2000; USGS 2000; Long et al. 2001). Chapters 8 and 9 in this volume are devoted to a discussion of the use of porewater tests in SQT studies.

Workshop Objectives

The workshop followed the format of previous SETAC Technical Workshops. As such, a wide array of specialists in different aspects of porewater testing research and applications were invited to participate. In total, 30 participants from 7 countries (USA, Canada, Mexico, United Kingdom, Italy, Australia, and New Zealand) assembled in Pensacola, Florida, USA, to engage in 4 days of intensive discussions. These participants, selected on the basis of their expertise and affiliation, included chemists, toxicologists, benthic ecologists, and regulators, representing universities, government research, regulatory agencies, mining and chemical industries, consulting agencies, as well as a nongovernmental organization (NGO).

The workshop began with a series of plenary talks. These presentations provided reviews of several porewater-related issues and identified key questions to be discussed by the participants during the workshop. The plenary topics and presenters were these:

- Comparison among porewater and solid-phase–based tests (Steven M. Bay, Southern California Coastal Water Research Project [SCCWRP])
- Sediment chemistry, porewater extraction methods: Methodological uncertainties (Bruce Williamson, National Institute for Water and Atmosphere [NIWA], New Zealand)
- Marine and freshwater sediment porewater toxicity testing: Methodological uncertainties and confounding factors (G. Allen Burton Jr., Wright State University)
- Use of porewater toxicity tests in sediment TIEs (Kay Ho, USEPA)

- Use of porewater toxicity tests in SQT studies (Edward L. Long, National Oceanic and Atmospheric Administration [NOAA])
- Use of porewater toxicity data for the development of sediment quality guidelines (SQGs) (R. Scott Carr, USGS)
- Regulatory applications of porewater toxicity testing (Linda Porebski, Environment Canada).

Following the opening talks, the participants were assigned to 1 of 5 workgroups as follows:

- Comparison of Porewater and Solid-Phase Sediment Toxicity Tests
- Porewater Chemistry: Effects of Sampling, Storage, Handling, and Toxicity Testing
- Issues and Recommendations for Porewater Toxicity Testing: Methodological Uncertainties, Confounding Factors, and TIE Procedures
- Uses of Porewater Toxicity Tests in SQT Studies
- Regulatory Applications of Porewater Toxicity Testing.

The chapters in this book include written versions of the plenary talks for each topic area, followed by a synopsis of the discussions from each workgroup for that topic. It is our hope that this volume will provide a useful summary of the current state of knowledge regarding studies with pore water and will serve as a basis for studies on this topic into the foreseeable future.

References

Adams DD. 1991. Sediment pore water sampling. In: Mudroch A, MacKnight SD, editors. Handbook of techniques for aquatic sediments sampling. Boca Raton FL, USA: CRC. Chapter 7.

Adams WJ, Kimerle RA, Mosher RG. 1985. Aquatic safety assessment of chemicals sorbed to sediments. In: Cardwell RD, Purdy R, Bahner RC, editors. Aquatic toxicology and hazard assessment: Seventh symposium. Philadelphia PA, USA: American Society of Testing and Materials (ASTM). ASTM STP 854. p 429–453.

Anderson BC. 1944. Toxicity thresholds of various substances found in industrial wastes as determined by the use of *Daphnia magna*. *Sewage Works J* 16:1156–1165.

[ASTM] American Society for Testing and Materials. 2000. Standard guide for conducting 10-d static sediment toxicity tests with marine and estuarine amphipods, Designation E 1367-92. Volume 11.05, ASTM annual book of standards. Philadelphia PA, USA: ASTM. 1758 p.

Bender M, Martin W, Hess J, Sayles F, Ball L, Lambert C. 1987. A whole-core squeezer for interfacial pore-water sampling. *Limnol Oceanogr* 32:1214–1225.

Burgess RM, Ho KT, Morrison GE, Chapman G, Denton DL. 1996. Marine toxicity identification evaluation (TIE). Phase I guidance document. Washington DC, USA: U.S. Environmental Protection Agency. EPA/600/R-96/054. 70 p.

Burgess RM, Schweitzer KA, McKinney RA, Phelps DK. 1993. Contaminated marine sediments: Water column and interstitial toxic effects. *Environ Toxicol Chem* 12:127–138.

Carignan R. 1984. Interstitial water sampling by dialysis: Methodological notes. *Limnol Oceanogr* 29:667–670.

Carignan R, Rapin F, Tessier A. 1985. Sediment porewater sampling for metal analysis: A comparison of techniques. *Geochim Cosmochim Acta* 49:2493–2497.

Carr RS. 1988. Development and evaluation of a sediment bioassessment technique using the polychaete *Dinophilus gyrociliatus*. Duxbury MA, USA: Battelle Ocean Sciences. Final report submitted to National Oceanic and Atmospheric Administration, Seattle WA. 60 p.

Carr RS. 1998. Marine and estuarine porewater toxicity testing. In: Wells PG, Lee K, Blaise C, editors. Microscale testing in aquatic toxicology: Advances, techniques and practice. Boca Raton FL, USA: CRC Lewis. p 523–538.

Carr RS, Biedenbach JM, Hooten RL. 2001. Sediment quality assessment survey and toxicity identification evaluation studies in Lavaca Bay, Texas, a marine Superfund site. *Environ Toxicol* 16:20–30.

Carr RS, Chapman DC. 1992. Comparison of solid-phase and pore-water approaches for assessing the quality of marine and estuarine sediments. *Chem Ecol* 7:19–30.

Carr RS, Chapman DC. 1995. Comparison of methods for conducting marine and estuarine sediment porewater toxicity tests: Extraction, storage, and handling techniques. *Arch Environ Contam Toxicol* 28:69–77.

Carr RS, Chapman DC, Howard CL, Biedenbach JM. 1996. Sediment Quality Triad assessment survey in the Galveston Bay Texas system. *Ecotoxicology* 5:341–364.

Carr RS, Chapman DC, Presley BJ, Biedenbach JM, Robertson L, Boothe P, Kilada R, Wade T, Montagna P. 1996. Sediment porewater toxicity assessment studies in the vicinity of offshore oil and gas production platforms in the Gulf of Mexico. *Can J Fish Aquat Sci* 53:2618–2628.

Carr RS, Long ER, Chapman DC, Thursby G, Biedenbach JM, Windom H, Sloane G, Wolfe DA. 1996. Toxicity assessment studies of contaminated sediments in Tampa Bay, Florida. *Environ Toxicol Chem* 15:1218–1231.

Carr RS, Montagna PA, Biedenbach JM, Kalke R, Kennicutt MC, Hooten R, Cripe G. 2000. Impact of storm-water outfalls on sediment quality in Corpus Christi Bay, Texas. *Environ Toxicol Chem* 19:561–574.

Carr RS, Nipper M. 1998. Preliminary survey of sediment toxicity in the vicinity of Honolulu, Hawaii. Corpus Christi TX, USA: U.S. Geological Survey. Final report prepared for USGS, Geological Division, Menlo Park CA. 18 p, 5 tables, 8 figures, 4 attachments.

Carr RS, Nipper M, Biedenbach JM, Hooten RL, Miller K, Saepoff S. 2001. Sediment toxicity identification evaluation (TIE) studies at marine sites suspected of ordnance contamination. *Arch Environ Contam Toxicol* 41:298–307.

Carr RS, Williams JW, Fragata CTB. 1989. Development and evaluation of a novel marine sediment pore water toxicity test with the polychaete *Dinophilus gyrociliatus*. *Environ Toxicol Chem* 8:533–543.

Chin Y, Gschwend PM. 1991. The abundance, distribution, and configuration of porewater organic colloids in recent sediment. *Geochim Cosmochim Acta* 55:1309–1317.

Di Toro DM, Mahoney JD, Hansen DJ, Scott KJ, Hicks MB, Mayr SM, Redmond SM. 1990. Toxicity of cadmium in sediments: The role of acid volatile sulfide. *Environ Toxicol Chem* 9:1487–1502.

Di Toro DM, Zarba CS, Hansen DJ, Berry WJ, Swartz RC, Cowan CE, Pavlou SP, Allen HE, Thomas NA, Paquin PR. 1991. Pre-draft technical basis for establishing sediment quality criteria for non-ionic organic chemicals using equilibrium partitioning. Washington DC, USA: U.S. Environmental Protection Agency (USEPA), Office of Water.

Edmunds WM, Bath AH. 1976. Centrifuge extraction and chemical analysis of interstitial waters, *Environ Sci Technol* 10:467–472.

Gannon JE, Beeton AM. 1971. Procedures for determining the effects of dredged sediments on biota-benthos viability and sediment selectivity tests. *J Water Pollut Control Fed* 43:392–398.

Giesy JP, Graney RL, Newsted JL, Rosui CJ, Benda A, Kreis RG, Horvath FJ. 1988. Comparison of three sediment bioassay methods using Detroit River sediments. *Environ Toxicol Chem* 7:483–498.

Giesy JP, Rosiu CJ, Graney RL, Henry MG. 1990. Benthic invertebrate bioassays with toxic sediment and pore water. *Environ Toxicol Chem* 9:233–248.

Grindley J. 1946. Toxicity to rainbow trout and minnows of some substances known to be present in waste water discharged to rivers. *Ann Appl Biol* 33:103–12.

Hoepner J. 1981. Design and use of a diffusion sampler for interstitial water from fine grained sediments. *Environ Technol Lett* 2:187–196.

Hooten RL, Carr RS. 1998. Development and application of a marine sediment pore-water toxicity test using *Ulva fasciata* zoospores. *Environ Toxicol Chem* 17:932–940.

Kennicutt II MC. 1995. Gulf of Mexico offshore operations monitoring experiment. Phase I: Sublethal responses to contaminant exposure. New Orleans LA, USA: U.S. Department of the Interior, Minerals Management Service, Gulf of Mexico OCS Regional Office. OCS Study MMS 94-0045. 709 p.

Knezovich AP, Harrison FL. 1987. A new method for determining the concentrations of volatile organic compounds in sediment interstitial water. *Bull Environ Contam Toxicol* 38:937–940.

Landrum PF, Nihart SR, Eadie BJ, Herche LR. 1987. Reduction in bioavailability of organic contaminants to the amphipod *Pontoporeia hoyi* by dissolved organic matter of sediment interstitial waters. *Environ Toxicol Chem* 6:11–20.

Long ER, Buchman MR, Bay SM, Breteler RJ, Carr RS, Chapman PM, Hose JE, Lissner AL, Scott J, Wolfe DA. 1990. Comparative evaluation of five toxicity tests with sediments from San Francisco Bay and Tomales Bay, California. *Environ Toxicol Chem* 9:1193–1214.

Long ER, Hameedi J, Robertson A, Dutch M, Aasen S, Ricci C, Welch K, Carr RS, Johnson T, Biedenbach J, Scott KJ, Mueller C, Anderson JW. 2001. Sediment quality in Puget Sound: Year 2 - Northern Puget Sound. Silver Spring MD, USA: National Oceanic and Atmospheric Administration and Washington State Dept. of Ecology. NOAA NOS NCCOS CCMA Technical Memo No. 00-03-055. 347 p.

Long ER, Hameedi J, Robertson A, Dutch M, Aasen S, Ricci C, Welch K, Kammin W, Carr RS, Johnson T, Biedenbach J, Scott KJ, Mueller C, Anderson JW. 1999. Sediment quality in Puget Sound: Year 1 - Northern Puget Sound. Silver Spring MD, USA: National Oceanic and Atmospheric Administration and Washington State Dept. of Ecology. NOAA NOS NCCOS CCMA Technical Memo No. 139 and WSDOE Pub. No. 99-347. 221 p, 8 appendices.

Long ER, Sloane GM, Carr RS, Johnson T, Biedenbach J, Scott KJ, Thursby GB, Crecelius E, Peven C, Windom HL, Smith RD, Loganathon R. 1997. Magnitude extent of sediment toxicity in four bays of the Florida panhandle: Pensacola, Choctawhatchee, St. Andrew and Apalachicola. Silver Spring MD, USA: National Oceanic and Atmospheric Administration, Coastal Monitoring and Bioeffects Assessment Division. NOAA Technical Memorandum NOS ORCA 117. 219 p.

Long ER, Sloane GM, Carr RS, Scott KJ, Thursby GB, Wade T. 1995. Sediment toxicity in Boston Harbor: Magnitude, extent, and relationships with chemical toxicants. Silver Spring MD, USA: National Oceanic and Atmospheric Administration, Coastal Monitoring and Bioeffects Assessment Division. NOAA Technical Memorandum NOS ORCA 96. 85 p, 31 figures, 4 appendices.

Long ER, Sloane GM, Scott GI, Thompson B, Carr RS, Biedenbach J, Wade TL, Presley BJ, Scott KJ, Mueller C, Brecken-Fols G, Albrecht B, Anderson JW, Chandler GT. 1999. Magnitude and extent of chemical contamination and toxicity in sediments of Biscayne Bay and vicinity. Silver Spring MD, USA: National Oceanic and Atmospheric Administration. NOAA Technical Memorandum NOS NCCOS CCMA 141. 174 p.

Long ER, Wolfe DA, Carr RS, Scott KJ, Thursby GB, Windom HL, Lee R, Calder FD, Sloane GM, Seal T. 1995. Magnitude and extent of sediment toxicity in Tampa Bay, Florida. Silver Spring MD, USA: National Oceanic and Atmospheric Administration, Coastal Monitoring and Bioeffects Assessment Division. NOAA Technical Memorandum NOS ORCA 78. 84 p, 2 appendices.

McGreer ER. 1979. Sublethal effects of heavy metal contaminated sediments on the bivalve *Macoma balthica* (L.). *Mar Pollut Bull* 10:259–262.

Mount DI, Stephan CE. 1967. A method for establishing acceptable toxicant limits for fish – Malathion and 2,4-D. *Trans Am Fish Soc* 96:185–193.

Nipper MG, Badaró-Pedroso C, José VF, Prósperi VA. 1990. Marine bioassays and their applications in coastal management and biological monitoring. II Simpósio sobre Ecossistemas da Costa Sul e Sudeste Brasileira: Estrutura, Função e Manejo, Proc. Vol. 1. São Paulo SP: Academia de CiLncias do Estado de São Paulo. p 160–168.

Nipper M, Carr RS. 2000. Toxicity testing of sediment pore water from the Flower Garden Banks, Gulf of Mexico. Corpus Christi TX, USA: Texas A&M University - Corpus Christi. TAMU-CC-0002-CCS. 7 p, 3 tables, 3 figures, 3 attachments.

Nipper M, Carr RS. 2001. Porewater toxicity testing: A novel approach for assessing contaminant impacts in the vicinity of coral reefs. *Bull Mar Sci* 69:407–420.

Presley BJ, Brooks RR, Kappel HM. 1967. A simple squeezer for removal of interstitial water from ocean sediments. *J Mar Res* 25:355–357.

Reeburgh WS. 1967. An improved interstitial water sampler. *Limnol Oceanogr* 12:163–165.

Sarda N, Burton Jr GA. 1995. Ammonia variation in sediments: Spatial, temporal and method-related effects. *Environ Toxicol Chem* 14:1499–1506.

Sayles FL, Mangelsdorf Jr PC, Wilson TRS, Hume DN. 1976. A sampler for the in situ collection of marine sedimentary pore waters. *Deep-Sea Res* 23:259–264.

Schults DW, Ferraro SP, Smith LM, Roberts FA, Poindexter CK. 1992. A comparison of methods for collecting interstitial water for trace organic compounds and metals analyses. *Water Res* 26:989–995.

Swartz RC, DeBen WA, Cole FA. 1979. A bioassay for the toxicity of sediment to marine macrobenthos. *J Water Pollut Cont Fed* 51: 944–950.

[USEPA] U.S. Environmental Protection Agency. 1991. Toxicity Identification Evaluation: Characterization of chronically toxic effluents, Phase I. Duluth MN, USA: USEPA, Environmental Research Laboratory. EPA/600/6-91/003.

[USEPA] U.S. Environmental Protection Agency. 1993a. Methods for aquatic Toxicity Identification Evaluations: Phase II toxicity identification procedures for samples exhibiting acute and chronic toxicity. Duluth MN, USA: USEPA, Environmental Research Laboratory. EPA/600/R-92/080.

[USEPA] U.S. Environmental Protection Agency. 1993b. Methods for aquatic Toxicity Identification Evaluations: Phase III toxicity identification procedures for samples exhibiting acute and chronic toxicity. Duluth MN, USA: USEPA, Environmental Research Laboratory. EPA/600/R-92/081.

[USEPA] U.S. Environmental Protection Agency. 2000. Technical basis for the derivation of equilibrium partitioning sediment quality guidelines (ESGs) for the protection of benthic organisms. Washington DC, USA: USEPA, Office of Science and Technology. EPA-822-R-00-001.

[USEPA/USACE] U.S. Environmental Protection Agency/U.S. Army Corps of Engineers. 1977. Ecological evaluation of proposed discharge of dredged material into ocean waters. Vicksburg MS, USA: U.S. Army Engineer Waterways Experiment Station, Environmental Effects Laboratory.

[USEPA/USACE] U.S. Environmental Protection Agency/U.S. Army Corps of Engineers. 1991. Evaluation of dredged material proposed for ocean disposal. Washington DC, USA: USEPA. EPA-503/8-91/001.

[USEPA/USACE] U.S. Environmental Protection Agency/U.S. Army Corps of Engineers. 1998. Evaluation of dredged material proposed for discharge in waters of the U.S. Washington DC, USA: USEPA. EPA-823-B-98-004.

[USGS] U.S. Geological Survey. 1999. Toxicity testing of sediments from the New York Bight. Corpus Christi TX, USA: USGS, Marine Ecotoxicology Research Station. Final report prepared for the USGS Coastal and Marine Geological Team, Woods Hole MA, USA. 7 p, 7 tables, 1 figure, 4 attachments.

[USGS] U.S. Geological Survey. 2000. Toxicity testing of sediments from the BEST/EMAP estuary group monitoring study. Corpus Christi TX, USA: USGS, Marine Ecotoxicology Research Station. Final report prepared for the USGS BEST program, Madison WI. 10 p, 22 tables, 3 figures, 4 attachments.

Winger PV, Lasier PJ. 1991. A vacuum-operated pore-water extractor for estuarine and freshwater sediments. *Arch Environ Contam Toxicol* 2:321–324.

Word JQ, Ward JA, Franklin LM, Cullinan VI, Kiesser SL. 1987. Evaluation of the equilibrium partitioning theory for estimating the toxicity of the nonpolar organic compound DDT to the sediment dwelling amphipod *Rhepoxynius abronius*. Sequim WA, USA: Battelle Marine Research Laboratory. Report prepared for U.S. Environmental Protection Agency, Criteria and Standards Division, Washington DC. 60 p.

Relative Performance of Porewater and Solid-Phase Toxicity Tests: Characteristics, Causes, and Consequences

Steven M Bay, Brian S Anderson, R Scott Carr

Porewater toxicity tests have been used to assess sediment quality for more than a decade. When this approach was used initially, there were little data to compare the relative sensitivity or precision of porewater tests with the more common whole-sediment tests. During the past decade, in numerous large-scale sediment quality assessment surveys, a suite of toxicity tests, including porewater tests, have been conducted concurrently. These databases, in conjunction with other studies available from the literature, provide sufficient information to compare solid-phase and porewater tests. The purpose of this chapter is to assemble these comparative toxicity databases from both marine and freshwater studies in order to examine the relative performance of the different tests.

Objectives

The objectives of this chapter are 3-fold:

1) Compare the relative sensitivity of porewater and solid-phase toxicity tests. This objective is accomplished through the analysis of marine and freshwater studies in which both types of tests were applied to the same samples.

2) Examine the importance of 3 categories of key factors potentially responsible for differences in test performance: a) specific differences in sensitivity to contaminants, b) differences in contaminant exposure between test matrices, and c) influence of naturally occurring toxicants.

3) Explore the relevance of porewater toxicity data for assessing sediment quality. Selected marine toxicity studies that address this objective are discussed.

Porewater Toxicity Testing: Biological, Chemical, and Ecological Considerations. R. Scott Carr and Marion Nipper, editors.
© 2003 Society of Environmental Toxicology and Chemistry (SETAC). ISBN 1-880611-65-1

Approach

A literature review was conducted in order to obtain data on the relative performance of porewater and solid-phase toxicity tests. Studies in which both types of tests were conducted on the same sediment samples were compiled from scientific journals and reports from local and national agencies in the United States and Canada.

Two criteria were used to identify studies containing suitable data for analysis. First, each study had to report test results for individual stations. Second, only data for tests of pore water and bulk sediment were summarized; tests of elutriates or sediment extracts were not included in the analysis.

The information obtained from each study included test species, test duration and endpoints, total number of samples tested, and number of stations in each of the following porewater and solid-phase results categories: toxic/toxic, toxic/not toxic, not toxic/toxic, and not toxic/not toxic. The number of stations in each category was expressed as a percentage of the total stations for each study and as a percentage of all marine and estuarine or freshwater samples.

Relative Performance of the Tests

General characteristics of the data

A total of 36 marine and freshwater studies met the criteria for inclusion (see Appendix 2-1, p 30). Substantial differences exist in the types of comparative test data available between marine and freshwater tests. A summary of the compiled marine and estuarine data (Table 2-1) shows that most porewater toxicity tests were conducted with early life stages of echinoderms or mollusks and measured sublethal responses such as fertilization or embryo development. These data were almost always compared to solid-phase tests that measured amphipod survival. Porewater test data for other taxa (e.g., polychaetes, fish, and algae) and the use of sublethal endpoints in the solid-phase tests represented less than 10% of the available data.

The number of test comparisons located for freshwater species (593) was less than 30% of the marine dataset (Appendix 2-1, p 30). The experimental design of these porewater tests also differed from that used in marine studies (Table 2-2). Most freshwater toxicity tests used juvenile or adult amphipods as the porewater test organism. Like their marine counterparts, most freshwater solid-phase tests used amphipod survival as the measure of toxicity. Taxonomic diversity was less in freshwater studies; only 3 metazoan phyla were represented in the studies, whereas the marine studies included 6 phyla.

Table 2-1 Summary of toxicity test comparisons for marine and estuarine samples[a]

	Porewater test (no. of species)							
	Embryo and/or larval					Juvenile and/or adult		
	Fertilization	Development				Survival	Growth and reproduction	
Sediment test (no. of species)	Echinoid (4)	Echinoid (3)	Mollusk (2)	Fish (1)	Algae (1)	Crustacean (2)	Polychaete (1)	Bacteria (1)
Survival								
Crustacean (8)	38	25	7	1	1	2	3	3
Growth and reproduction								
Crustacean (3)	3	3	—	—	—	—	—	—
Mollusk (2)	1	2	—	—	1	—	1	—
Echinoderm (1)	2	2	—	—	—	—	—	—

[a] Values indicate number of studies containing data for each combination of porewater and solid-phase test species. There are multiple entries for studies reporting data for more than 1 combination of species or test methods. A total of 1911 pairwise comparisons were evaluated.

Table 2-2 Summary of toxicity test comparisons for freshwater samples[a]

	Porewater test (no. of species)				
	Embryo and/or larval	Juvenile and/or adult			
	Survival	Survival			Growth
Sediment test (no. of species)	Insect (2)	Crustacean (3)	Oligochaete (2)	Fish (1)	Bacteria (1)
Survival					
Insect larvae (2)	3	—	—	—	—
Crustacean (2)	—	10	2	1	1
Oligochaete (2)	—	1	2	1	—
Fish (1)	—	1	1	1	—
Growth or feeding					
Insect larvae (1)	—	1	—	—	1
Crustacean (1)	—	2	—	—	1

[a] Values indicate the number of studies containing data for each combination of porewater and solid-phase test species. Studies containing data for more than 1 combination of species are listed multiple times. Data for a total of 593 pairwise comparisons were evaluated.

In spite of their differences, the marine and freshwater studies shared 2 overarching characteristics: Most of the comparative data for porewater and solid-phase tests were based on a relatively small number of test types, and few of the solid-phase toxicity tests measured sublethal effects such as growth, reproduction, or feeding.

Overall correspondence between tests

The results of pairwise comparisons of the matched toxicity data were summarized to indicate the percentage agreement between the porewater and solid-phase methods in detecting toxicity. Porewater and solid-phase test results were in agreement the majority of the time (Figure 2-1). For marine tests, the same classification (either toxic or nontoxic) was obtained for 54% of the samples tested. Freshwater tests had a higher level of agreement, with 69% of the samples producing similar results for either test matrix. This difference is likely due to the difference between the species and life stages routinely employed in marine and freshwater porewater tests.

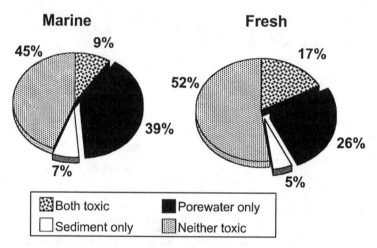

Figure 2-1 Comparative response of porewater and solid-phase toxicity tests of the same sediment sample for marine or freshwater studies. The results summarize the response of several different test species, as listed in Appendix 2-1 (p 30).

Most of the agreement among the test results was due to the absence of toxicity in the samples (Figure 2-1). Much less agreement was present when only toxic samples were examined. Agreement between test matrices (both tests classified the sample as toxic) was present in only 16% and 35% of the samples for marine and freshwater tests, respectively. Porewater toxicity tests were much more likely than solid-phase tests to detect the toxicity.

Patterns among test types

The influence of variations in test methods on relative test performance was evaluated by comparing the test results of individual studies. Test performance was summarized in terms of the amount of unique toxicity information produced by each test method; this was defined as the percentage of the total number of toxic samples (detected by either test) that were classified as toxic only by 1 test method (either porewater or solid-phase test).

Marine porewater toxicity tests provided most of the unique toxicity data, regardless of the test method or species used (Figure 2-2). In most studies, the solid-phase test did not detect any toxic samples that were not also identified as toxic on the basis of the porewater test results. This pattern of test performance was similar for most embryo and/or larval tests, regardless of whether fertilization or embryo development was used as the endpoint. The best performance among the solid-phase tests was reported by studies that measured sublethal responses (growth or reproduction); these tests often provided a larger percentage of unique toxicity data than the corresponding porewater tests. None of the marine studies examined reported a 100% correspondence between test types.

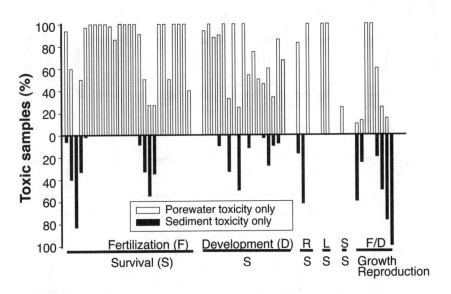

Figure 2-2 Relative amount of unique toxicity information provided by either porewater or solid-phase tests of marine samples. Each bar represents the percentage of the total number of toxic stations in a single study that was identified as toxic by only 1 test method. Results for similar test method combinations are grouped, as shown by the lines below the graph (upper and lower headings indicate test type for pore water or sediment, respectively). R = reproduction, L = luminescence.

Analysis of the relative performance data for freshwater tests produced a pattern that was similar to the marine dataset (Figure 2-3). Porewater tests provided most or all of the unique toxicity data regardless of the species tested. This pattern was often reversed for solid-phase tests incorporating chronic exposures or sublethal endpoints, as was observed for the marine tests. There were 4 comparisons that yielded 100% agreement. Three of these comparisons included chronic and/or sublethal solid-phase test endpoints, including amphipod feeding rate (Winger et al. 1993) and oligochaete chronic survival (Sibley et al. 1997). There were 13 sets of comparisons that used the same species for the porewater and short-term solid-phase tests. The agreement among these comparisons was poor and similar in nature to that obtained with mixed species (e.g., the porewater test was usually more sensitive).

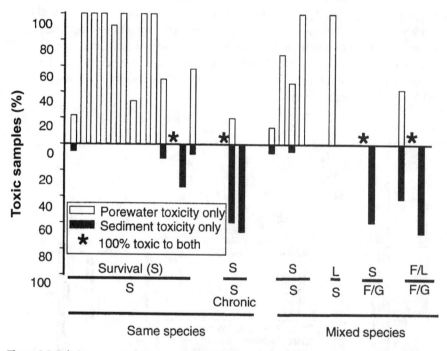

Figure 2-3 Relative amount of unique toxicity information provided by either porewater or solid-phase tests of freshwater samples. Each bar represents the percentage of the total number of toxic stations in a single study that was identified as toxic by only 1 test method. Results for similar test method combinations are grouped, as shown by the lines below the graph (upper and lower headings indicate test type for pore water or sediment, respectively). Asterisks indicate cases where complete agreement (0% unique toxicity information) was reported. L = luminescence, F = feeding activity, G = growth.

Factors Modifying Test Response

The ability of toxicity tests to respond to many different factors, including differences in contaminant bioavailability, unmeasured contaminants, mixtures of chemicals, and naturally occurring substances, is both an advantage and a limitation. These characteristics give toxicity tests the power to provide an integrated assessment of exposure, but they also complicate the identification of the cause of a toxic response.

The differential sensitivity of porewater and solid-phase test procedures to multiple contaminants and natural factors may account for the observed differences in relative performance among test types. These potentially modifying factors can be grouped into 3 categories:

1) differential sensitivity to contaminants among test species,

2) variations in contaminant exposure or bioavailability related to the test method, and

3) influence of naturally occurring toxicants (confounding factors).

The potential contribution of each of these factors is discussed in the following sections.

Differential sensitivity to contaminants among test species

Many marine porewater toxicity tests were developed with the objective of increasing sensitivity to contaminants. Therefore, differential sensitivity is expected to be a characteristic of porewater and solid-phase tests. In spite of many years of testing, there are relatively few contaminants for which dose–response data are available for both types of test methods. While differential sensitivity exists between test methods for selected chemicals, no consistent pattern is evident (Figure 2-4). Porewater toxicity tests representing a variety of taxa (echinoderms, mollusks, algae) often show greater sensitivity to copper and zinc, compared to amphipod or polychaete survival. Solid-phase test species appear to have a similar sensitivity to other chemicals, however, such as cadmium and pentachlorophenol. Embryo and/or larval tests are often more sensitive to ammonia and hydrogen sulfide, 2 naturally occurring substances produced by the decomposition of organic matter.

The data summarized in this paper do not provide a definitive illustration of relative test sensitivity for 2 reasons. First, limited data are available for some species and chemicals. The lack of comparable dose–response data for polycyclic aromatic hydrocarbons (PAHs) is especially pronounced. In addition, some of the studies summarized here did not report measured values of the toxicants. Consequently, the actual exposure concentrations of some chemicals (especially nonpolar organics) are uncertain.

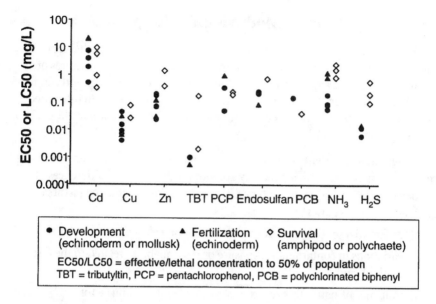

Figure 2-4 Sensitivity of marine porewater (development or fertilization) and solid-phase (survival) test species to selected chemicals. Data are EC50 (closed symbols) or LC50 (open symbols). Sources for the data are listed in Appendix 2-2 (p 34).

Though differential toxicant sensitivity exists among test methods, it is not the major factor producing the heightened sensitivity of porewater toxicity tests. Evidence for this conclusion is shown by the analysis of the freshwater test comparisons (Figure 2-3). Many freshwater porewater toxicity tests use the same species and a similar (or shorter) exposure period compared to solid-phase tests, which eliminates toxicant sensitivity as a modifying factor. Yet these porewater tests are almost always more responsive to the samples than are the corresponding solid-phase tests.

Variations in contaminant exposure or bioavailability

Exposure of the test organism to sediment-associated contaminants can be influenced by several aspects of porewater toxicity tests. First, the isolation of pore water from the sediment disrupts concentration gradients established by equilibrium partitioning (EqP) processes with sediment particles. The concentration of porewater constituents tends to decline because of sorption onto container surfaces soon after the pore water is extracted from the sediment and because of reduced solubility resulting from changes in porewater oxidation state (see Chapter 4).

Test organism exposure to contaminants in porewater tests may also be reduced because of shortened exposure time and lack of sediment ingestion. The typical duration of porewater toxicity tests is 1 to 4 days, which may be insufficient to allow some contaminants to attain steady-state tissue concentrations. The absence of sediment particles also eliminates contaminant bioaccumulation from sediment ingestion. Recent studies have shown that sediment ingestion is an important route of contaminant uptake for contaminants in some organisms (Harkey et al. 1994).

When we compare different tests, we know that there will be some differences in exposure related to the different test conditions, but these differences cannot account for the differences that we observe in test sensitivity. These factors tend to decrease contaminant exposure, which would decrease the relative sensitivity of porewater tests. The opposite situation (i.e., increased sensitivity of porewater tests) was often observed in the studies examined (Figures 2-2 and 2-3).

The relatively homogeneous exposure matrix found in a porewater test does have the potential to increase contaminant exposure, as compared with solid-phase tests, however. Sediment-dwelling organisms such as amphipods and annelids can reduce exposure to toxic pore water through behavioral changes such as sediment avoidance or altered ventilation rates (Keilty et al. 1988; Landrum and Robbins 1990). Porewater tests do not provide the test organisms with an opportunity to reduce exposure by avoiding toxic substances. Thus, the test organisms in a porewater test may receive a greater overall exposure to porewater constituents than do some solid-phase test species. Conversely, porewater tests may underestimate the exposure to nonionic organic chemicals with log octanol–water partition coefficient (K_{OW}) values >3.5 because of adsorption to exposure chambers and/or uptake by the test organisms, a problem common to all static exposure designs with high log K_{OW} chemicals.

Influence of naturally occurring toxicants

Naturally occurring toxicants such as ammonia and sulfide have been identified as confounding factors in toxicity tests (Ankley et al. 1990; Kohn et al. 1994; Knezovich et al. 1996). The heightened sensitivity of many marine porewater test species to these compounds (Figure 2-4) may be a significant factor in the relative sensitivity of porewater toxicity tests. Ammonia was identified as the dominant cause of toxicity to sea urchin embryos exposed to pore water from coastal sediments in southern California, USA (Bay 1996). Hydrogen sulfide, though highly toxic, is rarely present in porewater samples because of oxidation during sample processing. However, when procedures to minimize sample oxidation are used with porewater samples extracted from organically enriched sediments, toxic concentrations of hydrogen sulfide may be present (Southern California Coastal Water Resource Project [SCCWRP] 1995).

The influence of ammonia or sulfide can be accounted for by comparing the concentrations of these substances to known toxic thresholds (Bay 1996). Such analyses for ammonia indicate that this confounding factor is significant but not the only factor responsible for increased porewater test sensitivity. This situation is illustrated by data for sediment samples from Boston Harbor, Massachusetts, USA (Long, Sloane et al. 1996). Sea urchin embryo development was strongly affected by virtually every porewater sample tested, while effects on amphipod survival were infrequently observed (Figure 2-5). Most of the porewater samples contained toxic levels of ammonia. Dilution of porewater samples to a 25% concentration reduced ammonia concentrations to nontoxic levels for most of the samples. However, reduced embryo development was still observed in many samples (Figure 2-5), indicating that additional factors were responsible for the greater relative sensitivity of the sea urchin embryo test.

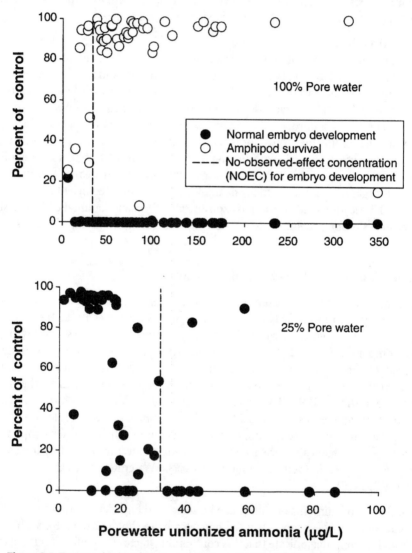

Figure 2-5 Toxicity of sediment and porewater samples from Boston Harbor, Massachusetts, USA. Sea urchin (*Arbacia punctulata*) embryo development in 100% and 25% pore water and corresponding amphipod (*Ampelisca abdita*) survival results are plotted against porewater ammonia concentration. Data from Long, Sloane et al. 1996.

Not enough is known about the influence of other potentially confounding factors on porewater or solid-phase toxicity test species. Embryo and/or larval tests are sensitive to many seawater constituents, many of which may vary in pore water. For example, the activity, viability, and fertilizing ability of sea urchin sperm is affected by many naturally occurring constituents, such as calcium, potassium, and bicarbonate (Table 2-3). It is likely that some of these constituents also affect other test species, but little data are available to document these effects. Lasier et al. (2000) determined that natural variation in porewater manganese concentration is sufficient to cause toxicity in *Hyalella azteca* and *Ceriodaphnia dubia*.

Table 2-3 Constituents known to influence sea urchin egg fertilization and vary in pore water

Naturally occurring constituent	Affects fertilization[a]	Porewater concentration[b]	
		Elevated	Reduced
pH	√		√
Ca^{2+}	√		√
Na^+	√	√	
K^+	√	√	
Mg^{2+}	√	√	
Zn^{2+}	√	√	
$Fe(OH)_3$	√	√	
Co^{2+}	√	√	
Ni^{2+}	√	√	
NH^{4+}	√	√	
HS^-	√	√	
CO_2	√	√	
CH_4	?	√	
PO_4^{3-}	?	√	
SiO_4^{2-}	?	√	
O_2	?		√
NO^{3-}	?		√
Fe^{3+}	?		√
Mn^{2+}	?		√
HCO^{3-}	√		√
SO_4^{2-}	?		√
Dissolved organic matter (DOM)	?	√	

[a] √ = shown to produce an effect on sea urchin fertilization in laboratory experiments; ? = not enough information.

[b] Typical change in concentration relative to overlying water; relative concentration may vary among different habitat types or depths.

Variation in pH is an important factor that is not consistently addressed in the design or interpretation of porewater toxicity tests. Marine test species are generally less tolerant of pH variations than are freshwater species. Early life stages of marine species, such as sea urchins and mollusks, are adversely impacted by variations in pH that occur in pore water. The pH of marine pore water is often lower than seawater, with typical values ranging from 7.0 to 8.4 (Figure 2-6). The sperm and embryos of the purple sea urchin can be adversely affected by pH values less than 7.4 or greater than 8.3 (Figure 2-6). Approximately 40% of porewater samples obtained from sediments along the Pacific coast of the U.S. had a pH of less than 7.4, yet pH data are often not included in toxicity test data reports.

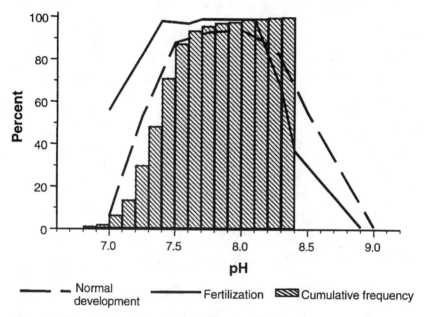

Figure 2-6 Effect of pH variation on purple sea urchin (*Strongylocentrotus purpuratus*) egg fertilization and embryo development. Bars indicate the cumulative percentage of samples having a porewater pH equal to or below the value indicated (*n* = 407).

Consequences for Sediment Quality Assessment

The high relative sensitivity of porewater tests prompts questions about the ecological relevance of the data. Two issues are of principal concern: the use of water column organisms not adapted to pore water and the use of sublethal endpoints that may not correspond to reduced survival of benthic organisms.

Because water column species are similar to benthic species in their sensitivity to contaminants, the use of water column species in porewater tests is not a significant cause for concern (U.S. Environmental Protection Agency [USEPA] 2000). The response of water column and meiobenthic species to pore water was examined by Carr, Chapman, Presley et al. (1996). The response of sea urchin (*Arbacia punctulata*) embryos exposed to pore water from metal-enriched sediments was similar to the response of 2 infaunal meiobenthic invertebrates, the polychaete *Dinophilus gyrociliatus* and nauplii of the copepod *Longipedia americana* (Figure 2-7).

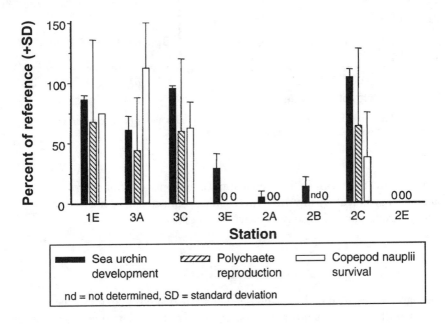

Figure 2-7 Comparative response of sea urchin (*A. punctulata*) embryos and meiobenthic taxa to pore water. Data from Carr, Chapman, Presley et al. 1996.

Comparison of porewater and solid-phase test results to benthic community changes provides perhaps the best measure of ecological relevance because community change reflects long-term and sublethal impacts. Limited comparisons have been made, so it is premature to draw conclusions regarding ecological relevance. But some comparisons between toxicity tests and benthic communities indicate that the data provide a useful measure of ecological impacts. For example, Anderson et al. (1998) observed that porewater toxicity to abalone embryos was significantly correlated to infaunal abundance (Figure 2-8). The correlation, though relatively low, was similar in magnitude to the correlation between solid-phase toxicity (amphipod survival) and the number of infaunal species in the sample.

Multivariate analyses conducted by Carr et al. (2000) showed that porewater toxicity was significantly correlated with adverse impacts on benthic communities in Corpus Christi Bay, Texas, USA. A highly significant correlation between sea urchin porewa-

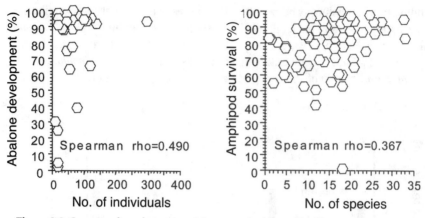

Figure 2-8 Correspondence between toxicity test endpoints and benthic community parameters. Data from Anderson et al. 1998.

ter toxicity and a decrease in the presence and abundance of echinoderms were found in an Sediment Quality Triad (SQT) study conducted in northern Puget Sound, Washington, USA (Long, Hameedi et al. 1999).

Conclusions

The analyses presented in this chapter demonstrate that porewater tests detect toxicity much more frequently than do solid-phase tests. This pattern is remarkably similar regardless of the salinity of the samples or species tested.

A variety of biological and chemical factors are probably responsible for this difference in sensitivity, and there are insufficient data to fully evaluate the significance of each factor. But some conclusions appear evident. First, differences that are species specific to anthropogenic contaminants appear to play a minor role because porewater tests show greater sensitivity than solid-phase tests even when the same species is used in both. Differences in exposure that result from chemical changes and lack of sediment ingestion, while present in many cases, would tend to reduce the sensitivity of porewater tests and are therefore minor factors. However, increased contaminant exposure may result from the inability of porewater test organisms to avoid toxicants and may play a significant role in differential test responses.

The influence of naturally occurring toxicants, such as ammonia and sulfide, certainly plays an important role in differential test sensitivity. While the importance of several of these potentially confounding factors has been established, more research is needed to identify additional factors, especially when porewater tests are conducted with gametes or early life stages. Additional studies are needed that examine the relationship between porewater and solid-phase toxicity tests and the effectiveness of these tests for predicting impacts on benthic communities.

References

[ASTM] American Society for Testing and Materials. 1993. Standard guide for conducting 10-d static sediment toxicity tests with marine and estuarine amphipods, Designation E 1367-92. Volume 11.04, ASTM annual book of standards. Philadelphia PA, USA: ASTM. p 1138–1164.

Anderson B, Hunt J, Phillips B, Newman J, Tjeerdena R, Wilson CJ, Kapahi G, Sapudar RA, Stephenson M, Puckett M, Fairey R, Oakden J, Lyons M, Birosik S. 1998. Sediment chemistry, toxicity, and benthic community conditions in selected water bodies of the Los Angeles region. Sacramento CA, USA: California State Water Resources Control Board, Division of Water Quality. Final report. 232 p plus appendices.

Anderson B, Hunt J, Tudor S, Newman J, Tjeerdema R, Fairey R, Oakden J, Bretz C, Wilson C, LaCaro F, Kapahi G, Stephenson M, Puckett M, Anderson J, Long E, Fleming T, Summers K. 1997. Chemistry, toxicity and benthic community conditions in sediments of selected southern California bays and estuaries. Sacramento CA, USA: Report to California State Water Resources Control Board. 141 p plus appendices.

Ankley GT, Katko A, Arthur JW. 1990. Identification of ammonia as an important sediment–associated toxicant in the lower Fox River and Green Bay, Wisconsin. *Environ Toxicol Chem* 9:313–322.

Ankley GT, Schubauer-Berigan MK, Dierkes JR. 1991. Predicting the toxicity of bulk sediments to aquatic organisms with aqueous test fractions: Pore water vs. elutriate. *Environ Toxicol Chem* 10:1359–1366.

Bay SM. 1996. Sediment toxicity on the mainland shelf of the Southern California Bight in 1994. In: Allen MJ, Francisco C, Hallock D, editors. Westminster CA, USA: Southern California Coastal Water Research Project. Annual report 1994–95. p 128–136.

Bay S, Burgess R, Nacci D. 1993. Status and applications of echinoid (Phylum Echinodermata) toxicity test methods. In: Landis WG, Hughes JS, Lewis MA, editors. Environmental toxicology and risk assessment. Philadelphia PA, USA: American Society for Testing and Materials. STM STP 1179. p 281–302.

Bay S, Jones BH, Schiff K, Washburn L. 1997. Study of the impact of stormwater discharge on the beneficial uses of Santa Monica Bay. Westminster CA, USA: Southern California Coastal Water Research Project. Second year progress report.

Bay SM, Greenstein DJ, Jirik AW, Brown JS. 1998. Southern California Bight 1994 Pilot Project: VI. Sediment toxicity. Westminster CA, USA: Southern California Coastal Water Research Project. Technical report 309. 56 p.

Brix KV, Cardwell RD, Sweeney FP, Farr CH. 2002. Effects of tributyl tin on reproduction of two west coast species: The giant kelp (*Macrocystis pyrifera*) and the sand dollar (*Dendraster excentricus*). Bremerton WA, USA: Parametrix, unpublished data.

Carr RS, Biedenbach JM. 1996. Influence of ammonia on sea urchin porewater toxicity tests. Proceedings, 23rd Annual Aquatic Toxicity Workshop; 1996 Oct 6–9; Calgary AB, Canada.

Carr RS, Chapman DC. 1992. Comparison of solid-phase and pore-water approaches for assessing the quality of marine and estuarine sediments. *Chem Ecol* 7:19–30.

Carr RS, Chapman DC, Howard CL, Biedenbach JM. 1996. Sediment Quality Triad assessment survey of the Galveston Bay, Texas system. *Ecotoxicology* 5:341–364.

Carr RS, Chapman DC, Presley BJ, Biedenbach JM, Robertson L, Boothe L, Kilada R, Wade T, Montagna P. 1996. Sediment porewater toxicity assessment studies in the vicinity of offshore oil and gas production platforms in the Gulf of Mexico. *Can J Fish Aquatic Sci* 53:2618–2628.

Carr RS, Long ER, Chapman DC, Thursby G, Biedenbach JM, Windom H, Sloane G, Wolfe DA. 1996. Toxicity assessment studies of contaminated sediments in Tampa Bay, Florida. *Environ Toxicol Chem* 15:1218–1231.

Carr RS, Montagna PA, Biedendach JM, Kalke R, Kennicutt C, Hooten R, Cripe G. 2000. Impact of storm-water outfalls on sediment quality in Corpus Christi Bay, Texas, USA. *Environ Toxicol Chem* 19:561–574.

DeWitt TH, Swartz RC, Lamberson JO. 1989. Measuring the toxicity of estuarine sediment. *Environ Toxicol Chem* 8:1035–1048.

Dillon TM, Moore DW, Gibson AB. 1993. Development of a chronic sublethal bioassay for evaluating contaminated sediment with the marine polychaete worm *Nereis (Neanthes) arenaceodentata*. *Environ Toxicol Chem* 12:589–605.

Dimick RE, Breese WP. 1965. Bay Mussel embryo bioassay. Proceedings of the 12th Pacific Northwest Industrial Waste Conference. Seattle WA, USA: Univ Washington, College of Engineering. p 165–175.

Dinnel PA. 1991. Toxicity testing with oyster and mussel embryos. In: Chapman GA, editor. Culture and toxicity testing of west coast marine organisms. Newport OR, USA: U.S. Environmental Protection Agency, Environmental Research Laboratory-Narragansett. Technical report N147.

Dinnel PA, Link JM, Stober QJ, Letourneau MW, Roberts WE. 1989. Comparative sensitivity of sea urchin sperm bioassays to metals and pesticides. *Arch Environ Contam Toxicol* 18:748–755.

Environment Canada. 1996. Suivi environnemental au site d'immersion CM-7 des déblais de dragage du havre de Cap-aux-Meules, Îles-de-la Madeleine, Québec. Montréal QC, Canada: Environment Canada – Région du Québec. 67 p plus appendices.

Fairey R, Roberts C, Jacobi M, Lamerdin S, Clark R, Downing J, Long E, Hunt J, Anderson B, Newman J, Tjeerdema R, Stephenson M, Wilson C. 1998. Assessment of sediment toxicity and chemical concentrations in the San Diego Bay region, California, USA. *Environ Toxicol Chem* 17:1570–1581.

Giesy JP, Graney RL, Newsted JL, Rosiu CJ, Benda A, Kreis Jr RG, Horvath F. 1988. Comparison of three sediment bioassay methods using Detroit River sediments. *Environ Toxicol Chem* 7:483–498.

Green AS, Chandler GT, Blood ER. 1993. Aqueous-, pore-water, and sediment-phase cadmium: Toxicity relationships for a meiobenthic copepod. *Environ Toxicol Chem* 12:1497–1506.

Greenstein DJ, Alzadjali S, Bay SM. 1996. Toxicity of ammonia to Pacific purple sea urchin (*Strongylocentrotus purpuratus*) embryos. In: Allen MJ, Francisco C, Hallock D, editors. Southern California Coastal Water Research Project. Westminster CA, USA: Annual report 1994–95. p 72–77.

Harkey GA, Landrum PF, Klaine SJ. 1994. Comparison of whole-sediment, elutriate and pore-water exposures for use in assessing sediment-associated organic contaminants in bioassays. *Environ Toxicol Chem* 13:1315–1329.

Hartwell SI, Hameedi J, Harmon M. 2001. Magnitude and extent of contaminated sediment and toxicity in Delaware Bay. Silver Spring MD, USA: National Oceanic and Atmospheric Administration, National Ocean Service. NOAA Technical Memorandum NOS/NCCOS/CCMA 148.

Ho KT, McKinney RA, Kuhn A, Pelletier MC, Burgess RM. 1997. Identification of acute toxicants in New Bedford Harbor sediments. *Environ Toxicol Chem* 16:551–558.

Hooten RL, Carr RS. 1998. Development and application of a marine sediment pore-water toxicity test using *Ulva fasciata* zoospores. *Environ Toxicol Chem* 17:932–940.

Hunt JW, Anderson BS. 1989. Sublethal effects of zinc and municipal sewage effluents on larvae of the red abalone *Haliotis rufescens*. *Mar Biol* 101:545–552.

Hunt JW, Anderson BS, Englund M, Phillips BM. 1997. Recent advances in microscale toxicity testing with marine mollusks. In: Wells PG, Lee K, Blaise C, editors. Microscale aquatic toxicology: Advances, techniques and practice. Boca Raton FL, USA: Lewis. p 423–436.

Hunt, JW, Anderson BS, Phillips BM, Newman J, Tjeerdema RS, Taberski K, Wilson CJ, Stephenson M, Puckett HM, Fairey R, Oakden J. 1998. Sediment quality and biological effects in San Francisco Bay. Sacramento CA, USA: California State Water Resources Control Board. Bay Protection and Toxic Cleanup Program final technical report. 188 p plus appendices.

Jones BH, Washburn L, Bay S, Schiff K. 1996. Study of the impact of stormwater discharge on the beneficial uses of Santa Monica Bay. Alhambra CA, USA: Report to Los Angeles County Department of Public Works.

Keilty, TJ, White DS, Landrum PF. 1988. Short-term lethality and sediment avoidance assays with endrin-contaminated sediment and two oligochaetes from Lake Michigan. *Arch Environ Contam Toxicol* 17:95–101.

Knezovich JP, Steichen DJ, Jelinski JA, Anderson SL. 1996. Sulfide tolerance of four marine species used to evaluate sediment and pore-water toxicity. *Bull Environ Contam Toxicol* 57:450–457.

Kohn NP, Word JQ, Niyogi DK, Ross LT, Dillon TM, Moore DW. 1994. Acute toxicity of ammonia to four species of marine amphipod. *Mar Environ Res* 38:1–15.

Landrum PF, Robbins JA. 1990. Bioavailability of sediment-associated contaminants to benthic invertebrates. In: Baudo R, Giesy J, Muntau H, editors. Sediments: Chemistry and toxicity of in-place pollutants. Ann Arbor MI, USA: Lewis. p 237–263.

Lasier PJ, Winger PV, Bogenrieder KJ. 2000. Toxicity of manganese to *Ceriodaphnia dubia* and *Hyalella azteca*. *Arch Environ Contam Toxicol* 38:298–304.

Long ER, Buchman MF, Bay SM, Breteler RJ, Carr RS, Chapman PM, Hose JE, Lissner AL, Scott J, Wolfe DA. 1990. Comparative evaluation of five toxicity tests with sediments from San Francisco Bay and Tomales Bay, California. *Environ Toxicol Chem* 9:1193–1214.

Long ER, Hameedi J, Robertson A, Aasen S, Dutch M, Ricci C, Welch K, Kammin W, Carr RS, Johnson T, Biedenbach J, Scott KJ, Mueller C, Anderson JW. 1999. Survey of sediment quality in Puget Sound, Year 1 - Northern Puget Sound. Silver Spring MD, USA: National Oceanic and Atmospheric Administration. NOAA Technical Memorandum NOS NCCOS CCMA 139. 249 p plus 11 appendices.

Long ER, Robertson A, Wolfe DA, Hameedi J, Sloane GM. 1996. Estimates of the spatial extent of sediment toxicity in major U.S. estuaries. *Environ Sci Technol* 30:3585–3592.

Long ER, Scott GI, Fulton M, Kucklick J, Thompson B, Carr RS, Biedenbach J, Scott KJ, Thursby GB, Chandler GT, Anderson JW, Sloane GM. 1998. Magnitude and extent of sediment toxicity in selected estuaries of South Carolina and Georgia. Silver Spring MD, USA: National Oceanic and Atmospheric Administration. NOAA Technical Memorandum NOS ORCA 128.

Long ER, Sloane GM, Carr RS, Scott KJ, Thursby GB, Wade TL. 1996. Sediment toxicity in Boston Harbor: Magnitude extent, and relationships with chemical toxicants. Silver Spring MD, USA: National Oceanic and Atmospheric Administration. NOAA Technical Memorandum NOS ORCA 96. 133 p.

Long ER, Sloane GM, Scott GI, Thompson B, Carr RS, Biedenbach J, Wade TL, Presley BJ, Scott KJ, Mueller C, Brecken-Fols G, Albrecht B, Anderson JW, Chandler GT. 1999. Magnitude and extent of sediment contamination and toxicity in Biscayne Bay and vicinity. Silver Spring MD, USA: National Oceanic and Atmospheric Administration. NOAA Technical Memorandum NOS NCCOS CCMA 141.

Martin M, Hunt JW, Anderson BS, Espinosa L, Palmer FH. 1986. Marine Bioassay Project: Second Report: Acute toxicity tests with red abalone, mysid shrimp, and giant kelp. Sacramento CA, USA: California State Water Resources Control Board. 65 p.

Martin M, Osborn KE, Billig P, Glickstein N. 1981. Toxicities of ten metals to *Crassostrea gigas* and *Mytilus edulis* embryos and *Cancer magister* larvae. *Mar Pollut Bull* 12:305–308.

Meador JP. 1993. The effect of laboratory holding on the toxicity response of marine infaunal amphipods to cadmium and tributyltin. *J Exp Mar Biol Ecol* 174: 227–242.

Nacci D, Jackim E, Walsh R. 1986. Comparative evaluation of three rapid marine toxicity tests: Sea urchin early embryo growth test, sea urchin sperm cell toxicity test and Microtox. *Environ Toxicol Chem* 5:521–525.

Nicely PA, Hunt JW, Anderson BS, Palmer FA, Carley S. 1999. Tolerance of several marine toxicity test organisms to ammonia and artificial salts. Proceedings, Society of Environmental Toxicology and Chemistry (SETAC) 20th Annual Meeting; 1999 Nov 14–18; Philadelphia, PA. Pensacola FL, USA: SETAC.

Nipper MG, Roper DS, Williams EK, Martin ML, Van Dam LF, Mills GN. 1998. Sediment toxicity and benthic communities in mildly contaminated mudflats. *Environ Toxicol Chem* 17:502–510.

Pesch CE, Morgan D. 1978. Influence of sediment in copper toxicity tests with the polychaete *Neanthes arenaceodentata. Water Res* 13:747–751.

Phillips B, Anderson B, Hunt J, Newman J, Tjeerdema R, Wilson CJ, Long ER, Stephenson M, Puckett M, Fairey R, Oakden J, Dawson S, Smythe H. 1998. Sediment chemistry, toxicity, and benthic community conditions in selected water bodies of the Santa Ana region. Sacramento CA, USA: California State Water Resources Control Board, Division of Water Quality. 105 p plus appendices.

Phillips BM, Nicely PA, Hunt JW, Anderson BS, Tjeerdema RS, Palmer SE, Palmer FH, Puckett HM. 2002. Toxicity of metal mixtures to larval purple sea urchins (*Strongylocentrotus purpuratus*) and bay mussels (*Mytilus galloprovincialis*). *Bull Environ Contam Chem.* Forthcoming.

Porebski LM, Doe KG, Zajdlic BA, Lee D, Pocklington P, Osborne JM. 1999. Evaluating the techniques for a tiered testing approach to dredged sediment assessment – A study over a metal concentration gradient. *Environ Toxicol Chem* 11:2600–2610.

Reish DJ, Gerlinger TV. 1984. The effect of cadmium, lead, and zinc in the polychaetous annelid *Neanthes arenaceodentata. Proc Linnean Soc N S W* (Special Volume):383–389.

Roberts MH. 1987. Acute toxicity of tributyltin chloride to embryos and larvae of two bivalve mollusks, *Crassostrea virginica* and *Mercenaria mercenaria. Bull Environ Contam Toxicol* 39:1012–1019.

Rossi SS, Neff JM. 1978. Toxicity of polynuclear aromatic hydrocarbons to the polychaete, *Neanthes arenaceodentata. Mar Pollut Bull* 9:220–223.

Sasson-Brickson G, Burton GA. 1991. In situ and laboratory sediment toxicity testing with *Ceriodaphnia dubia. Environ Toxicol Chem* 10:201–207.

[SCCWRP] Southern California Coastal Water Research Project. 1986. PCB metabolites similar to parent PCBs in toxicity to sea urchin embryos. Westminster CA, USA: SCCWRP. Annual Report 1986. p 29–30.

[SCCWRP] Southern California Coastal Water Research Project. 1995. Toxicity of sediments on the Palos Verdes shelf. In: Cross JN, Francisco C, Hallock D, editors. Westminster CA, USA: SCCWRP. Annual Report 1993–94. p 79–90.

Schiff K, Bay S, Stransky C. 2002. Characterization of stormwater toxicants from an urban watershed to freshwater and marine organisms. *Urban Water* 4:215–227.

Sibley PK, Legler J, Dixon DG, Barton DR. 1997. Environmental health assessment of the benthic habitat adjacent to a pulp mill discharge. 1. Acute and chronic toxicity of sediments to benthic macroinvertebrates. *Arch Environ Contam Toxicol* 32:274–284.

Sullivan DL. 1996. Ucluelet Inlet survey. North Vancouver, BC, Canada: Environment Canada, Environmental Protection, Pacific and Yukon Region. 23 p.

Swartz RC, DeBen WA, Jones JKP, Lamberson JO, Cole FA. 1985. Phoxocephalid amphipod bioassay for marine sediment toxicity. In: Cardwell RD, Purdy R, Bahner RC, editors. Aquatic toxicity and hazard assessment, Seventh Symposium. Philadelphia PA, USA: American Society for Testing and Materials. ASTM STP 854. p 284–307.

Swartz RC, Schults DW, DeWitt TH, Ditsworth GR, Lamberson JO. 1990. Toxicity of fluoranthene in sediment to marine amphipods: A test of the equilibrium partitioning approach to sediment quality criteria. *Environ Toxicol Chem* 9:1071–1080.

Swartz RC, Cole FA, Lamberson JO, Ferraro SP, Schults DW, Deben WA, Lee II H, Ozretich RJ. 1994. Sediment toxicity, contamination and amphipod abundance at a DDT- and dieldrin-contaminated site in San Francisco Bay. *Environ Toxicol Chem* 13:949–962.

Tang A, Kalocai JG, Santos S, Jamil B, Stewart J. 1997. Sensitivity of blue mussel and purple sea urchin larvae to ammonia. Proceedings, Society of Environmental Toxicology and Chemistry (SETAC) 18th Annual Meeting; 1997 Nov 16–20; San Francisco CA. Pensacola FL, USA: SETAC.

[USEPA] U.S. Environmental Protection Agency. 1980. Ambient water quality criteria for endosulfan. Washington DC, USA: USEPA, Office of Water. EPA/440/5-80-046.

[USEPA] U.S. Environmental Protection Agency. 1986. Ambient water quality criteria for pentachlorophenol. Washington DC, USA: USEPA, Office of Water. EPA/440/5-86-009.

[USEPA] U.S. Environmental Protection Agency. 1995a. Ambient water quality-copper. Addendum. Washington DC, USA: USEPA, Office of Water. Draft.

[USEPA] U.S. Environmental Protection Agency. 1995b. Ambient water quality-zinc. Addendum. Washington DC, USA: USEPA, Office of Water. Draft.

[USEPA] U.S. Environmental Protection Agency. 1995c. Short-term methods for estimating the chronic toxicity of effluents and receiving waters to west coast marine and estuarine organisms. Washington DC, USA: USEPA, Office of Research and Development. EPA/600/R-95/136.

[USEPA] U.S. Environmental Protection Agency. 2000. Technical basis for the derivation of equilibrium partitioning sediment guidelines (ESGs) for the protection of benthic organisms: Nonionic organics. Washington DC, USA: USEPA, Office of Science and Technology. EPA-822-R-00-002.

Vismann B. 1990. Sulfide detoxification and tolerance in *Nereis (Hediste) diversicolor* and *Nereis (Neanthes) virens* (Annelida: Polychaeta). *Mar Ecol Prog Ser* 59: 229–238.

Werner I, Nagel R. 1997. Stress proteins HSP60 and HSP70 in three species of amphipods exposed to cadmium, diazinon, dieldrin, and fluoranthene. *Environ Toxicol Chem* 16:293–2403.

Winger PV, Lasier PJ. 1995. Sediment toxicity in Savannah Harbor. *Arch Environ Contam Toxicol* 28:357–365.

Winger PV, Lasier PJ. 1998. Toxicity of sediment collected upriver and downriver of major cities along the lower Mississippi River. *Arch Environ Contam Toxicol* 35:213–217.

Winger PV, Lasier PJ, Geitner H. 1993. Toxicity of sediments and pore water from Brunswick Estuary, Georgia. *Arch Environ Contam Toxicol* 25:371–376.

Winger PV, Lasier PJ, White DH, Seginak JT. 2000. Effects of contaminants in dredge material from the lower Savannah River. *Arch Environ Contam Toxicol* 38:128–136.

Zajdlik BA, Doe KG, Porebski LM. 2000. Report on biological toxicity tests using pollution gradient studies: Sydney Harbour. Ottawa, ON, Canada: Environment Canada, Marine Environment Division. EPS 3/AT/2.

Appendix 2-1 Comparative toxicity data for porewater and solid-phase toxicity tests[a]

Pore water		Solid phase		Porewater or solid-phase toxicity classification[c]				Reference
Method	Result[b]	Method	Result[b]	T/T	T/N	N/T	N/N	
Marine tests								
Arbacia punctulata 60 min	F	*Ampelisca abdita* 10 d	S	0	17	1	42	Long et al. 1998
	F	*A. abdita* 10 d	S	0	6	4	70	Hartwell et al. 2001
	F	*A. abdita* 10 d	S	1	0	5	49	Long, Sloane et al. 1996
	F	*A. abdita* 10 d	S	8	23	15	59	Long, Sloane et al. 1999
	F	*A. abdita* 10 d	S	1	56	1	63	Long, Sloane et al. 1999
	F	*A. abdita* 10 d	S	0	38	0	45	Long et al. 1998
	F	*A. abdita* 10 d	S	0	17	0	22	Long, Robertson et al. 1996
	F	*A. abdita* 10 d	S	0	1	0	30	Long, Robertson et al. 1996
	F	*A. abdita* 10 d	S	0	4	0	62	Long, Robertson et al. 1996
	F	*A. abdita* 10 d	S	3	117	0	45	Carr, Long et al. 1996
	F	*A. abdita* 10 d	S	1	6	0	29	Carr et al. 2000
	F	*Grandidierella japonica* 10 d	S	0	6	0	18	Carr, Chapman, Howard, Biedenbach 1996
	F	*G. japonica* 10 d	S	0	3	0	0	Carr and Chapman 1992
Strongylocentrotus purpuratus 40 min	F	*Rhepoxynius abronius* 10 d	S	0	2	0	3	SCCWRP 1995
	F	*R. abronius* 10 d	S	0	1	0	9	Jones et al. 1996
	F	*R. abronius* 10 d	S	0	10	1	5	Bay et al. 1997
	F	*A. abdita* 10 d	S	1	3	2	6	Phillips et al. 1998
	F	*R. abronius* 10 d	S	2	3	6	7	Phillips et al. 1998
	F	*R. abronius* 10 d	S	10	7	9	4	Anderson et al. 1997
	F	*Ampelisca virginiana* 10 d	S	0	3	0	3	Porebski et al. 1999
	S	*Hyalella azteca* 10 d	S	0	23	0	18	Winger et al. 2000
Lytechinus pictus 20 min	F	*A. virginiana* 10 d	S	0	5	0	1	Porebski et al. 1999
	F	*A. virginiana* 10 d	S	3	3	3	0	Zajdlik et al. 2000
	F	*R. abronius* 10 d	S	0	21	0	2	Environment Canada 1996

Appendix 2-1 *(cont'd.)*

Pore water		Solid phase		Porewater or solid-phase toxicity classification[c]				Reference
Method	Result[b]	Method	Result[b]	T/T	T/N	N/T	N/N	
Marine tests *(cont'd.)*								
Dendraster excentricus 20 min								
	F	*A. virginiana* 10 d	S	0	6	0	0	Porebski et al. 1999
	F	*A. virginiana* 10 d	S	0	5	0	4	Sullivan 1996
	F	*A. virginiana* 10 d	S	3	2	0	1	Zajdlik et al. 2000
A. punctulata 48 h								
	D	*A. abdita* 10 d	S	1	17	0	18	Carr et al. 2000
	D	*G. japonica* 10 d	S	0	12	0	12	Carr, Chapman, Howard, Biedenbach 1996
	D	*A. abdita* 10 d	S	3	21	0	0	Long, Sloane et al. 1996
	D	*A. abdita* 10 d	S	0	9	1	2	Long et al. 1998
	D	*A. abdita* 10 d	S	0	37	0	27	Long, Sloane et al. 1999
	D	*A. abdita* 10 d	S	1	1	1	6	Long, Sloane et al. 1999
	D	*G. japonica* 10 d	S	0	2	0	1	Carr and Chapman 1992
Fellaster zelandiae 38 h								
	D	*Chaetocorophium lucasi* 28 d	S	2	2	4	0	Nipper et al. 1998
S. purpuratus 72 h								
	D	*A. abdita* 10 d	S	0	15	0	38	Bay et al. 1998
S. purpuratus 96 h								
	D	*R. abronius* 10 d	S	57	88	19	32	Fairey et al. 1998
	D	*A. abdita* 10 d	S	3	9	0	1	Phillips et al. 1998
	D	*Eohaustorius estuarius* 10 d	S	1	1	0	0	Phillips et al. 1998
	D	*R. abronius* 10d	S	19	17	1	2	Phillips et al. 1998
	D	*E. estuarius* 10 d	S	6	28	13	62	Hunt et al. 1998
	D	*R. abronius* 10 d	S	16	10	3	1	Anderson et al. 1997
Haliotis rufescens 48 h								
	D	*R. abronius* 10 d	S	4	55	5	12	Anderson et al. 1998
	D	*R. abronius* 10 d	S	5	10	0	0	Phillips et al. 1998
Dinophilus gyrociliatus 7 d								
	R	*A. abdita* 10 d	S	0	5	1	9	Long et al. 1990
	R	*R. abronius* 10 d	S	5	0	8	2	Long et al. 1990
	R	*G. japonica* 10 d	S	0	3	0	0	Carr and Chapman 1992

Appendix 2-1 *(cont'd.)*

Pore water		Solid phase		Porewater or solid-phase toxicity classification[c]				Reference
Method	Result[b]	Method	Result[b]	T/T	T/N	N/T	N/N	
Marine tests *(cont'd.)*								
Microtox 15 min	L	*A. virginiana* 10 d	S	0	0	0	6	Porebski et al. 1999
	L	*R. abronius* 10 d	S	0	2	0	7	Sullivan 1996
	L	*R. abronius* 10 d	S	0	23	0	0	Environment Canada 1996
Amphiascus tenuiremis 7 d	S	*A. tenuiremis* 7 d	S	3	1	0	1	Green et al. 1993
A. punctulata 60 min	F	*Americamysis bahia* 7 d	G	6	2	12	16	Carr et al. 2000
S. purpuratus 40 min	F	*L. pictus* 35 d	G	5	1	2	3	SCCWRP 1995
	F	*L. pictus* 35 d	G	0	5	0	1	Bay et al. 1997
F. zelandiae 38 h	D	*C. lucasi* 28 d	G	0	4	0	4	Nipper et al. 1998
	D	*Macomona liliana* 28 d	G	1	3	1	3	Nipper et al. 1998
A. punctulata 60 min	F	*A. tenuiremis* 7 d	R	3	3	6	2	Long et al. 1998
	F	*A. tenuiremis* 7 d	R	1	2	10	2	Long, Sloane et al. 1999
	D	*A. tenuiremis* 7 d	R	0	0	2	1	Long, Sloane et al. 1999
Freshwater tests								
Ceriodaphnia dubia 48 h	S	*H. azteca* 96 h	S	12	2	1	5	Ankley et al. 1991
	S	*Lumbriculus variegatus* 96 h	S	5	11	0	9	Ankley et al. 1991
	S	*Pimephales promelas* 96 h	S	8	8	1	8	Ankley et al. 1991
Tubifex tubifex 48 h	S	*H. azteca* 48 h	S	0	1	0	11	Sibley et al. 1997
Microtox 15 min	L	*H. azteca* 10 d	S	0	2	0	0	Winger et al. 1993
H. azteca 96 h	S	*H. azteca* 96 h	S	13	4	1	4	Ankley et al. 1991
H. azteca 48 h	S	*H. azteca* 48 h	S	0	2	0	10	Sibley et al. 1997
H. azteca 96 h	S	*H. azteca* 10 d	S	0	5	0	35	Winger and Lasier 1998
	S	*H. azteca* 10 d	S	0	23	0	18	Winger et al. 2000
H. azteca 10 d	S	*H. azteca* 10 d	S	1	10	0	15	Winger and Lasier 1995
	S	*H. azteca* 10 d	S	0	2	0	1	Winger et al. 1993

Appendix 2-1 *(cont'd.)*

| Pore water | | Solid phase | | Porewater or solid-phase toxicity classification[c] | | | | Reference |
Method	Result[b]	Method	Result[b]	T/T	T/N	N/T	N/N	
Freshwater tests *(cont'd.)*								
Daphnia magna 48 h	S	D. magna 48 h	S	2	1	0	9	Sibley et al. 1997
Chironomus riparius 48 h	S	C. riparius 48 h	S	0	3	0	9	Sibley et al. 1997
		C. riparius 10 d	S	0	3	0	9	Sibley et al. 1997
P. promelas 48 h	S	P. promelas 48 h	S	7	9	2	8	Ankley et al. 1991
C. dubia 48 h	S	C. dubia 48 h	S	1	0	0	1	Sasson-Brickson and Burton 1991
Hexagenia spp. 96 h	S	Hexagenia spp. 96 h	S	4	0	2	5	Sibley et al. 1997
L. variegatus 48 h	S	L. variegatus 48 h	S	4	7	1	15	Ankley et al. 1991
T. tubifex 48 h	S	T. tubifex 30 d	S	1	0	0	11	Sibley et al. 1997
H. azteca 48 h	S	H. azteca 28 d	S	1	1	3	6	Sibley et al. 1997
D. magna 48 h	S	D. magna 30 d	S	3	0	6	1	Sibley et al. 1997
H. azteca 10 d	BF	H. azteca 10 d	BF	2	5	5	14	Winger and Lasier 1995
		H. azteca 10 d	BF	2	0	0	1	Winger et al. 1993
D. magna 48 h	S	Chironomus tentans 10 d	G	10	0	15	5	Giesy et al. 1988
		H. azteca 10 d	BF	2	0	0	0	Winger et al. 1993
Microtox 15 min	L	C. tentans 10 d	G	8	0	17	5	Giesy et al. 1988

[a] Data are arranged by method types and test species.

[b] S = survival, G = growth (weight or length), L = luminescence, BF = behavior or feeding, F = fertilization, D = embryo or larva development; R = reproduction

[c] T/T = toxic pore water and toxic solid phase; T/N = toxic pore water and nontoxic solid phase; N/T = nontoxic pore water and toxic solid phase; N/N = nontoxic pore water and nontoxic solid phase.

Appendix 2-2 Toxicity of selected chemicals (dissolved in seawater) to marine species frequently used in solid-phase or porewater tests

Species	Group	Toxicant	Exposure	Endpoint	EC50/LC50[a] (mg/L)	Reference
S. purpuratus	Sea urchin	Cd	96 h	Development	0.510	Dinnel et al. 1989
	Sea urchin		80 min	Fertilization	18.400	Dinnel et al. 1989
A. punctulata	Sea urchin	Cd	4 h	Development	13.900	Nacci et al. 1986
	Sea urchin		80 min	Fertilization	38.000	Nacci et al. 1986
	Sea urchin		48 h	Development	7.380	Carr, Chapman, Presley et al. 1996
	Sea urchin		60 min	Fertilization	20.100	Carr, Chapman, Presley et al. 1996
Mytilus edulis	Mussel	Cd	48 h	Development	3.890	Phillips et al. 2002
A. abdita	Amphipod	Cd	96 h	Survival	0.330	ASTM 1993
E. estuarius	Amphipod	Cd	96 h	Survival	9.330	DeWitt et al. 1989
R. abronius	Amphipod	Cd	96 h	Survival	0.920	Swartz et al. 1985
Neanthes arenaceodentata	Polychaete	Cd	96 h	Survival	5.600	Reish and Gerlinger 1984
Ulva fasciata	Alga	Cd	96 h	Germination	1.930	Hooten and Carr 1998
S. purpuratus	Sea urchin	Cu	72 h	Development	0.016	USEPA 1995c
	Sea urchin		40 min	Fertilization	0.026	USEPA 1995c
A. punctulata	Sea urchin	Cu	4 h	Development	0.014	Nacci et al. 1986
	Sea urchin		80 min	Fertilization	0.028	Nacci et al. 1986
	Sea urchin		48 h	Development	0.038	Hooten and Carr 1998
	Sea urchin		48 h	Development	0.004	Carr, Chapman, Presley et al. 1996
	Sea urchin		60 min	Fertilization	0.006	Carr, Chapman, Presley et al. 1996
M. edulis	Mussel	Cu	48 h	Development	0.007	Phillips et al. 2002
H. rufescens	Abalone	Cu	48 h	Development	0.009	Hunt and Anderson 1989
A. abdita	Amphipod	Cu	96 h	Survival	0.026	USEPA 1995a
N. arenaceodentata	Polychaete	Cu	96 h	Survival	0.077	Pesch and Morgan 1978
U. fasciata	Alga	Cu	96 h	Germination	0.044	Hooten and Carr 1998

Appendix 2-2 *(cont'd.)*

Species	Group	Toxicant	Exposure	Endpoint	EC50/LC50[a] (mg/L)	Reference
S. purpuratus	Sea urchin	Zn	96 h	Development	0.023	Dinnel et al. 1989
	Sea urchin	Zn	40 min	Fertilization	0.029	Schiff et al. 2002
A. punctulata	Sea urchin	Zn	4 h	Development	0.205	Nacci et al. 1986
	Sea urchin		80 min	Fertilization	0.121	Nacci et al. 1986
	Sea urchin	Zn	48 h	Development	0.073	Carr, Chapman, Presley et al. 1996
	Sea urchin		60 min	Fertilization	0.112	Carr, Chapman, Presley et al. 1996
M. edulis	Mussel	Zn	48 h	Development	0.175	Martin et al. 1981
H. rufescens	Abalone	Zn	48 h	Development	0.068	Hunt and Anderson 1989
A. abdita	Amphipod	Zn	96 h	Survival	0.390	USEPA 1995b
N. arenaceodentata	Polychaete	Zn	96 h	Survival	1.400	Reish and Gerlinger 1984
U. fasciata	Alga	Zn	96 h	Germination	0.202	Hooten and Carr 1998
A. punctulata	Sea urchin	PCP[c]	80 min	Fertilization	0.900	USEPA 1986
M. edulis	Mussel	PCP	48 h	Development	0.345	Dimick and Breese 1965
H. rufescens	Abalone	PCP	48 h	Development	0.050	Hunt et al. 1997
E. estuarius	Amphipod	PCP	96 h	Survival	0.243	Toxscan unpublished data[d]
R. abronius	Amphipod	PCP	96 h	Survival	0.191	Toxscan unpublished data[d]
S. purpuratus	Sea urchin	Diazinon	40 min	Fertilization	>12.0	Bailey personal communication[e]
A. punctulata	Sea urchin	Diazinon	48 h	Development	>9.600	Bay et al. 1993
A. abdita	Amphipod	Diazinon	48 h	Survival	0.010	Werner and Nagel 1997
R. abronius	Amphipod	Diazinon	96 h	Survival	0.009	Werner and Nagel 1997
S. purpuratus	Sea urchin	Endosulfan	96 h	Development	0.227	Dinnel et al. 1989
	Sea urchin	Endosulfan	60 min	Fertilization	0.081	Dinnel et al. 1989
M. edulis	Mussel	Endosulfan	48 h	Development	0.212	Dinnel 1991
H. rufescens	Abalone	Endosulfan	48 h	Development	0.252	Martin et al. 1986
N. arenaceodentata	Polychaete	Endosulfan	96 h	Survival	0.73	USEPA 1980
S. purpuratus	Sea urchin	DDT	96 h	Development	>0.008	Dinnel et al. 1989
	Sea urchin	DDT	80 min	Fertilization	<0.001	Dinnel et al. 1989
E. estuarius	Amphipod	DDT	10 d	Survival	0.003	Swartz et al. 1994

Appendix 2-2 *(cont'd.)*

Species	Group	Toxicant	Exposure	Endpoint	EC50/LC50[a] (mg/L)	Reference
D. excentricus	Sand dollar	TBT[f]	60 min	Fertilization	0.0005	Brix et al. unpublished data[g]
M. edulis	Mussel	TBT	48 h	Development	0.001	Roberts 1987
E. estuarius	Amphipod	TBT	96 h	Survival	0.002	Meador 1993
R. abronius	Amphipod	TBT	96 h	Survival	0.173	Meador 1993
A. abdita	Amphipod	Fluoranthene	96 h	Survival	0.010	ASTM 1993
R. abronius	Amphipod	Fluoranthene	10 d	Survival	0.024	Swartz et al. 1990
N. arenaceodentata	Polychaete	Fluoranthene	96 h	Survival	0.500	Rossi and Neff 1978
S. purpuratus	Sea urchin	NH_3	72 h	Development	0.057	Greenstein et al. 1996
S. purpuratus	Sea urchin	NH_3	40 min	Fertilization	1.150	Bay unpublished[b]
A. punctulata	Sea urchin	NH_3	48 h	Development	0.060	Carr, Chapman, Presley et al. 1996
S. purpuratus	Sea urchin	NH_3	60 min	Fertilization	0.600	Carr, Chapman, Presley et al. 1996
M. edulis	Mussel	NH_3	48 h	Development	0.190	Tang et al. 1997
H. rufescens	Abalone	NH_3	48 h	Development	0.082	Nicely et al. 1999
A. abdita	Amphipod	NH_3	96 h	Survival	0.830	Kohn et al. 1994
E. estuarius	Amphipod	NH_3	96 h	Survival	2.490	Kohn et al. 1994
R. abronius	Amphipod	NH_3	96 h	Survival	1.590	Kohn et al. 1994
N. arenaceodentata	Polychaete	NH_3	21 d	Survival	0.83	Dillon et al. 1993
S. purpuratus	Sea urchin	H_2S	96 h	Development	0.012	Knezovich et al. 1996
S. purpuratus	Sea urchin	H_2S	40 min	Fertilization	0.014	SCCWRP 1995
M. edulis	Mussel	H_2S	48 h	Development	0.006	Knezovich et al. 1996
E. estuarius	Amphipod	H_2S	48 h	Survival	0.203	Knezovich et al. 1996
R. abronius	Amphipod	H_2S	48 h	Survival	0.097	Knezovich et al. 1996
N. arenaceodentata	Polychaete	H_2S	24 d	Survival	0.576	Vismann 1990
S. purpuratus	Sea urchin	Aroclor 1254	96 h	Development	0.148	SCCWRP 1986
A. abdita	Amphipod	Aroclor 1254	96 h	Survival	0.040	Ho et al. 1997

[a] EC50 = effective concentration to 50% of test population; LC50 = lethal concentration to 50% of test population.
[b] Steven Bay, Southern CA Coastal Water Reserve Project, Westminster CA, USA.
[c] PCP = pentachlorophenol.
[d] David B. Lewis, Director, Bioassay Division, ToxScan, Inc., Watsonville CA, USA.
[e] Howard Bailey, EVS Environmental Consultants, Seattle WA, USA.
[f] TBT = tributyltin.
[g] Kevin Brix, EcoTox, North Bend WA, USA.

CHAPTER 3

Comparison of Porewater and Solid-Phase Sediment Toxicity Tests

Parley V Winger (Workgroup Leader), Barbara Albrecht, Brian S Anderson, Steven M Bay, Francesca Bona, Gladys L Stephenson

Sediment is a natural particulate material that has been transported and deposited in aquatic ecosystems and is normally found below the water surface (American Society for Testing and Materials [ASTM] 1999). As such, sediments include a solid phase (particulate material) and a porewater phase (water in the interstitial spaces between particles). Sediments of both freshwater and marine environments provide habitats that support populations of benthic and infaunal organisms. These invertebrate and microbial communities reflect the quality and the "health" of an ecosystem. Although sediment quality assessment programs generally use infaunal invertebrate species to measure solid-phase toxicity, recent studies have included porewater toxicity because use of the aqueous phase of sediments has several advantages (Adams et al. 1985; Carr et al. 1989; Di Toro et al. 1991; Ankley et al. 1994; Carr, Long et al. 1996; Whiteman et al. 1996).

Advantages and Disadvantages of Porewater Tests

Porewater toxicity tests were developed relatively quickly. Many procedures that were already established for water-only exposures were applied directly, with or without procedural modifications, while some tests specific to pore water were developed. Many of the organisms used in the aqueous tests were also incorporated into porewater testing procedures, but additional test species have also been used. Consequently, there is considerable variation in the type of porewater assays available and the test species used (Burton 1998; see Chapters 6 and 7). To some extent, the lack of standardized methods or protocols for porewater tests is responsible for the reluctance of government agencies to use porewater testing for regulatory purposes (Chapters 10 and 12). Consequently, these tests are primarily used in regional surveys of sediment quality and research to elucidate exposure pathways and contaminant availability.

There are several advantages to using pore water to describe sediment quality. For some test species, sediment pore water represents the primary route of exposure for nonpolar organic compounds (Adams et al. 1985; Knezovich and Harrison 1987) and for metals (Swartz et al. 1985; Ankley et al. 1991; Di Toro et al. 1991). Porewater tests are frequently more likely to detect toxicity than are solid-phase tests in side-by-side comparisons of the same sediment (Winger and Lasier 1995; Carr, Chapman et al. 1996; Winger, Lasier, White, Seginak 2000). Porewater tests are rapid (short exposure durations: hours or days) and often less expensive to conduct than solid-phase tests that last weeks or months (Burton 1998). Another advantage of using porewater tests is the direct relationship between porewater chemistry and toxicity test results (Ankley et al. 1991). Porewater testing allows comparison with water quality standards and criteria (Giesy et al. 1990). Often, there is a good correlation with porewater chemistry and the toxicological responses observed in test organisms for short-term porewater toxicity tests (Ankley et al. 1991; Burton 1998).

Porewater toxicity tests also have several practical advantages. The test durations are usually short and generally require only small volumes of pore water (Carr et al. 2000; Winger, Lasier, White, Seginak 2000). Porewater tests are not confounded by predators or other indigenous organisms found in the solid phase of sediments (Reynoldson et al. 1994). Potentially confounding geochemical factors such as particle size and total organic carbon (TOC) are considered to be less important in porewater tests. It is also much easier to conduct a dilution series test with pore water than with whole sediments and to describe the subsequent concentration–response relationship (Giesy et al. 1990). These characteristics make pore water particularly amenable for toxicity identification evaluation (TIE) assessments of factors influencing toxicity (Ankley et al. 1996; see Chapters 6 and 7). Some variability encountered in solid-phase testing may be due to avoidance of the sediment by the test organisms during the exposure period. Porewater exposures obviate this avoidance response that could increase the probability of a false negative in some solid-phase tests (Hoke et al. 1995).

Despite the advantages discussed above, there are limitations to porewater toxicity tests that must be considered in planning their use. Extraction of pore water from sediments is often time-consuming, making it difficult to collect sufficient quantities of pore water necessary for conducting toxicity and chemical tests. The relatively large volumes of pore water required for chemical analyses can be prohibitive (Burton 1998; Forbes et al. 1998). Changes in the physical or chemical composition of pore water might occur during collection, extraction, and manipulation of porewater samples, but the influence of these methodological artifacts remains poorly understood (Forbes et al. 1998; see Chapters 4 through 7).

The interpretation of porewater toxicity tests is often hindered by uncertainty regarding the accuracy and representativeness of the results. The accuracy of porewater toxicity test results may be compromised by the use of species or life stages that are not ecologically relevant for predicting effects in sediments (i.e., use of pelagic or epibenthic species versus benthic or infaunal species). Porewater tests exclude consideration of other routes of exposure (e.g., those organisms that ingest sediment or prey on other infaunal species) that might be more relevant to infaunal test species (Boese et al. 1990; Giesy et al. 1990; Forbes et al. 1998; Lee et al. 2000). Porewater tests have the potential to "artificially" increase the sensitivity of benthic organisms to contaminants because infaunal test species might be stressed in the absence of sediment (Giesy et al. 1990; Ankley et al. 1991; Ingersoll 1995). Although porewater toxicity tests have the advantage of being rapid, they are particularly limited in freshwater studies to short-term exposures with lethality (depending upon the species or life stage exposed) as the primary endpoint; tests with a longer duration would require feeding and renewal or replacement of water. The application of porewater toxicity tests has been limited by the lack of standardized protocols for collection, manipulation, extraction, and toxicity testing.

Objectives of Porewater Tests

Variability in test methods and the sensitivity of pore water to manipulation artifacts creates uncertainty about the validity of porewater toxicity data and hence their suitability for describing sediment quality. One way to evaluate the performance of porewater toxicity tests is to examine the relationship between the test results and impacts on benthic organisms or communities. This has been done to a limited extent, particularly in habitat evaluations using the Sediment Quality Triad (SQT) (Carr et al. 2000; see Chapters 8 and 9). However, comparisons are not always possible because of resource limitations, environmental variability, and/or the lack of interpretive tools to identify benthic effects.

An alternative approach to understanding the performance of porewater toxicity tests is to compare the results with solid-phase toxicity tests. The extent of agreement ("concordance") or disagreement ("discordance") between these 2 testing approaches provides a tool to evaluate the relative performance of porewater toxicity tests.

Concordance, for the purpose of this discussion, occurs when both the porewater and the solid-phase tests identify a toxic response (significant difference from the control), or no toxicity is detected with either test. Discordance occurs when one matrix (e.g., pore water) is toxic but not the other (e.g., solid-phase). By identifying the causes of agreement or disagreement between tests, much can be learned about the factors that have the greatest impact on porewater test results. Therefore, the objectives of this chapter are to

1) identify factors that have an important effect on the relative performance of porewater tests,

2) provide a conceptual framework for interpreting the significance of a lack of concordance between solid-phase and porewater tests,

3) recommend procedures that would enhance interpretation of porewater tests, and

4) identify information needs that will improve the application and interpretation of porewater toxicity tests.

Concordance between Porewater and Solid-Phase Tests

There are various ways to examine the degree to which test results concur. Statistical approaches, such as Kendall's coefficient of concordance (Kendall 1970), provide a measure of statistical significance but do not describe the overall magnitude of the differences. A more direct approach may be the examination of the percentage of agreement and the visual comparison of the magnitude of responses between samples. This approach retains much of the information in the results and permits distinguishing between statistically significant (yet minor) differences and cases where large differences exist.

It is important to realize that complete agreement between solid-phase and porewater toxicity tests is not necessarily desirable. Porewater toxicity tests are often intended to complement solid-phase sediment quality assessments by providing information on more sensitive species or alternative exposure conditions. Consequently, variable levels of agreement between test methods are to be expected, especially in situations where toxicity is due to moderate or mild levels of contamination.

The response of an organism to a toxicant is influenced by 2 primary factors: the dose received (a function of toxicant concentration, exposure duration, and exposure route or bioavailability) and the sensitivity of organisms to toxicants. The interaction of these factors, therefore, has a strong influence on the concordance of toxicity tests (Table 3-1). High concordance would be expected with extreme levels (high or low) of contamination. Under these conditions, differences in species sensitivity or exposure conditions play a minor role. Under more typical or moderate levels of contamination, concordance is dependent upon the test methods, the bioavailability of the contaminants to the organism, and the sensitivity of the test organisms.

Table 3-1 Factors expected to result in high or low concordance between toxicity tests

High concordance	Low concordance
High sediment contamination	Confounding factors present
Low sediment contamination	Alteration of porewater characteristics
Primary exposure route is water	Exposure route is sediment related
Similar sensitivities of test species	Different sensitivities of test species
Both tests insensitive	

Low concordance is an expected result among tests that use organisms with markedly different sensitivities to contaminants (Table 3-1). Test results are also expected to disagree when the primary route of toxicant exposure is not the aqueous phase (e.g., sediment ingestion). In this case, the ability of a porewater test to discriminate toxicity should be less than that of the solid-phase test. Lack of concordance might also be caused by factors not related to the presence of contaminants. For example, natural constituents (e.g., ammonia) might be present at sufficiently high levels to mask the influence of contaminants (confounding factors). Alterations in porewater characteristics that are due to sediment handling and porewater extraction procedures might also affect the contaminant dose.

There is no single acceptable level of concordance between tests. Rather, the degree of concordance should be a function of the study objectives and the sample characteristics (Table 3-1). Where the goal of a study is to provide a high level of confidence in the detection of toxicity (low probability of a false positive), high concordance may be desirable. If the goal is to provide a highly protective measure for sediment quality to a diversity of species (low probability of false negatives), then a variety of sediment toxicity tests using species with different sensitivities to contaminants might be employed, with a corresponding increase in the potential for discordant results.

An analysis of porewater and solid-phase toxicity tests conducted on the same samples showed that concordant results were obtained 57% of the time (Chapter 2). Agreement between tests was similar for comparisons in marine (54%) and freshwater (69%) environments, despite the use of different porewater test strategies.

Compared to solid-phase tests, porewater tests detect toxic sediments more often and thus provide the majority of toxicity information obtained in surveys of field sediments. For marine samples, the porewater tests were the only responsive toxicity indicator in 38% of the studies examined (each study tested multiple sediment samples). Relatively low concordance would be expected for comparisons of tests with marine sediments because the porewater test methods often use highly sensitive, early life-history stages (e.g., embryos and larvae). The frequent use of the same species in freshwater porewater and solid-phase tests does not improve the concordance between methods (Chapter 2). Porewater tests using freshwater species were the only responsive indicator in 31% of the field studies (and in 64% of the samples tested) where the same species was used in each type of test.

The lack of concordance between porewater and solid-phase tests may be influenced by the duration of test exposure in the solid-phase tests. Exposure times for most short-term lethality tests with solid-phase sediments ranged from 10 to 14 days, but few of the sediment tests demonstrated toxicity. However, solid-phase tests that incorporated longer periods of exposure (chronic tests) and included more sensitive sublethal endpoints (e.g., growth) generally showed a higher level of concordance with porewater tests (Chapter 2).

High concordance studies

There are situations in which porewater and solid-phase toxicity tests would be expected to give similar results. These include studies of sediments from highly contaminated sites, studies of reference sites, studies of situations in which pore water is the primary route of exposure, and studies in which the same species are tested in both porewater and solid-phase matrices.

A number of examples of synoptic porewater and solid-phase assessments at highly contaminated sites show similar results. Anderson et al. (1998) found concordance between the 2 test procedures in a study conducted on a former naval base in Holmen, Denmark. Amphipod 10-day solid-phase acute tests with *Corophium volutator* and porewater tests with the alga *Skeletonema costatum* and the copepod *Acartia tonsa* all detected toxicity at a number of stations. Anderson et al. (2001) found significant acute mortality at a site (Consolidated Slip) in inner Los Angeles Harbor, California, USA, using a 10-day solid-phase sediment test with the amphipod *Rhepoxynius abronius* and a porewater test with red abalone (*Haliotis rufescens*) embryos (Table 3-2). The sediments were highly polluted by a number of organic and trace metal contaminants, including chlordane, polychlorinated biphenyls (PCBs), high molecular weight polycyclic aromatic hydrocarbons (PAHs), zinc, and mercury. Using sea urchin (*Strongylocentrotus purpuratus*) embryos in porewater and amphipod (*Eohaustorius estuarius*) survival in solid-phase tests, Hunt, Anderson, Phillips, Newman, Tjeerdema, Taberski et al. (1998) found significant toxicity to both species in samples from Islais Creek in San Francisco Bay, California, USA (Table 3-2). Sediments at this site were highly contaminated by a mixture of organic chemicals, hydrogen sulfide, and ammonia.

Although there are differences in the ability of the solid-phase and porewater tests to characterize sediment quality, results from a number of studies have demonstrated high concordance between porewater and solid-phase tests at both marine and freshwater reference sites. Anderson et al. (2001) also found high amphipod survival in solid-phase exposures and minimal toxicity to abalone embryos in pore water at a Terminal Island reference site in Los Angeles Harbor. Hunt, Anderson Phillips, Newman, Tjeerdema, Stephenson et al. (1998) reported minimal toxicity to amphipods (*E. estuarius*, 10-day solid-phase) and sea urchin embryos (*S. purpuratus*, 96-hour pore water) in samples collected from a relatively uncontaminated reference site in San Francisco Bay (Paradise Cove) over a 2-year period (Table 3-2).

Concordance between porewater and solid-phase exposures can also be evaluated using the same species exposed to both test matrices. Winger, Lasier, Bodenreider et al. (2000) investigated sediment toxicity in Hollis Creek, Mississippi, USA, using the porewater and solid-phase exposures with the amphipod *Hyalella azteca*. Sediments from a contaminated site on this stream contained metals and PAH compounds. Discordance between porewater and solid-phase tests was found at this site; there was significant amphipod mortality in the 96-hour porewater test (32% survival),

Table 3-2 Examples of high concordance between porewater and solid-phase toxicity tests

Station name	Water type[a]	Chemicals of concern	TOC (%)	Grain size % fines	Solid-phase survival (%)	Porewater embryo or larval development or amphipod survival (%)	Reference
Consolidated Slip (Los Angeles Harbor CA)	SW	DDE[b], Chlordane, PCBs, Zn, Hg, PAHs	4.6	91	58 *R. abronius* amphipod	0 *H. rufescens* embryo	Anderson et al. 2001
Islais Creek, (San Francisco Bay CA)	SW	Chlordane, PCBs, dieldrin, PAHs	3.9	39	57 *E. estuarius* amphipod	0 *S. purpuratus* embryo	Hunt, Anderson, Phillips, Newman, Tjeerdema, Taberski et al. 1998
Paradise Cove (San Francisco Bay CA)	SW	Ni, Cr	1.1	93	81 *E. estuarius* amphipod	95 *S. purpuratus* embryo	Hunt, Anderson, Phillips, Newman, Tjeerdema, Stephenson et al. 1998
Hollis Creek SC	FW	PAHs, metals	4.8	32	85% growth[c] *H. azteca* amphipod	32 *H. azteca* amphipod	Winger, Lasier, White, Seginak 2000
Reference stations							
Terminal Island CA	SW	DDE	0.6	91	80 *R. abronius* amphipod	95 *H. rufescens* embryo	Anderson et al. 2001
Savannah River SC	FW		16.6	36	100 *H. azteca* amphipod	96 *H. azteca* amphipod	Winger, Lasier, Bogenreider 2000

[a] SW = seawater; FW = freshwater.
[b] DDE = dichlorodiphenyldichloroethylene.
[c] Growth is reported as the percent of growth in the control.

but survival was 86% in a 28-day solid-phase exposure. However, there was concordance when amphipod growth in a chronic, 28-day, solid-phase test was considered. These results demonstrated that concordance was improved when exposure time was increased and a more sensitive sublethal endpoint was included.

In a study on the effects of bleached kraft mill effluent on freshwater benthic species, Sibley et al. (1997) also found that concordance between solid-phase and porewater results improved when bulk-sediment exposure durations were increased to include chronic effects. In this case, survival of amphipods (*H. azteca*) was considerably less in porewater exposures after 96 hours (0% survival) than in 10-day solid-phase exposures (100% survival). Amphipod mortality in solid-phase exposures approached that observed in the porewater tests when the duration of solid-phase exposures was increased to 28 days (33% survival).

In another study, using the same species exposed to porewater and solid-phase samples that had been spiked with cadmium, Green et al. (1993) compared survival of copepods (*Amphiascus tenuiremus*) after 96 hours. Although survival was somewhat lower in pore water (LC50 [lethal concentration to 50% of a test population] = 608 mg Cd/L) than in solid-phase exposures (LC50 = 860 mg Cd/L), the differences in LC50s were attributed to the dissimilarity in responses at the highest and lowest porewater concentrations. This study demonstrated that uptake via pore water was the primary route of cadmium exposure for this copepod, which also explained the relatively close agreement between tests conducted in the different sediment matrices.

Low concordance studies

Although similar results between porewater and solid-phase tests may be expected, low concordance is commonly observed (see Table 3-1). For example, Carr, Long et al. (1996) reported data from Tampa Bay, Florida, USA, where toxicity observed in the amphipod survival tests occurred in approximately 2% of the samples. Porewater tests performed as 1-hour sea urchin fertilization tests showed toxicity in 73% of the samples. Concordance between porewater and solid-phase tests on the same sediments was generally poor. As discussed by Bay et al. (Chapter 2, Figure 2-4), there are considerable differences in the relative sensitivities of the different species used in sediment assessments, especially when sensitivity is considered in terms of water-only exposures. This is one of the primary reasons a test battery with multiple species, varying endpoints, and different exposure pathways is recommended for toxicity assessments. Species-specific differences might, in some cases, account for observed differences in porewater and solid-phase tests at moderately contaminated sites.

Test protocols used for porewater assessments may also differ in their ability to discriminate toxicity of specific classes of contaminants. In some cases, the discriminatory power of the test also differs for naturally occurring chemicals. For example, there are relatively large differences in sensitivity between the sea urchin (*Arbacia punctulata*) fertilization and embryo development endpoints to unionized ammonia (Figure 3-1). This may lead to variations in concordance between porewater and solid-phase test results with amphipods such as *Ampelisca abdita*, which are relatively insensitive to this constituent (Figure 3-1). Therefore, in the absence of the influence of other contaminants, porewater samples that have unionized ammonia concentrations above the embryo lowest observed effects concentration (LOEC) of 90 µg/L may show less concordance between embryo development and amphipod survival than between sea urchin fertilization and amphipod survival. The impact of such differences in ammonia sensitivity can be substantial; data from throughout the United States show that approximately 35% of undiluted marine porewater samples contain ammonia concentrations above the LOEC for *A. punctulata* embryo development (Figure 3-1).

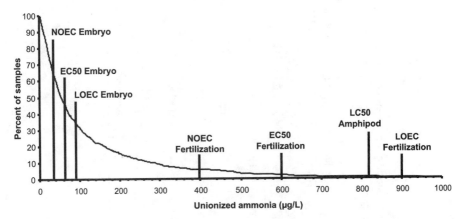

Figure 3-1 Concentration of unionized ammonia in porewater samples obtained from 865 coastal locations within the United States (Carr and Beidenbach 1996). The vertical reference lines indicate the median lethal concentration (LC50) for the amphipod *A. abdita* (Kohn et al. 1994) and the median effects concentration (EC50), the no observed effects concentration (NOEC), and the lowest observed effects concentration (LOEC) for embryo development and fertilization test with the sea urchin *A. punctulata* (Carr and Beidenbach 1996; Carr, Chapman et al. 1996).

The embryonic stages of some marine invertebrates (e.g., echinoids, mollusks) used in porewater tests are more sensitive to some trace metals than the amphipod species used in solid-phase exposures (Chapter 2), and this may explain some differences in responses to sediments observed in field assessments. Hunt, Anderson, Phillips, Newman, Tjeerdema, Taberski et al. (1998) compared sea urchin (*S. purpuratus*) embryos exposed in pore water to amphipod (*E. estuarius*) survival in 10-day solid-phase tests and found relatively large site-specific differences in their responses. In many instances, this was due to the greater sensitivity of sea urchin

Table 3-3 Examples of low concordance between porewater and solid-phase toxicity tests

Station name	Water type[a]	Chemicals of concern	TOC (%)	Grain size % fines	Solid-phase survival (%)	Porewater embryo or larval development or amphipod survival (%)	Reference
Guadalupe Slough (San Francisco Bay CA)	SW	H₂S, Ni, Zn, chlordane	2.2	76	85 *E. estuarius* amphipod	0 *S. purpuratus* embryo	Hunt, Anderson, Phillips, Newman, Tjeerdema, Taberski et al. 1998
Castro Cove (San Francisco Bay CA)	SW	PAHs	2.9	99	33 *E. estuarius* amphipod	96 *S. purpuratus* embryo	Hunt et al. 1998a
Back River (Savannah SC)	FW	As, Zn	9.0	53	88 *H. azteca* amphipod	66 *H. azteca* amphipod	Winger, Lasier, Bogenreider 2000
Mississippi River (Memphis TN)	FW	Cr, Pb, Cu, Ni, Zn	3.7	62	90 *H. azteca* amphipod	68 *H. azteca* amphipod	Winger and Lasier 1998

[a] SW = seawater; FW = freshwater.

embryos to elevated concentrations of ammonia in pore water. At the Guadalupe Slough site in San Francisco Bay, sea urchin development was considerably lower in the porewater exposures (0%) when compared to amphipod survival in the bulk phase (85%; Table 3-3). In this case, porewater TIEs were used to demonstrate that embryo development improved to 97% when ethylenediaminetetraacetic acid (EDTA) was added to pore water, suggesting that divalent cations were the source of embryo toxicity. Bulk-phase sediment analyses at this site indicated moderate concentrations of chlordane, nickel, and to a lesser extent, zinc.

Lack of concordance between porewater and solid-phase results may sometimes be caused by greater sensitivity of species used in solid-phase exposures. Hunt, Anderson, Phillips, Newman, Tjeerdema, Taberski et al. (1998) found survival of the amphipod *E. estuarius* was 33% in sediment from Castro Cove in San Francisco Bay, but sea urchin development in both porewater and solid-phase exposures was greater than 95% (Table 3-3). This site was highly contaminated by PAH compounds, which have been shown to be toxic to this amphipod species (DeWitt et al. 1989). Although the relative sensitivity of sea urchin embryos to PAHs has not been reported, the lack of concordance between the porewater and solid-phase results at this site may be due to variability in species sensitivity. It is also possible that the PAH concentrations in pore water were not elevated and that the difference in observed toxicity may be explained by variations in exposure (i.e., sediment ingestion by *E. estuarius*). The route of exposure to sea urchin embryos is membrane uptake of solubilized chemicals in pore water.

Sediment ingestion has been shown to be the primary route of exposure for a number of contaminants in a variety of species, and concordance between porewater and solid-phase tests will be influenced by this consideration. Lee et al. (2000) used variations in sediment concentrations of acid volatile sulfide (AVS) in laboratory experiments with metal-spiked sediments to show that sediment ingestion was the primary route of exposure to metals for a number of infaunal species, including the filter-feeding clam *Potamocorbula amurensis*, the deposit-feeding clam *Macoma balthica*, and the deposit-feeding polychaetes *Neanthes arenaceodentata* and *Heteromastis filiformis*. Similarly, Forbes et al. (1998) used a combination of feeding selectivity experiments and reaction-diffusion models to show that sediment ingestion was the primary route of exposure for the deposit-feeding polychaete *Capitella* sp. These studies demonstrate that, depending on the species used, contaminant uptake may occur through routes of exposure other than pore water, and this characteristic can influence concordance between test results.

The evaluation of spiked sediment using a variety of porewater and solid-phase tests illustrates the influence of variations in contaminant sensitivity on concordance. Mearns et al. (1995) used a variety of toxicity tests and variably aged sediments to investigate temporal variations in the toxicity of oiled sediments. In these experiments, toxicity of pore water from oiled sediments was evaluated using 3 tests: sea urchin fertilization, Microtox bioluminescence, and grass shrimp embryo develop-

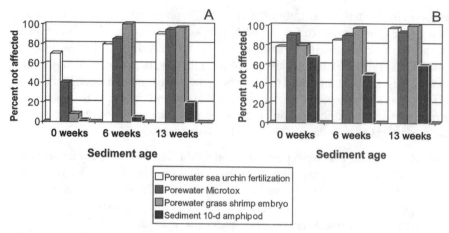

Figure 3-2 A) Toxicity of pore water and sediment composites from oiled sediment; **B)** Toxicity of pore water and sediment composites from unoiled sediment (from Mearns et al. 1995; reprinted with permission)

ment. Toxicity of solid-phase sediment samples was evaluated using the tube-building amphipod *Leptocheirus plumulosus* in 10-day acute exposures. The results were variable depending on the toxicity test used and the age of the sediments (Figure 3-2). Amphipod survival was low in all samples regardless of sediment aging. None of the porewater samples were toxic to sea urchins, and toxicity to grass shrimp embryos and Microtox varied with sample age (Figure 3-2). These results demonstrated that toxicity can vary between solid-phase and porewater tests because of differences in the relative sensitivities of the species to oil.

Ecological relevance of porewater toxicity tests

A primary goal of porewater toxicity testing is to elucidate the impact of sediment contamination on the benthos. One way to assess test performance is to investigate the relationship between the test results and characteristics of the in situ benthic community structure. A number of studies using solid-phase toxicity tests have demonstrated relationships between acute effects measured in laboratory tests and declines in corresponding benthic community metrics. For example, Swartz et al. (1979, 1994) found lower numbers of amphipods in benthic communities at stations where laboratory toxicity tests detected acute mortality. Anderson et al. (2001) used multivariate and univariate statistical analyses to show that mortality of amphipods in laboratory toxicity tests was negatively correlated with a number of benthic community metrics, including the number of infaunal species and the number of crustacean species measured in these samples.

Comparisons between porewater and solid-phase test results with benthic community characteristics were made in a number of studies, and in some cases, these comparisons demonstrated a better relationship with porewater tests than with solid-phase tests. In a recent survey of sediment toxicity in Puget Sound, Washington, USA, Long et al. (1999), using the amphipod *A. abdita* solid-phase test, found no toxicity at 100 stations, but inhibition of fertilization of the sea urchin *S. purpuratus* in porewater samples occurred at a number of stations. In this study, multivariate statistical analyses demonstrated negative correlations between the presence of echinoid species in benthic community samples and inhibition of fertilization. Anderson et al. (2001) found similar negative correlations between inhibition of mollusk (*H. rufescens*) embryo development in laboratory porewater exposures and the number of mollusk species and individuals in the Los Angeles Harbor benthos. Carr et al. (2000) also found that inhibition of sea urchin embryos (*A. punctulata*) was negatively correlated with benthic community metrics in samples where amphipod mortality was minimal. These comparisons demonstrate that porewater toxicity tests may provide ecologically meaningful information and that sometimes this information is unique to the porewater test species. This may be particularly useful in situations where contamination effects on solid-phase test species are not observed.

The correspondence between porewater toxicity and benthic community effects is not consistent, however. Porewater toxicity to sea urchin embryos was detected in 15 samples during a 1994 survey of sediment quality on the southern California shelf (Bay et al. 1998), but there was no correspondence with the occurrence of benthic community changes.

Interpretation of Concordance or Discordance

As outlined in the previous sections, multiple factors can affect the concordance between solid-phase and porewater toxicity tests. Consequently, there is a need for tools that help interpret the significance of discordance between solid-phase and porewater toxicity results. Effective tools for interpreting test data include the integration of either a TIE or an SQT approach into the testing framework. In many circumstances, however, the paucity of data does not allow the full application of such procedures, and a substantial amount of scientific judgment must be used to evaluate the results.

A conceptual framework for the interpretation of concordance and discordance from sediment and porewater toxicity tests is given in Figure 3-3. This flowchart was designed to provide guidance in interpreting test results by identifying factors most likely to influence test results. The framework classifies comparative test results on the basis of 3 key characteristics:

1) presence or absence of concordance between solid-phase and porewater tests,

2) presence or absence of toxicity, or

3) presence or absence of contamination.

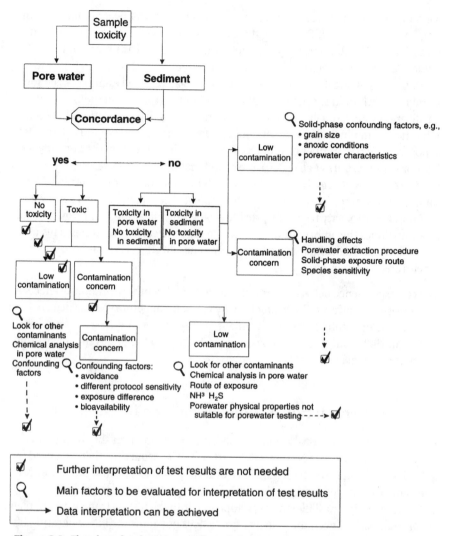

Figure 3-3 Flowchart for the interpretation of the concordance or discordance between porewater and solid-phase sediment toxicity tests

In some instances, an explanation of discordance may not be readily apparent. In such cases, further evaluation (e.g., chemical analyses) or retesting may be warranted. However, given that toxicity occurred in one of the matrices, the quality of the sediment is suspect and a conservative interpretation should be applied.

The conceptual framework (Figure 3-3) is not intended to present a complete strategy framework for assessment of sediment quality, and it does not take into consideration additional relevant ecological measures such as characteristics of the benthic community structure. However, it provides a simplified approach for interpreting the 2 most commonly used sediment toxicity tests. The application of the framework uses a simple set of assumptions regarding the study design: 2 types of sediment tests (porewater and solid-phase) and sediment contamination in 2 categories (high, above threshold effects, and low, similar to reference areas). The thresholds between the 2 levels of contamination must be determined by the user and may be based on sediment quality guidelines (SQGs) (Long and Morgan 1990; Long 1992; Long and MacDonald 1992; Long et al. 1995), depending on the purpose of the study.

Classification of the degree of test concordance is the first step in using the conceptual framework. When both porewater and solid-phase sediment tests show no toxicity, further interpretation from a toxicological perspective might not be required because the porewater toxicity test results confirm the results of the solid-phase test. When both toxicity tests indicate toxicity and this is consistent with the contaminant concentrations in the sediment, no further information is needed to interpret the results. If both tests indicate toxicity and the sediment contamination is apparently below all threshold effect levels, the presence of confounding factors or unmeasured contaminants should be investigated. Supplementary chemical analyses of the sediment and pore water are needed to determine whether confounding factors (Table 3-4) or unsuspected contaminants are influencing the test results.

Four categories of discordance are included in the conceptual framework (Figure 3-3), each of which may be the result of different factors. Most discordant data result from the greater sensitivity of the porewater test (Chapter 2). Caution should be used in attributing this discordance to anthropogenic chemicals just because sediment contamination is elevated (first category). This pattern could also be related to confounding factors that co-occur with contamination and to artifacts created by sample handling (e.g., change in contaminant bioavailability because of sample oxidation).

The presence of confounding factors and artifacts is even more likely in the second category of discordance when porewater tests indicate toxicity in the absence of high contamination. A thorough evaluation of the results is critical in this situation because the probability of a false positive (i.e., indication of a significant effect when none exists) is high. Measurement of contaminants and other constituents in pore water is recommended. There is the possibility that contaminants are present in the interstitial water at concentrations that exceed the threshold effect level. If porewater chemistry does not account for the test results, further investigation of causality (e.g., the presence of other toxicants in the sediment) is necessary. In some cases, porewater toxicity test species are more susceptible to natural variations in the physical or chemical properties of pore water than those in solid-phase tests. For example, in freshwater assays, it has been observed that typical concentrations of nonanthropogenic constituents such as ammonia, manganese, and alkalinity often

Table 3-4 Examples of physicochemical factors influencing concordance between porewater and solid-phase toxicity tests

Factors	Nature of effects[a]		Reference
Ammonia	Toxicity		Schubauer-Berigan and Ankley 1991; Fairey et al. 1998
Alkalinity	Toxicity (freshwater species)		Lasier et al. 1997
Salinity or conductivity	Bioavailability changes toxicity Toxicity		Flegal et al. 1994 Wiese et al. 1997
Hydrogen sulfide	Toxicity		Brouwer and Murphy 1995
	Metal bioavailability		Ankley et al. 1996; Wang and Chapman 1999
Dissolved organic carbon (DOC)	Bioavailability		Green et al. 1993; Van Ginneken et al. 1999
pH	Chemical speciation or bioavailability	X	Bay et al. 1993
Alkalinity	Ion toxicity	X	Lasier et al. 1997
Ferric or manganese oxides	Metal bioavailability	X	Bufflap and Allen 1995
Particle size distribution	Factor influencing toxicity	X	DeWitt et al. 1989
Biological factors	Predation or indigenous organisms	X	Reynoldson et al. 1994
Species sensitivity	Toxicity and exposure	X	Bay et al. Chapter 2
Avoidance behavior	Exposure modification	X	Oakden et al. 1984; Wang and Chapman 1999
Route of exposure (e.g., ingestion)	Exposure modification	X	Lee et al. 2000
Burrowing or tube-building behavior	Exposure modification	X	Krantzberg 1985
Exposure (test) durations	Differences affect toxicity concordance	X	Sibley et al. 1997

[a] Effects apply to both freshwater and marine tests, except where noted (X).

exceed tolerance levels of the test organisms, and concentrations of these compounds may be influenced by handling and porewater extraction procedures (Ankley et al. 1990; Lasier et al. 1997, 2000). A careful and methodological examination of the discordance in samples from apparently uncontaminated reference sites may be required to understand the causes of discordance.

The final 2 categories of discordance considered in the proposed framework arise when solid-phase sediment toxicity tests show effects that are not detected by porewater tests. The first step in this evaluation would again include a comparison of the toxicity results with the contaminant concentrations in both whole sediment and pore water. In the case of low contamination, the interpretation of discordance between the test results should focus on potential interferences associated with the solid-phase sediments. For example, the particle size distribution might not be appropriate for the selected test species, and/or anoxic conditions in the sediment or overlying water during the exposure could influence the survival and the health of organisms. These confounding factors are mainly related to the solid-phase testing procedure, and for this reason, they are implicated in causation when the porewater tests show no toxicity.

The last case considered in Figure 3-3 is one in which toxicity is detected only in the solid phase and there are contaminants present. This situation requires detailed examination of potential confounding factors related to the porewater procedure. The absence of porewater toxicity may be related to the fact that pore water is not a major pathway of exposure for certain highly hydrophobic contaminants, as pointed out by Forbes et al. (1998). Alternatively, the greater susceptibility of the test organisms in the solid-phase tests might be due to species-specific differences in contaminant sensitivity or reduced bioavailability of the contaminant in the pore water. This seems to be the case for sediment-ingesting organisms or for highly hydrophobic contaminants. Under such circumstances, the use of porewater bioassays might not be appropriate, and greater weight should be given to the results from the solid-phase sediment tests. Research is needed to better understand the (equilibrium) partitioning of contaminants between sediment and pore water. Possible causes for this type of discrepancy may also be related to artifacts caused by sediment sample storage, handling, and porewater extraction procedures (e.g., sorption of contaminants to particulates; see Chapters 5 and 7). Future research in this field should be directed toward a higher degree of methodological standardization.

Recommendations to Improve Interpretation of Porewater Toxicity Tests

Porewater testing provides additional insight into the assessment of sediment quality and has certain advantages over solid-phase exposures (short exposure duration, low volumes needed for testing, direct contact with test matrix, a primary route of exposure, etc.). For these reasons, pore water has been used as an alternative to or as an additional testing procedure to solid-phase exposures. But understanding and resolving differences in response between solid-phase and porewater tests may require additional evaluations. The added scrutiny needed to account for the discordance shown between these tests, however, may lead to a better understanding of the sediment quality and identification of factors that are environmentally important. The following recommendations are provided to enhance the assessment of sediment quality.

1) Whenever possible, test both solid-phase sediment and pore water in the assessment of sediment quality.

   Use of both testing procedures would ensure that potential discrepancies associated with contaminant exposure routes (water and sediments) would be accounted for in the exposures. Information from the 2 testing procedures would also contribute to the weight-of-evidence approach that would enhance the ability to ascertain sediment quality. Additionally, the level of discordance between the 2 tests may be useful in identifying causes of toxicity by employing a systematic evaluation of factors potentially responsible for the discrepancies between the tests.

2) Measure and identify potential confounding factors in pore water.

   Several potentially confounding factors inherent in some sediments and pore waters have been identified (Table 3-4). Steps to alleviate or minimize the influence of selected confounding factors have already been implemented in some sediment testing procedures through standardization. For example, daily renewal of overlying water in solid-phase tests with freshwater sediment alleviates problems associated with elevated concentrations of ammonia and other basic water quality parameters (ASTM 1995). The presence and quantification of these confounding factors will aid in the interpretation of test results and may account for discordance between or among tests. Information on confounding factors would also contribute directly to any TIE procedures that are undertaken.

   The use of alternative test methods or test species may enhance the resolution in testing of sediment pore water and reduce uncertainty in the interpretation of test results that are due to confounding factors. Selection of alternative test species may aid data interpretation by the use of endemic organisms that provide increased ecological relevance or through the selection of species that are tolerant to confounding factors. For example, zoospores of *Ulva fasciata* are tolerant of ammonia and may be a suitable alternative to more sensitive species for porewater testing (Hooten and Carr 1998).

3) When possible, use a battery of tests (including pore water) to evaluate sediment quality.

   The use of multiple porewater and sediment tests that include different species or life stages increases the probability that contamination will be detected. Data from a test battery allow the use of a weight-of-evidence approach to provide greater confidence in the results. The multispecies approach provides a better opportunity for at least one of the species or life stages to be more sensitive than the others to specific contaminants. For example, using different life stages and species enhanced the identification of contaminated sites in Tampa Bay, Florida and Corpus Christi Bay, Texas, USA (Carr, Long et al. 1996; Carr et al. 2000). A battery of tests maximizes information on exposure routes and sensitivities among species and life stages, and the information can be used to describe environmental quality.

4) Measure contaminants in pore water as well as in solid-phase sediments.

   Chemical analyses of pore water for basic chemistry and for contaminants are needed. Once identified, common confounding factors in pore water, such as ammonia, alkalinity, and manganese, can be accounted for and the interpretation of test results improved (Schubauer-Berigan and Ankley 1991; Lasier et al. 1997, 2000). Measurement of contaminants in the pore water also provides an estimate of the contaminants in equilibrium between the sediment and the pore water and of the bioavailable fraction to which the test organisms are exposed during the tests (Di Toro et al. 1991). Enhanced porewater chemistry analyses will also facilitate the incorporation of TIEs into more studies, which may provide information on the cause of toxicity (Mount and Anderson-Carnahan 1988; Ankley et al. 1995; Burgess et al. 1996).

Research Needs and Information Gaps

A lack of concordance between toxicity tests often occurs and has been shown to be caused by 4 conditions:

1) Differential sensitivity of test organisms to anthropogenic contaminants (Mearns et al. 1995; Anderson et al. 2001)

2) Differential sensitivity to naturally occurring toxic constituents (e.g., ammonia toxicity)

3) Altered exposure conditions (e.g., alkalinity and manganese toxicity) (Lasier et al. 1997, 2000)

4) Unidentified porewater or solid-phase sediment characteristics (Bay et al. 1998).

The occurrence of each of these situations influences the reliability and interpretation of porewater toxicity data. A conceptual framework was developed to assist in the identification of the causes of low concordance (Figure 3-3), but implementation of this approach is limited because of factors associated with porewater testing. The principal limitations are these:

1) The understanding of the types and significance of noncontaminant-related confounding factors is lacking.

2) The applicability of porewater tests to certain contamination or exposure scenarios is uncertain.

3) The sensitivity of porewater and solid-phase tests to important contaminants is often unknown.

4) Methods for measuring porewater chemistry are limited.

5) The ecological relevance of porewater toxicity tests is unknown.

Addressing the following research needs and information gaps should help resolve these limitations.

1) Identify naturally occurring porewater characteristics that influence toxicity test results.

While some porewater constituents such as ammonia and pH have been identified as important factors, other factors (unidentified) may also be important. Sufficient data and analytical tools are currently unavailable for the evaluation of the effects of nonanthropogenic factors. Toxicity at reference sites is often attributed to confounding factors, but there is a lack of consistency on how these data are interpreted and used to assess toxicity in field samples. Guidance is needed to insure consistency in these analyses.

At a minimum, the influence of known confounding factors (e.g., ammonia, sulfide, pH, alkalinity, manganese) should be determined for porewater test species and the concentrations of these factors measured in all toxicity tests. In addition, research should be conducted to identify additional confounding factors of significance. One approach to accomplishing this objective would be to identify the cause of porewater toxicity at uncontaminated reference sites. Laboratory studies should be conducted to determine the dose–response relationship of these confounding factors for porewater test species. The dose–response data should be used to identify thresholds for toxic effects or to develop other tools (e.g., models) to facilitate normalization or interpretation of test results influenced by these factors.

2) Clarify the role of porewater toxicity tests when the aqueous phase is not the primary route of exposure.

Pore water may not be an appropriate test medium when contaminants of concern are tightly sorbed onto particles (high octanol–water partition coefficient [K_{OW}] compounds) or the bioavailability of dissolved forms is reduced by porewater constituents such as dissolved organic carbon (DOC). In these cases, contaminant exposure may be due to sediment ingestion or bioaccumulation, and therefore, contaminant effects would be underestimated by short-term porewater toxicity tests.

Spiked sediment studies using high K_{OW} compounds should be conducted to determine whether porewater is an effective medium for conducting toxicity tests when the aqueous phase is not the primary route of exposure.

3) Determine the comparative sensitivity to contaminants for porewater and solid-phase test species.

The sensitivity of porewater test species to model toxicants is an important aid to selecting the appropriate test organism, interpreting the results of comparisons between solid-phase and porewater tests, and identifying the cause of toxicity. The types and quality of data available are highly variable among species. Few data are readily available for some species, and little comparable toxicity data for porewater and solid-phase test species are available for some contaminant groups (e.g., PAHs).

Efforts to eliminate these deficiencies should first be focused upon compiling existing data that are not accessible from conventional data sources. Following evaluation of these data, a select group of model toxicants should be identified for testing with selected species. These tests should be conducted using consistent methodologies (i.e., similar duration and formulation) so that the results will be comparable.

4) Improve analytical methods for chemical analysis of pore water.

Knowledge of porewater chemical composition is often essential to understanding the lack of concordance (Figure 3-3). Yet, contaminant analyses of porewater are rarely conducted because of cost considerations and limitations in the amount of sample available. Even when funds and adequate sample volumes are available, there are no guidelines for the types of analytes and the methods to be used for pore water.

The recommendations of geochemists and toxicologists regarding important porewater analytes (Chapter 5) should be incorporated into porewater toxicity test designs. In addition, cost-effective, microscale methods for the quantification of organic constituents at toxicologically significant concentrations should be developed.

5) Validate and improve porewater toxicity test methods.

Many existing porewater test methods are relatively sensitive, reliable, and cost-effective. Yet, confidence in the relevance of the results compared to solid-phase tests is often reduced for these reasons:

a) Standardized test methods for porewater evaluation often are not used, thereby reducing the comparability of results among studies.

b) The apparent sensitivity of some species to nonanthropogenic variations in porewater characteristics confounds the identification of contaminant-related toxicity.

c) The relevance of porewater toxicity test responses for estimating impacts to benthic communities or populations is uncertain.

Standard methods for the experimental design, sample collection, and sample handling of interstitial water should be applied whenever possible. Existing freshwater and marine test methods that measure sublethal effects, yet are relatively insensitive to natural variations in porewater characteristics, should be used when feasible. Finally, the ecological relevance of specific porewater toxicity test methods needs to be established through the analysis of existing SQT studies and additional in situ and laboratory comparative studies.

References

Adams WJ, Kimerle RA, Mosher RG. 1985. Aquatic safety assessment of chemicals sorbed to sediments. In: Cardwell RD, Purdy R, Bahner RC, editors. Aquatic toxicology and hazard assessment: Seventh Symposium. Philadelphia PA, USA: American Society for Testing and Materials. ASTM STP 854. p 429–453.

[ASTM] American Society for Testing and Materials. 1995. Standard test for measuring the toxicity of sediment-associated contaminants with freshwater invertebrates, Designation E 1706-95a. Volume 11.05, ASTM annual book of standards. Philadelphia PA, USA: ASTM. 1358 p.

[ASTM] American Society for Testing and Materials. 1999. Standard terminology relating to biological effects and environmental fate, Designation E 943-97b. Volume 11.05, ASTM annual book of standards. Philadelphia PA, USA: ASTM. 1586 p.

Anderson BS, Hunt JW, Phillips BM, Fairey R, Roberts CA, Oakden JM, Puckett HM, Stephenson M, Tjeerdema RS, Long ER, Wilson CJ, Lyons JM. 2001. Sediment quality in Los Angeles Harbor: A triad assessment. *Environ Toxicol Chem* 20:359–370.

Anderson HV, Kjolholt J, Poll C, Dahl SO, Stuer-Lauridsen F, Pedersen F, Bjornestad E. 1998. Environmental risk assessment of surface water and sediments in Copenhagen Harbour. *Water Sci Technol* 37:263–272.

Ankley GT, Kato K, Arthur JW. 1990. Identification of ammonia as an important sediment associated toxicant in the lower Fox River and Green Bay, Wisconsin. *Environ Toxicol Chem* 9:313–322.

Ankley GT, Schubauer-Berigan MK, Dierkes JR. 1991. Predicting the toxicity of bulk sediments to aquatic organisms with aqueous test fractions: Pore water vs. elutriate. *Environ Toxicol Chem* 10:1359–1366.

Ankley GT, Schubauer-Berigan MK, Dierkes JR. 1996. Application of toxicity identification evaluation techniques to pore water from Buffalo River sediments. *J Great Lakes Res* 22:534–544.

Ankley GT, Thomas NA, Di Toro DM, Hansen DJ, Mahony JD, Berry WJ, Swartz RC, Hoke RA, Garrison AW, Allen HE, Zarba CS. 1994. Assessing potential bioavailability of metals in sediments: A proposed approach. *Environ Manag* 18:331–337.

Bay S, Burgess R, Nacci D. 1993. Status and applications of echinoid (phylum Echinodermata) toxicity test methods. In: Landis WG, Hughes JS, Lewis MA, editors. Environmental toxicology and risk assessment. Philadelphia PA, USA: American Society for Testing and Materials. ASTM STP 1179. p 281–302.

Bay SM, Greenstein DJ, Jirik AW, Brown JS. 1998. Southern California Bight 1994 Pilot Project: VI. Sediment toxicity. Westminster CA, USA: Southern California Coastal Water Research Project. Technical Report 309.

Boese BL, Lee H, Sprecht DT, Randall RC, Winsor MH. 1990. Comparison of aqueous and solid-phase uptake for hexachlorobenzene in tellinid clam *Macoma nasuta* (Conrad): Mass balance approach. *Environ Toxicol Chem* 9:221–231.

Brouwer H, Murphy T. 1995. Volatile sulfides and their toxicity in freshwater sediments. *Environ Contam Toxicol* 14:203–208.

Bufflap SE, Allen HE. 1995. Sediment pore water collection methods for trace metal analysis: A review. *Water Res* 29:165–177.

Burgess RM, Ho KT, Morrison GE, Chapman G, Denton DL. 1996. Marine toxicity identification evaluation (TIE). Phase I guidance document. Washington DC, USA: U.S. Environmental Protection Agency. EPA/600/R-96/054. 70 p.

Burton Jr GA. 1998. Assessing aquatic ecosystems using pore waters and sediment chemistry. Ottawa ON, Canada: Natural Resources Canada (CANMET), Aquatic Effects Technology Evaluation Program. NRCan 97-0083.

Carr RS, Chapman DC, Howard CL, Biedenbach JM. 1996. Sediment Quality Triad assessment survey of the Galveston Bay, Texas system. *Ecotoxicology* 5:341–364.

Carr RS, Long ER, Windom HL, Chapman DC, Thursby G, Sloane GM, Wolfe DA. 1996. Sediment quality assessment studies of Tampa Bay, Florida. *Environ Toxicol Chem* 15:1218–1231.

Carr RS, Montagna PA, Biedenbach JM, Kalke R, Kennicutt MC, Hooten R, Cripe G. 2000. Impact of storm-water outfalls on sediment quality in Corpus Christi Bay, Texas. *Environ Toxicol Chem* 19:561–574.

Carr RS, Williams JW, Fragata CTB. 1989. Development and evaluation of a novel marine sediment porewater toxicity test with the polychaete *Dinophilus gyrociliatus*. *Environ Toxicol Chem* 8:533–543.

DeWitt TH, Swartz RC, Lamberson JO. 1989. Measuring the acute toxicity of estuarine sediments. *Environ Toxicol Chem* 8:1035–1048.

Di Toro DM, Hansen DJ, Berry WJ, Swartz RC, Cowan CE, Pavlou SP, Allen HE, Thomas NA, Paquin PR. 1991. Technical basis for establishing sediment quality criteria for nonionic organic chemicals using equilibrium partitioning. *Environ Toxicol Chem* 10:1541–1583.

Fairey R, Roberts C, Jacobi M, Lamerdin S, Clark R, Downing J, Long E, Hunt J, Anderson B, Newman J, Stephenson M, Wilson CJ. 1998. Assessment of sediment toxicity and chemical concentrations in the San Diego Bay region, California. *Environ Toxicol Chem* 17:1570–1581.

Flegal AR, Risebrough RW, Anderson B, Hunt J, Anderson S, Oliver J, Stephenson M, Pickard R. 1994. San Francisco estuary pilot regional monitoring program: Sediment studies. Oakland CA, USA: San Francisco Bay Regional Water Quality Control Board/State Water Resources Control Board.

Forbes TL, Giessing A, Hansen R, Kure LK. 1998. Relative role of pore water versus ingested sediment in bioavailability of organic contaminants in marine sediments. *Environ Toxicol Chem* 17:2453–2462.

Giesy JP, Rosiu CJ, Graney RL, Henry MG. 1990. Benthic invertebrate bioassays with toxic sediment and pore water. *Environ Toxicol Chem* 9:233–248.

Green AS, Chandler GT, Blood ER. 1993. Aqueous-, pore-water-, and sediment-phase cadmium toxicity relationships for meiobenthic copepod. *Environ Toxicol Chem* 12:1497–1506.

Hoke RA, Kosian PA, Ankley GT, Cotter AM, Vandermeiden FM, Phipps GL, Durhon EJ. 1995. Check studies with *Hyalella azteca* and *Chironomus tentans* in support of the development of a sediment quality criterion for dieldrin. *Environ Toxicol Chem* 4:435–443.

Hooten RL, Carr RS. 1998. Development and application of a marine sediment porewater toxicity test using *Ulva fasciata* and *U. lactuca* zoospores. *Environ Toxicol Chem* 17:932–940.

Hunt JW, Anderson BS, Phillips BM, Newman J, Tjeerdema RS, Stephenson M, Pucket HM, Fairey R, Smith RW, Taberski K. 1998. Evaluation and use of sediment reference sites and toxicity tests in San Francisco Bay. Sacramento CA, USA: State Water Resources Control Board. Bay Protection and Toxic Cleanup Program final technical report.

Hunt JW, Anderson BS, Phillips BM, Newman J, Tjeerdema RS, Taberski K, Wilson CJ, Stephenson M, Puckett HM, Fairey R, Oakden J. 1998. Sediment quality and biological effects in San Francisco Bay. Sacramento CA, USA: State Water Resources Control Board. Bay Protection and Toxic Cleanup Program final technical report.

Ingersoll CG. 1995. Sediment tests. In: Rand GM, editor. Fundamentals of aquatic toxicology. 2nd ed. Washington DC, USA: Taylor & Francis. p. 231–255.

Kendall MG. 1970. Rank correlation method. 3rd ed. London, UK: Charles Griffin.

Knezovich JF, Harrison FL. 1987. A new method for determining the concentration of volatile organic compounds in sediment interstitial water. *Bull Environ Contam Toxicol* 38:937–940.

Kranzberg G. 1985. The influence of bioturbation on physical, chemical and biological parameters in aquatic environments: A review. *Environ Pollut* 39:99–122.

Lasier PJ, Winger PV, Bogenreider KJ. 2000. Toxicity of manganese to *Ceriodaphnia dubia* and *Hyalella azteca*. *Arch Environ Contam Toxicol* 38:298–304.

Lasier PJ, Winger PV, Reinert RE. 1997. Toxicity of alkalinity to *Hyalella azteca*. *Bull Environ Contam Toxicol* 59:807–814.

Lee BG, Griscom SB, Lee JS, Choi HJ, Koh CH, Luoma SN, Fisher NS. 2000. Influence of dietary uptake and acid-volatile sulfide on bioavailability of metals to sediment-dwelling organisms. *Science* 287:282–284.

Long ER. 1992. Ranges in chemical concentrations in sediments associated with adverse biological effects. *Mar Pollut Bull* 24:38–45.

Long, ER, Hameedi J, Robertson A, Dutch M, Aasen S, Ricci C, Welch K, Kammin W, Carr RS, Johnson T, Biedenbach J, Scott KJ, Mueller C, Anderson J. 1999. Sediment quality in Puget Sound, Year 1-Northern Puget Sound. Seattle WA, USA: National Oceanic and Atmospheric Administration. NOS NCCOS CCMA Technical Report No. 139.

Long ER, MacDonald DD. 1992. National status and trends program approach. In: Sediment classification methods compendium. Washington DC, USA: U.S. Environmental Protection Agency, Office of Water. p 14–18.

Long ER, MacDonald DD, Smith SL, Calder FD. 1995. Incidence of adverse biological effects within ranges of chemical concentrations in marine and estuarine sediments. *Environ Manag* 19:81–97.

Long ER, Morgan LG. 1990. The potential for biological effects of sediment-sorbed contaminants tested in the National Status and Trends Program. Seattle WA, USA: U.S. National Oceanic and Atmospheric Administration. NOAA Tech Memo NOS OMA 52.

Mearns A, Doe K, Fisher W, Hoff R, Lee K, Siron R, Mueller C, Venosa A. 1995. Toxicity trends during an oil spill bioremediation experiment on a sandy shoreline in Delaware, USA. Eighteenth Arctic and Marine Oilspill Program Technical Seminar, Proceedings. Alberta, Canada: Environment Canada, Environmental Protection Service.

Mount DI, Anderson-Carnahan L. 1988. Methods for aquatic toxicity identification evaluations: Phase I toxicity characterization procedures. Duluth MN, USA: U.S. Environmental Protection Agency. EPA 600/3-88-034.

Oakden JM, Oliver JS, Flegal AR. 1984. Behavioral responses of phoxocephalid amphipods to organic enrichment and trace metals in sediment. *Mar Ecol Prog Ser* 14:253–257.

Reynoldson TB, Day KE, Clark C, Milani D. 1994. Effects of indigenous animals on chronic endpoints in freshwater sediment toxicity tests. *Environ Toxicol Chem* 13:937–977.

Schubauer-Berigan MK, Ankley GT. 1991. The contribution of ammonia, metals, and nonpolar organic compounds to the toxicity of sediment interstitial water from an Illinois River tributary. *Environ Toxicol Chem* 10:925–940.

Sibley PK, Benoit DA, Ankley GT. 1997. The significance of growth in *Chironomus tentans* sediment toxicity tests: Relationship to reproduction and demographic endpoints. *Environ Toxicol Chem* 16:336–345.

Swartz RC, Cole FA, Lamberson JO, Ferraro SP, Schults DW, Deben WA, Lee II H, Ozretich RJ. 1994. Sediment toxicity, contamination and amphipod abundance at a DDT- and dieldrin-contaminated site in San Francisco Bay. *Environ Toxicol Chem* 13:949–962.

Swartz, RC, DeBen WA, Cole FA. 1979. A bioassay for the toxicity of sediment to marine macrobenthos. *J Water Pollut Contr Fed* 5:944–950.

Swartz RC, DeBen WA, Jones JKP, Lamberson JO, Cole FA. 1985. Phoxocephalid amphipod bioassay for marine sediment toxicity. In: Cardwell RD, Purdy R, Bahner RC, editors. Aquatic Toxicology and Hazard Assessment, Seventh Symposium. Philadelphia PA, USA: American Society for Testing and Materials. STP 854. p 284–307.

Van Ginneken L, Chowdhury MJ, Blust R. 1999. Bioavailability of cadmium and zinc to the common carp, *Cyprinus carpio*, in complexing environments: A test for the validity of the free ion activity model. *Environ Toxicol Chem* 18:2295–2304.

Wang F, Chapman PM. 1999. Biological implications of sulfide in sediment: A review focusing on sediment toxicity. *Environ Toxicol Chem* 18:2526–2532.

Whiteman FW, Ankley GT, Kahl MD, Rau DM, Bacer MD. 1996. Evaluation of interstitial water as a route of exposure for ammonia in sediment tests with benthic macroinvertebrates. *Environ Toxicol Chem* 15:794–801.

Wiese SB, MacLeod CL, Lester JN. 1997. Partitioning of metals between dissolved and particulate phases in the salt marshes of Essex and North Norfolk (UK). *Environ Technol* 18:399–407.

Winger PV, Lasier PJ. 1995. Sediment toxicity in Savannah Harbor. *Arch Environ Contam Toxicol* 28:357–365.

Winger PV, Lasier PJ. 1998. Toxicity of sediment collected upriver and down river of major cities along the lower Mississippi River. *Arch Environ Contam Toxicol* 35:213–217.

Winger PV, Lasier PJ, Bogenreider KJ. 2000. Bioassessment of Hollis Creek, Oktibbeha County. Vicksburg MS, USA: Preliminary report submitted to U.S. Fish and Wildlife Service, Ecological Services Office.

Winger PV, Lasier PJ, White DH, Seginak JT. 2000. Effects of contaminants in dredge material from the lower Savannah River. *Arch Environ Contam Toxicol* 38:128–136.

CHAPTER 4

Sediment and Porewater Chemistry

Bruce Williamson, Robert M Burgess

This chapter reviews sediment chemistry, its effect on porewater chemistry and how this chemistry changes from place to place. We focus on the overall chemical environment of the sediments, for which a great deal is known from studies on sediment diagenesis and from which some predictions can be made on the potential chemical changes that occur during porewater extraction. These changes are examined in more detail in Chapter 5. The fundamental questions we address here are these:

1) What are the dominant chemical characteristics of sediment pore water, and how do these vary from place to place?

2) How do these characteristics affect contaminant bioavailability in pore water?

3) What potential changes in these characteristics need to be considered when pore water is extracted?

Sediment Diagenesis and Redox Chemistry

Surface sediments, after being deposited, go through a series of chemical and physical changes termed "early diagenesis." By far the most important factor in early diagenesis is the biological oxidation of metabolizable forms of particulate organic carbon. Particulate organic matter is input from sedimentation of autochthonous and allocthonous material, filter feeding, and defecation, and it is broken down to a wide range of organic molecules and oxidized carbon (e.g., CO_2). In the process, dissolved organic matter (DOM) or carbon (DOC) is formed (see "Porewater Dissolved Organic Carbon").

Biological oxidation, together with the resulting anaerobic conditions, produces large changes to the form of iron, manganese, sulfur, and DOC, constituents that play key roles in "binding" contaminants in sediment and releasing them to the pore water. Biological oxidation of organic carbon occurs in the sequence O_2 > nitrate > Mn(IV) oxide > Fe(III) oxide > sulfate > CO_2 (Froelich et al. 1979). Figure 4-1 illustrates conceptual views of sediment chemistry in terms of this energetic sequence.

the diagenetic view

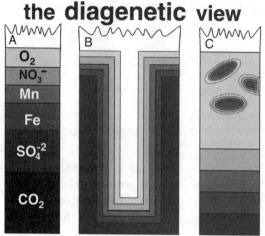

Figure 4-1 Conceptual views of sediment chemistry in terms of the energetic sequence O_2 > nitrate > Mn(IV) oxide > Fe(III) oxide > sulfate > CO_2 (from Aller 1982, reprinted with permission from Kluwer Academic/Plenum Publishers)

When organic matter (OM) decays in sediment, oxygen is used preferentially. When oxygen is depleted by aerobic metabolism in the surface of sediments, then anaerobic metabolism prevails and proceeds in order of most favorable energetics: The microbes use NO_3^-, MnO_x, FeOOH, SO_4^{2-}, and CO_2 as the terminal electron acceptor, in that preferred order. Aller (1982) found that this "linear" view is complicated by burrows, so that this diagenetic sequence is appropriately rearranged around the burrow (Figure 4-1B) and by microzones of decaying OM (Figure 4-1C), for example, fecal pellets. "The concept of biochemical successions, together with the view of sedimentary deposits as laterally homogeneous and essentially one-dimensional bodies, has led to the dogma that decomposition reactions are vertically stratified below the sediment–water interface" (Aller 1982). As well as the diagenetic sequence, many oxidation–reduction reactions occur abiotically; for example, Fe^{3+} and MnO_2 can oxidize (and be reduced by) HS^-. Further, the introduction of oxygen through burrows and burrowing results in oxygen reacting with reduced substances such as Fe^{2+}, Mn^{2+}, and HS^-. Consequently, much of the oxygen demand in sediment is due to nonrespiratory reactions rather than directly to OM breakdown (Jorgensen et al. 1990; Canfield, Jorgensen et al. 1993).

Actual sediment chemistry profiles can be more complex than the simple diagrams presented in Figure 4-1 because of the complexity of bioturbation (Aller 1978; Rhoads and Boyer 1982; Robbins 1982; Morrisey et al. 1999). Steep chemical gradients are found at surfaces (Revsbech et al. 1980; Davison et al. 1991, 1997) and throughout the sediment profile (Fones et al. 1998; Luther et al. 1998; Shuttleworth et al. 1999; Williamson et al. 1999). An example of sharp gradients in sediment chemistry is illustrated in Figure 4-2, which is a simplified diagram of the vertical profile photographed in an intertidal estuarine muddy sediment.

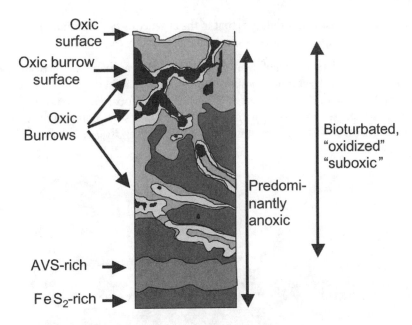

Figure 4-2 A stylized picture of the complex redox environment of an intertidal estuarine muddy sediment (AVS = acid volatile sulfide; based on data from Williamson et al. 1999)

While the diagenetic sequence is not linear, there is the general observation of a surface oxic layer underlain by anoxic sediment. Figure 4-2 illustrates many of the features of the redox structure of sediments. It shows the thin (2 to 5 mm) oxic surface layer, which extends down burrow walls. The surface layer can be viewed as being folded into sediment by burrows, thus creating a very complex 3D boundary between the oxic and anoxic sediments. The burrow wall chemistry does differ from the surface sediment chemistry, however (e.g., the oxygen profile may be thinner or thicker, and burrow walls often accumulate orange FeOOH [Williamson et al. 1999]). In addition, resident animals sometimes construct burrows from mucus-cemented grains. The areas between burrows contain no measurable oxygen and moderate amounts of acid volatile sulfide (AVS), which is dominantly FeS in uncontaminated sediments. High AVS concentrations accumulate below the bioturbated layer, and below this, pyrites accumulate because FeS is converted to FeS_2. Sometimes the sediment below the bioturbated zone is termed the "anoxic zone" because of the sharp boundary change to much darker sediments and sometimes because of the smell of sulfide. However, the interval between the thin oxic surface layer and the dark AVS zone is also largely anoxic. It is often termed the "suboxic" or "oxidized" zone because bioturbation results in the introduction of oxygen and the oxidation of reduced sediments, so the sediments are lighter ("oxidized"). Therefore, the redox condition in sediments is not a simple 2-layer oxic–anoxic system or a 3-layer system, but reflects the complexity described above for early diagenesis. The corollary is that in the absence of bioturbation, the sediment system usually reverts to a simple, 2-layer redox condition.

Although our understanding of sediment chemistry is reasonably sophisticated, we still often view it or treat it in very simplistic terms, such as in this simple 2D view of water and sediment. This simplistic view becomes very apparent, or tends to prevail, when we sample sediment — whether for chemical analysis or porewater toxicity testing — because we tend to treat it as a homogeneous entity, as a "black box" (Figure 4-3).

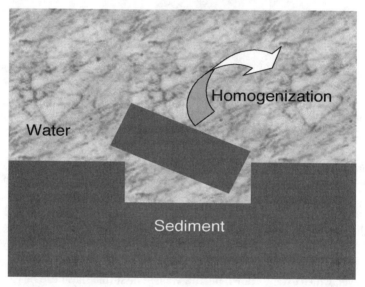

Figure 4-3 Treatment and sampling of sediment as a homogeneous matrix ("black box")

Factors that Control Sediment Chemistry

A great deal is known about the general chemistry of early diagenesis; a lot is known about the different major chemical cycles for organic carbon, sulfur, iron, and manganese (e.g., Berner and Westrich 1985; Burdige 1991; Canfield, Jorgensen et al. 1993; Canfield, Thamdrup, Hansen 1993; Davison 1993). However, the unknown, as far as general chemistry is concerned is "where," rather than "how" or "what." In other words, the complexity of interactions between bioturbation, other physical processes, mineralogy, sediment texture, and OM input can create a complex 3D profile that may vary markedly from place to place. While a lot is known about the major chemical processes in early diagenesis, there is a great deal yet to be learned about the chemistry of toxicants (Luther 1995; Santschi et al. 1997) and minor components of sediment diagenesis that are critical to the fate of contaminants, such as the complex nature of DOM (Santschi et al. 1997) (see "Porewater Dissolved Organic Carbon").

Some generalizations can be made about the variability of sediment chemistry in terms of the 4 major factors that largely determine the "where" of sediment chemistry: bioturbation, OM input, sediment texture, and overlying water hydraulics.

Bioturbation is highly dependent on the type and density of bioturbating organisms (McCall and Tevesz 1982; Rhoads and Boyer 1982). Figure 4-2 illustrates the strong effect that crabs have on sediment chemistry, through excavating to form burrows and by their irrigation of their burrows. Figure 4-4 illustrates the effects of another important bioturbator, the lug worm *Arenicola marina*, which also transports sediment to the surface by feeding at depth and defecating on the surface. Its feeding behavior also causes sediment to subside. It also irrigates its burrow and feeding pocket by pumping surface water into the feeding pocket, thus creating an oxic zone that sometimes extends up to the surface (Carney 1981; Meyers et al. 1987; Banta et al. 1999). Bioturbation can thus create a high degree of heterogeneity in redox conditions, both vertically and horizontally. This is especially important when the sampling scale is small compared to the scale of bioturbation features.

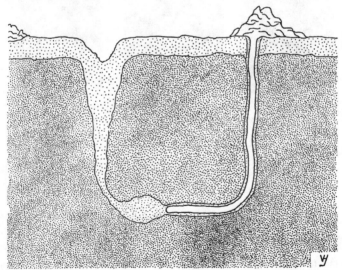

Figure 4-4 Transport of sediment and oxygenation of burrow and feeding pocket by lug worm *A. marina* (reprinted from Carney 1981, copyright 1981, with permission from Elsevier Science)

Organic matter inputs are the fuel for sediment chemistry. There is a basic pattern for the effects with organic enrichment (Figure 4-5), which follows the energetic pattern of aerobic and anaerobic metabolism (Figure 4-1). At low OM inputs, after dissolved oxygen (DO) is used up, NO_3^- and manganese oxides (a mixture of hydrous oxides of +3 and +4 oxidation states, represented by "MnO_x") are used in anaerobic metabolism. Little hydrous iron oxide (FeOOH) or SO_4^{2-} is used (or, the corollary, little dissolved Fe^{2+} or HS^- is formed). As the inputs of OM increase, the higher anaerobic metabolism demands shift to FeOOH, and Fe^{2+} concentrations increase in pore water. Any HS^- produced by sulfate reduction tends to be oxidized by FeOOH or precipitated as FeS. Thus, the relatively high concentrations of iron oxide present in sediment buffer against an increase in dissolved sulfide (Canfield 1989). As OM input continues to increase, the supply of FeOOH and MnO_x is essentially used up and sulfate reduction predominates, so a high concentration of dissolved sulfide can be formed.

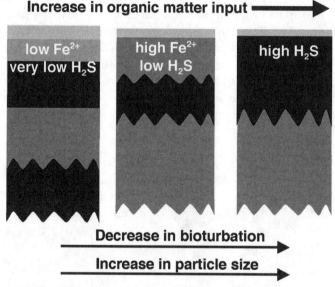

Figure 4-5 Basic pattern of effects with organic enrichment that follow the energetic pattern of aerobic and anaerobic metabolism (see Figure 4-1)

The effect of enrichment by OM is also very much dependent on bioturbation and sediment texture. Bioturbation introduces oxygen at depth, which oxidizes a large proportion of reduced species such as Fe^{2+} or HS^-. Texture determines mineralogy and FeOOH and MnO_x content. For sandy sediments, which usually contain much lower concentrations of FeOOH and MnO_x, sulfate reduction becomes significant at moderate levels of organic enrichment (Williamson et al. 1994).

Other major factors, such as flow and texture, are very closely linked. Both depend on the hydrodynamic energy of the water body. For example, strong currents (flow, waves) are usually associated with coarse sediments (sands, gravels, cobbles). Current flows (e.g., tidal currents) can induce large pressure changes over topographical features on the sediment bed. This can lead to oxygenated surface water being pumped down and reduced pore water being advected up and out of the sediment. Thus sands (and coarser sediments) have enhanced diffusion coefficients, whereas in mud, penetration is controlled by molecular diffusion. For example, oxygen penetrates flat muddy sand to 3.6 mm at 10 cm s^{-1} (Lohse et al. 1996). When the sand is mounded, a current of 10 cm s^{-1} advects O_2 into the mound to 20 mm. Furthermore, the pressure changes produce adjacent upwelling of reduced pore water because of the low-pressure field above the mound (a suction effect) (Ziebis et al. 1996; Huettel et al. 1998). More dramatic DO penetration can be found in fast-flowing rivers (Rutherford et al. 1993), and in fact, the subsurface (hyporheic) flow in rivers is a major process supplying O_2 and nutrients to an important ecological niche within the upland riverine ecosystem (Boulton et al. 1998; Figure 4-6). In slow-flowing lowland rivers, which usually contain fine sediments, porewater chemistry probably reflects sediment chemistry more closely than it reflects overlying water composition.

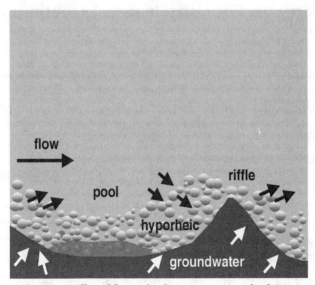

Figure 4-6 Effect of flow and sediment texture in upland rivers

We summarize the foregoing and show how this links to porewater chemistry in Figure 4-7.

Aerobic and anaerobic metabolism consume O_2, NO_3^-, FeOOH, MnO_x, SO_4^{2-}, and particulate organic matter and produce CO_2, Mn^{2+}, Fe^{2+}, HS^-, NH_3, and DOM. We have listed only a few of the products (there are many more), but these are the dominant ones implicated in toxicity, either directly (e.g., NH_3 toxicity to test organisms) or indirectly (e.g., CO_2 controlling the chemical environment — in this case, pH — which indirectly affects NH_3 toxicity).

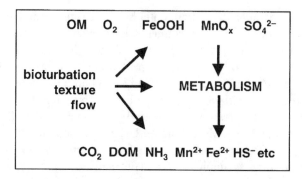

Figure 4-7 Summary of consumption and production of important sediment chemicals during aerobic and anaerobic metabolism

Porewater Chemistry: Inorganic Constituents

Oxygen and redox conditions in pore water

Dissolved oxygen levels in sediments define the oxic layer. It is usually a thin layer extending down any burrow walls and includes any burrow water. The oxic layer would be the best sediment environment to sample pore water for toxicity testing because then its redox environment would not need to be changed by oxygenating the water for testing. However, the oxic layer is usually very thin. Figure 4-8 shows some oxygen profiles in crab-bioturbated sediment under static conditions. Concentrations decrease to 0 within 2 to 3 mm, and this is probably true for most coastal and lake sediments (Meyers et al. 1987; Carlton and King 1990; Lohse et al. 1996; Burke 1999), even when the sediments are permeable sands. For lowland rivers, or rivers with large inputs of fine sediments, we suspect that the oxic layer would be of similar thickness. In upland rivers with coarse substrates, river water and oxygen are advected deep within the sediments.

The thickness of the oxic layer can vary with pressure fluctuations brought about by currents flowing over bed irregularities. It can also vary with temperature; at low temperatures, the metabolic reactions slow, and O_2 penetrates further. A reduction in temperature from 15 °C to 4 °C results in the oxic layer deepening from 2 mm to 10 mm in the estuarine muds depicted in Figure 4-8. Such effects may be important in sampling, (e.g., marked changes in the DO profile occurred after retrieving a core and allowing it to warm from 6 °C to 18 °C over 1 hour; Carlton and King 1990). When OM inputs are low, as in deeper oceanic or lake sediments, the oxic layer can be thicker, from 10 to 100 mm (Carlton and King 1990; Gehlen et al. 1997). Photosynthesis by benthic microalgae can raise O_2 concentrations in surface sediments above those in overlying waters and can deepen oxygen penetration into lake and coastal sediments (Carlton and King 1990; Burke 1999), especially below algal mats (Revsbech et al. 1983).

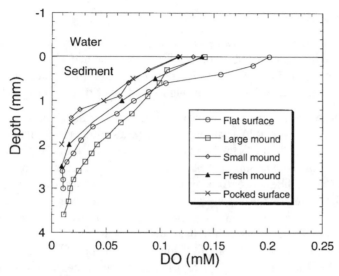

Figure 4-8 Oxygen profiles in crab-bioturbated sediment under static conditions. Mounds are crab excavation deposits; "pocked" surfaces are formed by bioturbation caused by intensive surface feeding

One of the implications of this is that samples for toxicity testing may contain a mixture of oxic and anoxic waters. We have calculated some approximate proportions from published work on burrow morphology in a range of crab-bioturbated sediment (Table 4-1). For the highest bioturbated mud, and assuming a 5 cm sample depth, there are significant proportions of oxic water (34%). For the least bioturbation, at a sandy site (less attractive to crabs), there is about 10% oxic and 90% anoxic water. Therefore, samples from strongly bioturbated sediments will probably contain a mixture of oxic and anoxic waters. Weakly bioturbated or weakly burrowed sediment samples probably yield mostly anoxic water, with only a few percent oxic water. In very heavily contaminated sites where the fauna is sparse, the sediments may also be anoxic.

Table 4-1 Proportion of oxic and anoxic waters (%) in the surface 5 cm of crab-bioturbated estuarine sediments[a]

Site	Sediment type	Oxic pore water (%)	Oxic burrow water (%)	Anoxic pore water (%)
1	Mud	5	29	66
2	Mud	5	21	74
3	Muddy sand	6	7	87
4	Sand	6	6	88

[a] Calculated from Morrisey et al. 1999.

Dissolved iron

Dissolved iron (mainly Fe^{2+}) is an example of a porewater characteristic that is reasonably well known, and some of its effects can be defined or calculated. Fe^{2+} is formed by the reduction of FeOOH, either directly as the electron acceptor in anaerobic metabolism or indirectly through oxidation of free sulfide by FeOOH.

In sedimentary environments with low OM inputs and/or high bioturbation, we observe low Fe^{2+} (Figure 4-5). Examples are deep sea sediments (Hammond et al. 1999; Skrabal et al. 2000). As the input of OM is increased, more FeOOH is reduced to Fe^{2+}. When the FeOOH supply is exhausted, the bacterial degradation switches to sulfate, and free sulfide concentrations increase and Fe^{2+} concentrations decrease (Figure 4-5). Therefore, the dissolved iron content can vary from low to high to low again as OM inputs increase, and pore water can contain a wide range of Fe^{2+} concentrations. These processes lead to a wide variability in Fe^{2+} concentrations from place to place and also to variation down a sediment profile, so the dissolved iron content of the extracted pore water will depend on the depth of sediment sampled. For example, in a profile measured in estuarine mud (Figure 4-9), concentrations range from 0 near the surface to over 100 µM at depth. Concentrations as high as 1000 µM are not uncommon. Texture also controls dissolved iron content because of its effect on O_2 penetration. For example, Williamson et al. (1994) found sandy sediments had lower dissolved iron and manganese concentrations than muddy sediments in the same area, despite similar OM inputs. Gel samplers and microelectrodes have allowed greater differentiation of Fe^{2+} (Mn^{2+} and S^{2-}) profiles in sediments, and reveal a great deal of vertical and horizontal complexity in their concentrations (Luther et al. 1998, 1999; Shuttleworth et al. 1999; Davison et al. 2000).

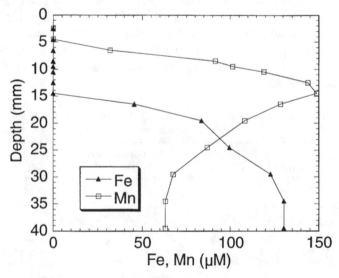

Figure 4-9 Dissolved iron and manganese profiles in estuarine mud

Porewater dissolved Fe^{2+} is rapidly oxidized by any oxygen introduced by natural processes (e.g., bioturbation) or during sampling and testing (Figure 4-10). The rate of oxidation has a half-life on the order of a few minutes (Davison 1993). Oxidation produces colloids of FeOOH, which can be very small in clean water (typically a few nm) and larger in water that contains other particles (Davison 1993). Rate of coagulation of these colloids is quite slow; it can take many hours, depending on the characteristics of the water. Therefore, even with high concentrations of Fe^{2+}, no yellow or orange FeOOH precipitates may be observed until many hours after oxidation. Typically, pore water that contains significant quantities of dissolved Fe^{2+} gradually turns straw-yellow on aeration, and this color intensifies to a brown or orange. One implication of Fe^{2+} precipitation may be its potential adverse effect to water column–based animals (e.g., from colloids clogging gills). Any such effect probably does not occur with sediment in-situ animals because presumably they have adapted to high dissolved Fe^{2+} concentrations.

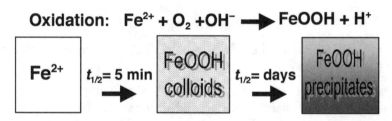

Oxidation: $Fe^{2+} + O_2 + OH^- \longrightarrow FeOOH + H^+$

| Fe^{2+} | $t_{1/2}$ = 5 min $\rightarrow$ | FeOOH colloids | $t_{1/2}$ = days $\rightarrow$ | FeOOH precipitates |

Figure 4-10 Summary of processes undergone by Fe^{2+} on aerating pore water

The implications of this oxidation and precipitation of Fe^{2+} can be examined from the point of view of the surface area formed or the adsorptive properties of the freshly precipitated FeOOH. Freshly precipitated FeOOH has a relatively high surface area per unit weight, as high as 600 m^2 per gram (Dzombak and Morel 1990). Figure 4-11A plots the theoretical surface areas for different iron concentrations in pore water. At low concentrations of iron, the surface area is small. At higher concentrations, the areas can be quite large, for example, 3 m^2 for a precipitate from 50 mg Fe/L (900 μM) in 100 ml of pore water. The adsorption properties of amorphous FeOOH for heavy metals and other substances are well known from early laboratory studies (Benjamin and Leckie 1978). A number of models have been derived to interpret the adsorption of heavy metals at the oxide surface; the most widely accepted one, the surface complexation model, is based on an extension of solution coordination chemistry to the description of the reaction of ions with reactive surface sites present on the solid–solution interface (Stumm et al. 1970). The adsorption model for FeOOH has been well developed, and computation methods are readily available for a wide range of different ions (Dzombak and Morel 1990).

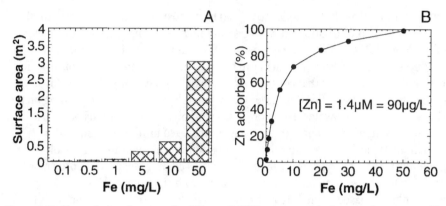

Figure 4-11 Theoretical surface areas for different FeOOH concentrations in aerated pore water (A). Proportion (%) of Zn^{2+} adsorbed by different FeOOH concentrations in aerated pore water (B).

$$\equiv Fe - OH + M^{z+} + mH_2O \rightarrow \; \equiv Fe - OM\,(OH)_m^{z-m-1} + (m+1)H^+,$$

where $\equiv Fe - OH$ is a surface site and $\equiv Fe - OM\,(OH)_m^{z-m-1}$ is a metal ion adsorbed to that site.

We have scoped the importance of adsorption by precipitated FeOOH in test solutions by considering a simple example with dissolved Zn^{2+} (Figure 4-11B). For the Zn^{2+} concentration, we used the U.S. Environmental Protection Agency (USEPA 1999) acute criteria for seawater, which is 90 µg/L (1.4 µM). We have used the surface complexation model of Dzombak and Morel (1990) in the chemical equilibrium program MINTEQA2 (USEPA 1991) to calculate the proportion of Zn^{2+} adsorbed. A high proportion of dissolved Zn^{2+} is adsorbed at the higher FeOOH concentrations, where Fe is >10 mg/L. Therefore, iron precipitation may represent a mechanism for removing toxic metals if pore water is oxidized. The corollary is that low concentrations of dissolved Fe^{2+} will not lead to the adsorption of significant quantities of toxic metals. Other heavy metals (e.g., Cd^{2+}, Cu^{2+}, Ni^{2+}, Pb^{2+}) are also strongly adsorbed by FeOOH and will undergo similar adsorption. Less well understood is the adsorption of DOM and other polar organics onto FeOOH (Davis 1982; Gu et al. 1994; Teermann and Jekel 1999). Hydrophobic organics may adsorb as well (Turner et al. 1999), especially through DOM intermediaries (Santschi et al. 1997). While it is well known that DOC affects the bioavailability of hydrophobic contaminants, it is not clear whether FeOOH alone will affect their bioavailability, and this requires further work.

Dissolved manganese

Dissolved manganese is formed through the reduction of MnO_x during anaerobic metabolism and is more commonly found in sediments than Fe^{2+} because its reduction is energetically more favorable than FeOOH reduction. However, MnO_x

concentrations are usually lower than FeOOH concentrations in sediments. Hence, the concentrations of Mn^{2+} and Fe^{2+} are often of similar orders of magnitude, although Mn^{2+} concentrations are often higher near the surface (Figure 4-9).

Like Fe^{2+}, Mn^{2+} is oxidized under aerobic conditions. The formation of high concentrations of MnO_x in oxic pore waters would affect contaminant bioavailability through surface adsorption reactions similar to those that occur on FeOOH. However, whereas FeOOH formation is virtually instantaneous, Mn^{2+} oxidizes slowly to form mixed oxides MnO_x (Stumm and Morgan 1981; Davison 1993). The rate of this reaction varies widely — half-lives for Mn^{2+} range from days to years — and depends on many factors, such as pH, water chemistry, and catalysis via surface adsorption onto MnO_x (self-catalysis) or FeOOH. At low FeOOH and MnO_x concentrations, the oxidation reaction is mediated by microbial activity, and half-lives can be considerably shortened compared to abiotic reactions (1 to 100 days in lakes) (Davison 1993). One consequence of this is that Mn^{2+} concentrations can be high in the pore water of surface sediments, despite the presence of oxygen and bioirrigation, whereas Fe^{2+} is rapidly oxidized under these conditions. In most natural waters, Mn^{2+} oxidation is unimportant over time scales of 1 to 4 days (the length of porewater toxicity tests). However, aerated pore water differs from many natural waters because it can contain high concentrations of Mn^{2+} and FeOOH. Therefore, we estimated half-lives for dissolved Mn^{2+} oxidation from known rate equations (Davison 1993) to determine whether this reaction could produce elevated MnO_x concentrations in a relatively short time. If Fe^{2+} concentration is high, FeOOH catalysis is likely because this oxide is formed immediately on aeration. Self-catalysis from particulate MnO_x is small because it is formed slowly. We varied pH and FeOOH concentrations to provide the estimates in Table 4-2. Microbial-mediated oxidation was also assumed to be insignificant in these calculations.

Table 4-2 Half-life for Mn^{2+} oxidation (days) in pore water containing FeOOH

[FeOOH]	[Mn] (total)	pH 7	pH 8	pH 9
1×10^{-6} M	1×10^{-5} M	2.5×10^5	2500	25
1×10^{-5} M	1×10^{-5} M	25000	250	2.5
1×10^{-4} M	1×10^{-5} M	2500	25	0.25
1×10^{-3} M	1×10^{-5} M	250	2.5	0.025
0	1×10^{-5} M	2.5×10^5	2500	25
0	1×10^{-4} M	18500	185	1.85
0	1×10^{-3} M	11700	117	0.12

Table 4-2 shows that Mn^{2+} oxidation will be very slow and insignificant in most situations. High particulate FeOOH concentrations catalyze the oxidation so that significant oxidation can occur over short time spans. However, concentrations of Fe^{2+} need to be very high (>10 mg/L) and the pH above 8 before significant abiotic oxidation and MnO_x precipitation will occur, and these conditions will be rare in most pore waters. We also examined the self-catalysis case, but because the rate of formation of MnO_x is slow at the start, its formation has little effect over a short time span. Exceptions are very high manganese concentrations and pH near or above 9. Lasier et al. (2000) found a decrease of <3% in Mn^{2+} concentrations during 4-day toxicity testing, which is entirely consistent with abiotic rates of oxidation.

Dissolved sulfide

High dissolved sulfide concentrations can be found with high OM inputs and /or coarse sediments (Figure 4-5). Under these conditions, concentrations will vary with depth. They may be high near the surface if there is a large input of OM, for example, under benthic algal mats (Visscher et al. 1991). They may be low at the surface and increase at mid-depths, then decrease again, approximating the diagenetic sequence (Figure 4-1B) (Williamson et al. 1994), or show a high variability throughout the profile because of burrows and microzones (Figure 4-1C). Free dissolved sulfide occurs mostly as HS^- at the pH of natural pore water with very small proportions of H_2S and S^{2-} depending on porewater pH. Free sulfide is important in pore water because, first, it can be toxic to animals (Wang and Chapman 1999) (Figure 4-12), and second, it reacts with trace metals to form insoluble precipitates (Davies-Colley et al. 1985) or sulfide complexes of unknown bioavailability (Emerson et al. 1983; Dyrssen 1985; Al-Farawati and van den Berg 1999). Free sulfide is oxidized by FeOOH and by any oxygen introduced by natural processes or during sampling and testing. With oxygen, dissolved sulfide is oxidized to $SO_4{}^{2+}$ or elemental sulfur, and this reaction has a half-life of approximately 1 to 24 hours (depending on the water chemistry) (Millero 1986; Yao and Millero 1996). Aeration will volatilize H_2S, and the rate of this process has a half-life similar to oxidation, depending on the aeration rate. Elemental sulfur precipitates, which often appear as white opalescence in oxidized porewater samples, may adsorb toxic constituents, thus affecting their bioavailability; this issue requires further study.

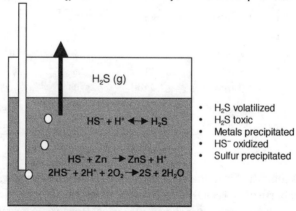

Figure 4-12 Summary of processes undergone by dissolved sulfide on aerating pore water

Porewater Dissolved Organic Carbon

Dissolved organic carbon is one of the more complex and important constituents of pore water. It is formed mainly during early diagenesis from the breakdown of particulate OM, but it can include toxic anthropogenic organic chemicals. While it has a complex chemistry, being made up of many different organic molecules and polymers, its importance to porewater toxicity is better appreciated than that of other constituents because of its well-known effect on toxicant bioavailability through complexation, partitioning, and other associations. Organic carbon in sediment pore waters exists in 2 forms: dissolved and colloidal. Conventionally, the 2 forms are discriminated from each other by some measurement of size. Consequently, several definitions of colloidal organic carbon (COC) and of DOC are based on size (Stumm and Morgan 1981; Buffle et al. 1992; Buffle and Leppard 1995). For example, Stumm and Morgan (1981) define colloidal particles as ranging between 10 and 10,000 nm in size. Any particles smaller than this range are considered truly dissolved, specifically individual, organic molecules, for example, carbohydrates, proteins, lipids, and contaminants. In contrast, COC consists of clusters of molecules. The most common clusters of organic molecules in pore waters are humic substances. Humic substances are unique because they do not have a specific chemical structure but are composed of complex aggregations of organic molecules and functional groups. These aggregations are commonly termed "polyfunctional geopolymers," which reflects their ability to be reactive while also showing hydrocarbon-like hydrophobicity (Leenheer 1985). Consequently, humic substances are able to interact with a variety of chemicals. It has been shown that organic acidic groups (e.g., carboxylic) are among the dominant functional groups lending some humic substances aqueous solubility despite large molecular mass (Thurman 1985).

The standardized use of macrofiltration to separate size classes of particles in environmental samples started in the 1950s. With the advent of ultrafiltration in the 1970s, it became possible to separate samples by molecular weight. Thus, porewater organic carbon could be separated into size classes as small as <500 molecular weight units (mwu). More recently, Gustafsson and Gschwend (1997) applied a more chemical-centric definition for COC: "any constituent that provides a molecular milieu into and onto which chemicals can escape from bulk aqueous solution, while its vertical movement is not significantly affected by gravitational settling." Here, the "bulk aqueous solution" would be pore water. While a size-based definition is useful in a practical sense, the chemical-centric definition is more relevant to issues discussed later in this section.

In the literature, porewater organic carbon is called by many names, including DOC and COC, but also "macromolecules," "humic substances," "colloids," and "nonsettling particles." For consistency in this section, the term DOC will be used to refer to both dissolved and colloidal carbon pools present in pore waters, unless otherwise noted.

Distributions, concentrations, and diagenesis of porewater DOC

Since the early 1970s, a good understanding has developed on the distribution of porewater DOC with depth in sediment (Krom and Sholkovitz 1977). Figure 4-13 shows the concentration of DOC with depth for several sites around the world. In general, we observe an increase in concentration with depth. In the upper sediment where overlying water oxygen is able to diffuse, or is introduced by bioturbation, there is relatively little accumulation of DOC. However, pore waters present in anoxic sediments show elevated concentrations of DOC that increase with increasing depth.

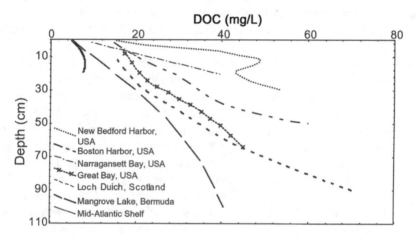

Figure 4-13 Concentration distribution of DOC with depth in sediment cores from around the world. Sites include Loch Duich, Scotland (Krom and Sholkovitz 1977); Mangrove Lake, Bermuda (Orem et al. 1986); Narragansett Bay, Rhode Island (Elderfield 1981); Mid-Atlantic Shelf (Burdige and Gardner 1998); New Bedford Harbor, Massachusetts (Brownawell and Farrington 1986); Boston Harbor, Massachusetts, (Chin and Gschwend 1991); and Great Bay, New Hampshire, USA (Orem and Gaudette 1984).

Concentrations of porewater DOC vary from approximately 5 mg/L to >400 mg/L (Thurman 1985). However, as shown in Figure 4-13, a more common range is 10 to 50 mg/L. Apart from the depth effect, DOC concentrations in pore waters depend on the input of organic material to the sediment system and the rate of its transformation to other forms (e.g., CO_2 and CH_4) (Valiela 1984). Consequently, coastal pore waters are frequently elevated in DOC, while pore waters from offshore sediments demonstrate low concentrations. For example, compare the Mid-Atlantic Shelf to other, more coastal locations (Figure 4-13). Another factor affecting porewater DOC concentrations is the presence of organic contaminants. In the sediments of interest for porewater toxicity testing, organic contaminants are often present, including polychlorinated biphenyls (PCBs), polycyclic aromatic hydrocarbons (PAHs), pesticides, and other organic molecules. These contaminants may contribute to the measured DOC concentration along with the naturally occurring organic carbon. Consequently, some contaminated sites may have unusual DOC concentration distributions, compared to pristine sites. Two sediment cores collected from the continental slope of North America demonstrate this point. One of the sites is "upstream" and unaffected by a deepwater dumpsite (DWDS), and shows increasing DOC with depth behavior. The other core is located near the dumpsite and shows the exact opposite trend with depth (Figure 4-14) (Martin and McCorkle 1993). The differences in the porewater DOC distributions may be attributed to the contamination present at the dumpsite.

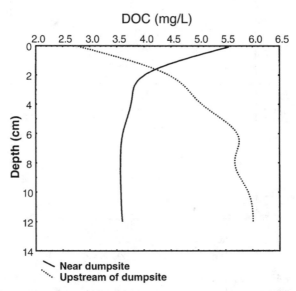

Figure 4-14 Concentration distribution of DOC with depth in sediment cores from 2 sites on the continental slope off of North America. One site is near a DWDS, while the other is upstream of the dumpsite (based on data from Martin and McCorkle 1993).

Figure 4-15 shows the distribution of porewater organic carbon separated into 2 specific molecular weight classes: <1000 mwu and 1000 to 50,000 mwu, as observed in samples collected from Great Bay, New Hampshire, USA (Orem and Gaudette 1984). The general trend is that the low molecular weight DOC (<1000) is fairly constant regardless of depth, while the high molecular weight DOC (1000 to 50,000) increases. Because of the common occurrence of this type of distribution, first observed in the 1970s, a model for the formation or diagenesis of porewater DOC evolved. Basically, this model proposes that water column particulate carbon enters the sediment and is transformed into the low molecular weight DOC molecules via abiotic and biotic processes. These in turn are polymerized to high molecular weight DOC that eventually condenses into sedimentary particulate organic carbon (Krom and Sholkovitz 1977; Krom and Westrich 1981). An alternative theory speculates that the high molecular weight DOC results from the selective preservation of the refractory components of the overlying water particulate organic carbon (Hatcher et al. 1983; Orem et al. 1986). Recent work on DOC profiles has devised variants on these mechanisms (Amon and Benner 1996; Burdige and Gardner 1998).

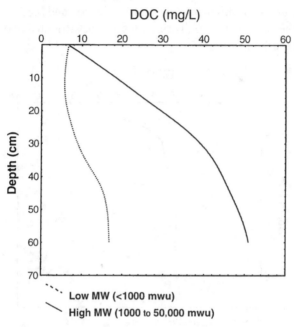

Figure 4-15 Concentration distribution of 2 size classes of DOC: low molecular weight DOC (<1000 mwu) and high molecular weight DOC (1000 to 50,000 mwu) (based on data from Orem and Gaudette 1984)

Interactions with porewater anthropogenic contaminants

The interaction of DOC and metals has long been known (i.e., since the 1920s) and used to help explain the behavior of metals in aqueous solution (Mantoura 1981). For example, early studies investigated the link between DOC and the supersaturation of iron in seawater, the bioavailability of iron to phytoplankton, the speciation and reactivity of trace metals, the occurrence of toxic algal blooms, and the reduction of metal toxicity (Mantoura 1981). Conversely, the interaction of DOC and organic contaminants has been known only since the late 1960s. This was first observed by Wershaw et al. (1969) when they reported the solubilization of DDT by a humic substance in natural waters. Since that study, this type of interaction has been observed for several classes of organic contaminants, including PCBs, pesticides, and PAHs (Sigleo and Means 1990; Suffet et al. 1994).

In recent years, several studies have also demonstrated the association of anthropogenic contaminants with porewater DOC. Table 4-3 presents a selection of studies in which the association of anthropogenic contaminants with DOC in pore water has been invoked, speculated, or demonstrated. A common consequence of the presence of DOC in pore waters is the apparent increase in solubility of organic contaminants above documented solubilities, giving the impression that pore waters are saturated with contaminants. For example, Chiou et al. (1986) reported the enhancement of the solubility of 2 pesticides, 2 PCBs, and a chlorinated benzene when DOC was present. Generally, they found apparent increases in solubility were positively correlated to increases in chemical octanol–water partition coefficient (K_{OW}) (Chiou et al. 1986). In other words, as organic contaminants become more insoluble, their affinity for DOC increases. In pore waters from New Bedford Harbor, Massachusetts, USA, Burgess et al. (1996) found total porewater concentrations of tetrachlorinated PCBs appeared to exceed solubility (Figure 4-16); however, when only the truly dissolved concentrations of these PCBs (and not the DOC-associated PCBs) were considered, the concentrations were actually below solubility.

While several of the early studies from the 1960s and 1970s demonstrated that in pore waters, metals such as copper and lead associated with DOC (Table 4-3), in recent years there has been little research. This may be because, although metals do interact with DOC, the level of this interaction is less significant than that of other environmental variables affecting their porewater distributions. For example, the presence of AVS has been demonstrated to strongly affect the distribution of many toxic metals in pore waters (Ankley et al. 1996).

The realization that contaminants associate with DOC forces us to consider how these pollutants exist in pore waters. Generally, the distribution of most contaminants in sediments has been described with a 2-phase model expressed as

$$C_T = C_P + C_D,$$

where C_T, C_P, and C_D are the total, particulate, and dissolved concentrations, respectively, of a given pollutant. Frequently, the partitioning of pollutants between the particulate and dissolved phases is expressed as the ratio of particulate to dissolved phases, known as the "partition coefficient" (K_P):

$$K_P = C_P / C_D.$$

Because of the importance of organic carbon in the partitioning of organic contaminants (Karickhoff et al. 1979) and potentially of metals (Mahony et al. 1996), the K_P is often normalized by the sediment organic carbon content (f_{oc}) to arrive at the organic carbon normalized partition coefficient (K_{OC}):

$$K_{OC} = K_P/f_{oc}.$$

Table 4-3 Selection of studies investigating the interaction of porewater DOC and anthropogenic contaminants

Contaminant	Location	Environment	Reference
Metal	Southern California, USA	Marine	Brooks et al. 1968
Metal	Saanich Inlet, Canada	Marine	Presley et al. 1972
Metal	Everglades, USA; Mobile Bay, USA	Marine	Lindberg and Harriss 1974
Metal	Saanich Inlet, Canada	Marine	Nissenbaum and Swaine 1976
Metal	Narragansett Bay, USA	Marine	Elderfield 1981
Metal	Gulf of California, Mexico	Marine	Brumsack and Gieskes 1983
Metal	Long Island Sound, USA	Marine	Lyons and Fitzgerald 1983
PCB, PAH, pesticide, phthalate	Great Lakes, USA	Freshwater	Landrum et al. 1985
PCB	New Bedford Harbor, USA	Marine	Brownawell and Farrington 1986
Metal	Narragansett Bay, USA	Marine	Douglas et al. 1986; Douglas and Quinn 1989
Metal	Bay of Horw, Switzerland	Freshwater	Piemontesi and Baccini 1986
PCB, PAH	Great Lakes, USA	Freshwater	Landrum et al. 1987
PAH, aliphatic hydrocarbon	Puget Sound, USA	Marine	Socha and Carpenter 1987
PCB, pesticide	Lake Superior, USA	Freshwater	Capel and Eisenreich 1990
PAH	Boston Harbor, USA	Marine	Chin and Gschwend 1992
PAH, pesticide	Lake Michigan, USA	Freshwater	Harkey et al. 1994
PCB, pesticide, PAH	Alsea Bay, USA	Marine	Ozretich et al. 1995
PCB	New Bedford Harbor, USA	Marine	Burgess et al. 1996; Burgess and McKinney 1997
Metal	Venice Lagoon, Italy	Marine	Bertolin et al. 1997
PCB	Brandywine River, USA	Freshwater	Hunchak-Kariouk et al. 1997
Metal	San Francisco Bay, USA	Marine	Rivera-Duarte and Flegal 1997
Metal	Gotland Deep, Baltic Sea	Marine	Brugmann et al. 1998
PCB	Southern California, USA	Marine	Pedersen et al. 1999
PAH	Elizabeth River, USA	Marine, freshwater	Mitra and Dickhut 1999
Metal	Chesapeake Bay, USA	Marine	Skrabal et al. 2000
Metal	Galveston Bay, USA	Marine	Warnken et al. 2001

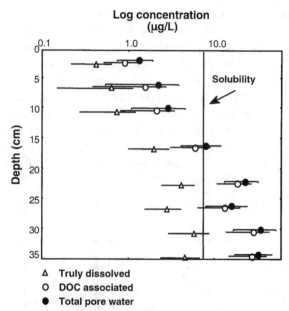

Figure 4-16 Concentration distribution of tetrachlorinated biphenyls in pore waters from New Bedford Harbor, USA. Tetrachlorinated biphenyls were measured in the truly dissolved, DOC-associated, and total porewater phases. The solid vertical line is the solubility of these PCBs (based on data from Burgess et al. 1996).

In pore waters, an additional phase may affect the results of toxicity tests and therefore must be considered. Addition of the DOC phase can be expressed as

$$C_T = C_P + C_D + C_{DOC},$$

where C_{DOC} is the concentration of a pollutant associated with DOC in the pore water. In this 3-phase model, K_P is now estimated as follows:

$$K_P = (C_P + C_{DOC}) / C_D.$$

Furthermore, because we are interested only in the pore water, the total porewater concentration (C_{PW}) can be presented as

$$C_{PW} = C_D + C_{DOC},$$

and the K_P can be more simply expressed as a porewater partition coefficient (K_{PW}),

$$K_{PW} = C_{DOC} / C_D,$$

and normalized for the effects of DOC,

$$K_{DOC} = K_{PW} / DOC,$$

where DOC is now equivalent to the concentration of DOC in the pore water. This last expression provides us with a useful tool for calculating how much of the organic contaminants present in pore water will be associated with DOC. However, measuring K_{PW} is quite difficult and time consuming, and it cannot be conducted routinely. An estimator of K_{DOC} is necessary in order to make this exercise practical. Recently, Burkhard (2000) performed a review of the literature relating K_{DOC} to K_{OW} for several types of hydrophobic organic pollutants. The review included several kinds of aquatic environments such as surface waters, sediment and soil pore waters, and groundwater (Figure 4-17). Based on his analysis of the relationship between K_{DOC} and K_{OW} in sediment pore waters, Burkhard (2000) generated the following linear relationship (Figure 4-17):

$$\log K_{DOC} = 0.99 \cdot \log K_{OW} - 0.88 \ (n = 396; \ r = 0.64).$$

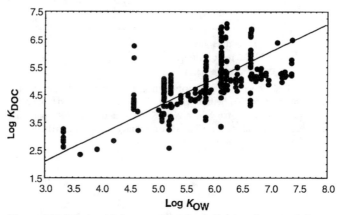

Figure 4-17 Relationship between $\log K_{DOC}$ and $\log K_{OW}$ for several classes of organic contaminants in freshwater and marine sediment pore waters (based on data from Burkhard 2000)

Using this expression, K_{DOC} can be calculated easily if we know the K_{OW} of the contaminant of interest. In practice, we can now estimate how much contaminant is truly dissolved,

$$C_D = C_{PW} / (K_{DOC} \cdot DOC + 1),$$

or the percent of the total porewater concentration that is truly dissolved,

$$\%C_D = 1 / (K_{DOC} \cdot DOC + 1) \cdot 100.$$

In the next section, we discuss how this information can be used to determine whether organic contaminants in pore waters will be expected to be bioavailable.

While several models exist for estimating the interaction of metals with organic carbon and specifically with DOC (Mantoura 1981; Honeyman and Santschi 1988; Santschi et al. 1997; Koelmans and Radovanovic 1998), much more research is needed in the area of porewater DOC and metal interactions. Furthermore, the interactions of ammonia (NH_4^+ and NH_3) and other porewater toxicants (e.g., $S^=$) with porewater DOC are largely unknown and require study.

Effects of porewater DOC on the bioavailability of porewater contaminants

We are concerned with the distribution, concentration, and interactions of DOC in pore waters because DOC has been shown to reduce the bioavailability of several classes of toxicants to aquatic organisms. Several early studies demonstrated that metal bioavailability was reduced when either synthetic or natural organic molecules (e.g., ethylenediaminetetraacetic acid [EDTA], nitrolotriacetic acid [NTA]) were present in solution containing toxic metals (Sunda et al. 1978). These studies also indicated the importance of the truly dissolved concentration of a metal as the direct cause of toxicity, compared to the total concentration. Similarly, Boehm and Quinn (1976) reported that DOC reduced the concentrations of hydrocarbons accumulated by a benthic bivalve. Reviews by McCarthy (1989), Sigleo and Means (1990), Suffet et al. (1994), and Haitzer et al. (1998) discuss the findings of studies conducted with DOC in natural systems, demonstrating reductions in the bioavailability of metals and organic contaminants.

For organic chemicals, the review by Haitzer et al. (1998) reported reductions in bioaccumulation ranging from 2% to 98% when DOC was present. Reductions in bioavailability are proposed to result from the formation of DOC-contaminant aggregates that are too large or too polar to be assimilated by organisms (McCarthy 1989; Haitzer et al. 1998). In rare instances and at relatively low DOC concentrations (≤ 10 mg/L), bioavailability was observed to increase by factors ranging from 2% to 303% (Haitzer et al. 1998). However, reductions in bioavailability were most likely to occur at environmentally relevant DOC concentrations, but the magnitude of the effect depended on the type of DOC (e.g., pore water versus overlying water). This last point is important in considering the effects of DOC on porewater contaminants. Because there are so few data on the effects of DOC on bioavailability in actual pore waters, there will be a tendency to assume that what occurs in overlying waters also occurs in pore waters. Until this is demonstrated, we must avoid generalizing.

The behavior of metals associated with DOC with regard to bioavailability is far less easy to summarize. The bioavailability of metals associated with DOC has been shown to increase and decrease depending on the species, the metal, the aqueous chemistry (e.g., hardness), and the DOC source (McCarthy 1989; Fent and Looser 1995; Hollis et al. 1996). Explanations for this level of variability range from changes in the truly dissolved metal concentration, resulting from altered interactions with organic and inorganic ligands, to competitive effects between the DOC and metal complex and biological membranes (McCarthy 1989). More research is required on the effects of DOC on metals bioavailability in pore waters.

Landrum et al. (1985, 1987) published the only studies directly investigating the effect of porewater DOC on the bioavailability of toxicants. In these studies, they exposed a freshwater amphipod to either control water, control water containing humic materials, or pore water from sediments collected from the Great Lakes, USA. The samples were amended with radiolabeled organic contaminants, including the PAHs phenanthrene, pyrene and benzo[*a*]pyrene, and a tetrachlorinated biphenyl. As predicted, the presence of DOC resulted in reductions in the amphipod uptake of the contaminants as a function of K_{OW} (Figure 4-18) (Landrum et al. 1987). For phenanthrene with a relatively low log K_{OW} of 4.16, the DOC caused very little change in accumulation, compared to the control water. Conversely, benzo[*a*]pyrene, log K_{OW} of 6.50, demonstrated an 85% reduction in uptake when DOC was present. The PCB and pyrene, with intermediate log K_{OW}s of 6.11 and 4.88, respectively, showed moderate reductions in accumulation. In a related study, Burgess and McKinney (1999) compared PCB distributions in several sedimentary exposure routes, including truly dissolved and DOC-associated, to the PCB distributions accumulated by 2 marine bivalve species, to determine which route was most bioavailable. A high level of covariance between exposure routes encumbered any definitive conclusions, but in the longer-term exposures, DOC-associated distributions were frequently the least correlated with PCB concentrations in the bivalves. Despite the paucity of information, the consensus in the literature is that porewater DOC will reduce the bioavailability of contaminants. For example, a recent document proposed for the regulation of contaminated sediments includes measures for considering the effects of porewater DOC on the bioavailability of organic contaminants (USEPA 2000). Clearly, more research is required to better understand and illustrate the effects of porewater DOC on the bioavailability of organic contaminants, metals, and ammonia.

As we have discussed earlier, sediment pore waters that contain elevated concentrations of DOC have been formed under anoxic conditions, and the DOC is in a "reduced" form. When aerated, this DOC may undergo oxidation. It is possible that any such oxidation will alter the interaction of the DOC with various contaminants (metals and hydrophobic organics) as well as with iron and manganese oxides. The little information available on this topic is discussed in Chapter 5.

Conclusions

Pore waters are complex, and their chemistry is highly variable. After sediments are laid down, they undergo a complex series of chemical and physical processes termed "early diagenesis." The dominant process is the breakdown of OM through aerobic and anaerobic metabolism, which consumes O_2, NO_3^-, FeOOH, MnO_x, SO_4^{2-}, and particulate organic matter and produces CO_2, Mn^{2+}, Fe^{2+}, HS^-, NH_3, and DOM (among many other products and reactants). The interaction of these processes with

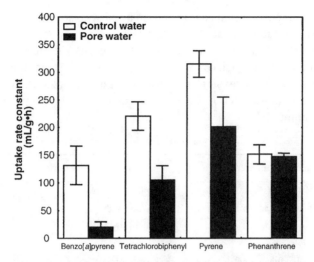

Figure 4-18 Uptake of several organic contaminants by a freshwater amphipod exposed in control water (no DOC present) and pore water. Log K_{OW}s for benzo[*a*]pyrene, the tetrachlorinated biphenyl, pyrene, and phenanthrene are 6.50, 6.11, 4.88, and 4.16, respectively (based on data from Landrum et al. 1987).

the rate of OM input, sediment texture, bioturbation, and overlying water flow produces a complex 3D pattern in the porewater chemical environment.

The chemical components of early diagenesis may indirectly or directly affect the bioavailability of contaminants. These components can play a large role in porewater toxicity, directly (e.g., NH_3 may be toxic to test organisms) and indirectly (e.g., DOM complexes toxicants). More important, some will be transformed when pore waters are oxygenated.

Oxidation of pore waters can result from natural processes (e.g., bioturbation) or the mixing of anoxic and oxic water, and it occurs if anoxic pore water is aerated during sampling or the ensuing toxicity test (Chapter 5). Oxidation–precipitation–adsorption processes involving dissolved iron, manganese, or sulfide may change contaminant bioavailability (e.g., adsorption of heavy metals, oxidation of H_2S) or introduce toxicity artifacts (e.g., possible toxicity of FeOOH colloids). These processes are especially important when early diagenesis has resulted in high concentrations of dissolved iron, manganese, or sulfide in the pore water. The corollary is that there are few of these complications when pore waters contain low concentrations of these substances, or contain high concentrations of contaminants, or the contaminants do not interact significantly with these inorganic porewater constituents. Oxidation may also affect DOC chemical characteristics and hence also contaminant bioavailability (Chapter 5).

One of the most important challenges porewater toxicity testing faces is the issue of realism in the chemical nature of test solutions because most pore water is anoxic in situ, while the tests are carried out in oxic water (Table 4-4). Anoxic water in the sediment (containing, e.g., high concentrations of NH_3 or H_2S) may not be in contact with animals that live there, but it is brought into contact with test organisms. Samples containing solely oxic pore waters do not suffer these problems, but these are less common.

Table 4-4 Artifacts inherent in porewater tests[a]

Sediment pore water	Porewater toxicity testing
Heterogeneous	Homogeneous
Intimate contact with sediment	Water only
Redox gradients	Saturated with oxygen
Mostly anoxic	Oxic
Many benthic animals are not in contact with anoxic pore water	Test animals are placed into pore water from anoxic regions
Reduced components stable	Reduced components oxidized

[a] For fine-grained sediments, which are not flushed by overlying flowing water, e.g., beach sand or permeable river sediments.

We know a great deal about some important porewater characteristics such as dissolved iron, manganese, sulfide, CO_2, NH_3, DO and the redox condition, DOC, and their cycles in sediments. In addition to their dependence on the rate of OM input to sediments, bioturbation, sediment texture, and overlying water flow, their concentration also varies with the depth of sampling. It is therefore very useful to understand the chemical nature of the sediments and their pore waters being sampled in order to provide guidance for the appropriate sampling strategy and for the assessment of the validity of the porewater toxicity results. There are some serious discrepancies in the way we sample sediments (treating it as a black box and sampling to a fixed depth) and the known complexity and structure of sediment and porewater chemistry. This theme is discussed more fully in Chapter 5.

References

Al-Farawati R, van den Berg CMG. 1999. Metal-sulfide complexation in seawater. *Mar Chem* 63:331–352.

Aller RC. 1978. Experimental studies of changes produced by deposit feeders on pore water, sediment and overlying water chemistry. *Am J Sci* 278:1185–1234.

Aller RC. 1982. The effects of macrobenthos on chemical properties of marine sediment and overlying water. In: McCall PL, Tevesz MJS, editors. Animal-sediment relations. New York NY, USA: Plenum. p 53–102.

Amon RMW, Benner R. 1996. Bacterial utilization of different size classes of dissolved organic matter. *Limnol Oceanogr* 41:41–51.

Ankley GT, Di Toro DM, Hansen DJ, Berry WJ. 1996. Technical basis and proposal for deriving sediment quality criteria for metals. *Environ Toxicol Chem* 15:2056–2066.

Banta GT, Holmer M, Jensen MH, Kristensen E. 1999. Effects of two polychaete worms, *Nereis diversicolor* and *Arenicola marina*, on aerobic and anaerobic decomposition in a sandy marine sediment. *Aquat Microb Ecol* 19:189–204.

Benjamin MM, Leckie JO. 1978. Competitive adsorption of Cd, Cu, Zn and Pb on amorphous iron oxyhydroxide. *J Colloid Interface Sci* 83:410–419.

Berner RA, Westrich JT. 1985. Bioturbation and the early diagenesis of carbon and sulfur. *Am J Sci* 285:193–206.

Bertolin A, Mazzocchin GA, Rudello D, Ugo P. 1997. Seasonal and depth variability of reduced sulfur species and metal ions in mud-flat pore-waters of the Venice lagoon. *Mar Chem* 59:127–140.

Boehm PD, Quinn JG. 1976. The effect of dissolved organic matter in sea water on the uptake of mixed individual hydrocarbons and number 2 fuel oil by a marine bivalve. *Estuar Coast Mar Sci* 4:93–105.

Boulton AJ, Findlay S, Marmonier P, Stanley EH, Valett HM. 1998. The functional significance of the hyporheic zone in streams and rivers. *Ann Rev Ecol Syst* 29:59–81.

Brooks RR, Presley BJ, Kaplan IR. 1968. Trace elements in the interstitial waters of marine sediments. *Geochim Cosmochim Acta* 32:397–414.

Brownawell BJ, Farrington JW. 1986. Biogeochemistry of PCBs in interstitial waters of a coastal marine sediment. *Geochim Cosmochim Acta* 50:167–169.

Brugmann L, Hallberg R., Larsson C, Loffler A. 1998. Trace metal speciation in sea and pore water of the Gotland Deep, Baltic Sea, 1994. *Appl Geochem* 13:359–368.

Brumsack HJ, Gieskes JM. 1983. Interstitial water trace-metal chemistry of laminated sediments from the Gulf of California, Mexico. *Mar Chem* 14:89–106.

Buffle J, Leppard GG. 1995. Characterization of aquatic colloids and macromolecules. 1. Structure and behavior of colloidal material. *Environ Sci Technol* 29:2169–2175.

Buffle J, Perret D, Newman M. 1992. The use of filtration and ultrafiltration for size fractionation of aquatic particles, colloids, and macromolecules. In: Buffle J, van Leeuwen HP, editors. Environmental particles. Boca Raton FL, USA: Lewis. p 171–230.

Burdige DJ. 1991. The kinetics of organic matter mineralization in anoxic marine sediments. *J Mar Res* 49:727–761.

Burdige DJ, Gardner KG. 1998. Molecular weight distribution of dissolved organic carbon in marine sediment pore waters. *Mar Chem* 62:45–64.

Burgess RM, McKinney RA. 1997. Effects of sediment homogenization on interstitial water PCB geochemistry. *Arch Environ Contam Toxicol* 33:125–129.

Burgess RM, McKinney RA. 1999. Importance of interstitial, aqueous and whole sediment exposures to bioaccumulation by marine bivalves. *Environ Pollut* 104:373–382.

Burgess RM, McKinney RA, Brown WA. 1996. Enrichment of marine sediment colloids with polychlorinated biphenyls (PCBs): Trends resulting from PCB solubility and chlorination. *Environ Sci Technol* 30:2556–2566.

Burke CM. 1999. Molecular diffusive fluxes of oxygen in sediments of Port Phillip Bay in southeastern Australia. *Mar Freshwat Res* 50:557–566.

Burkhard LP. 2000. Estimating dissolved organic carbon partition coefficients for nonionic organic chemicals. *Environ Sci Technol* 34:4663–4668.

Canfield DE. 1989. Reactive iron in marine sediments. *Geochim Cosmochim Acta* 53:619–632.

Canfield DE, Jorgensen BB, Fossing H, Glud R, Gundersen J, Ramsing NB, Thamdrup B, Hansen JW, Nielsen LP, Hall POJ. 1993. Pathways of organic carbon oxidation in three continental margin sediments. *Mar Geol* 113:27–40.

Canfield DE, Thamdrup B, Hansen JW. 1993. The anaerobic degradation of organic matter in Danish coastal sediments: Iron reduction, manganese reduction, and sulfate reduction. *Geochim Cosmochim Acta* 57:3867–3883.

Capel PD, Eisenreich SJ. 1990. Relationship between chlorinated hydrocarbons and organic carbon in sediment and porewater. *J Great Lakes Res* 16:245–257.

Carlton RG, King MJ. 1990. Spatial and temporal variations in microbial processes in aquatic sediments: Implication for the nutrient status of lakes. In: Baudo R, Giesy J, Muntau H, editors. Sediments: Chemistry and toxicity of in-place pollutants. Boca Raton FL, USA: Lewis. p 107–130.

Carney RS. 1981. Bioturbation and biodeposition. In: Boucot AJ, editor. Principles of benthic marine paleoecology. Sydney, Australia: Academic.

Chin Y-P, Gschwend PM. 1991. The abundance, distribution, and configuration of porewater organic colloids in recent sediments. *Geochim Cosmochim Acta* 55:1309–1317.

Chin Y-P, Gschwend PM. 1992. Partitioning of polycyclic aromatic hydrocarbons to marine porewater organic colloids. *Environ Sci Technol* 26:1621–1626.

Chiou CT, Malcolm RL, Brinton TI, Kile DE. 1986. Water solubility enhancement of some organic pollutants and pesticides by dissolved humic and fulvic acids. *Environ Sci Technol* 20:502–508.

Davies-Colley RJ, Nelson PO, Williamson KJ. 1985. Sulfide control of cadmium and copper concentrations in anaerobic estuarine sediments. *Mar Chem* 16:173–186.

Davis JA. 1982. Adsorption of natural dissolved organic matter at the oxide/water interface. *Geochim Cosmochim Acta* 46:2381–2393.

Davison W. 1993. Iron and manganese in lakes. *Earth-Sci Rev* 34:119–163.

Davison W, Fones GR, Grime GW. 1997. Dissolved metals in surface sediment and a microbial mat at 100-µm resolution. *Nature* 387:885–888.

Davison W, Fones G, Harper M, Teasdale P, Zhang H. 2000. Dialysis, DET and DGT: In situ diffusional techniques for studying water, sediments and soils. In: Buffle J, Horvai G, editors. In situ monitoring of aquatic systems: Chemical analysis and speciation. International Union of Pure and Applied Chemistry (IUPAC). Chichester, UK: J Wiley. p 495–569.

Davison W, Grime GW, Morgan JAW, Clarke K. 1991. Distribution of dissolved iron in sediment pore waters at submillimetre resolution. *Nature* 352: 323–326.

Douglas GS, Mills GL, Quinn JG. 1986. Organic copper and chromium complexes in the interstitial waters of Narragansett Bay sediments. *Mar Chem* 19:161–174.

Douglas GS, Quinn JG. 1989. Geochemistry of dissolved chromium-organic-matter complexes in Narragansett Bay interstitial waters. In: Suffet IH, MacCarthy P, editors. Aquatic humic substances: Influence on fate and treatment of pollutants. Washington DC, USA: American Chemical Soc. p 297–320.

Dyrssen D. 1985. Metal complex formation in sulphidic seawater. *Mar Chem* 15:284–293.

Dzombak DA, Morel FMM. 1990. Surface complexation models. Hydrous ferric oxide. London, UK: J Wiley.

Emerson S, Jacobs L, Tebo B. 1983. The behaviour of trace metals in marine anoxic waters: Solubilities at the oxygen-hydrogen sulfide interface. In: Wong CS, Boyle E, Bruland KW, Burton JD, Goldberg ED, editors. Trace metals in sea water. New York NY, USA: Plenum. p 579–608.

Elderfield H. 1981. Metal-organic associations in interstitial waters of Narragansett Bay sediments. *Am J Sci* 281:1184–1196.

Fent K, Looser PW. 1995. Bioaccumulation and bioavailability of tributyltin chloride: Influence of pH and humic acids. *Water Res* 29:1631–1637.

Fones GR, Davison W, Grime GW. 1998. Development of constrained DET for measurements of dissolved iron in surface sediments at sub-mm resolution. *Sci Total Environ* 221:127–137.

Froelich PN, Klinkhammer GP, Bender ML, Luedtke NA, Heath GR, Cullen D, Dauphin D, Hammond D, Hartman B, Maynard V. 1979. Early oxidation of organic matter in pelagic sediments of the eastern equatorial Atlantic: Suboxic diagenesis. *Geochim Cosmochim Acta* 43:1075–1090.

Gehlen M, Rabouille C, Ezat U, GuidiGuilvard LD. 1997. Drastic changes in deep-sea sediment porewater composition induced by episodic input of organic matter. *Limnol Oceanogr* 42:980–986.

Gu B, Schmitt J, Chen Z, Liang L, McCarthy JF. 1994. Adsorption and desorption of natural organic matter on iron oxide: Mechanisms and models. *Environ Sci Technol* 28:38–46.

Gustafsson O, Gschwend PM. 1997. Aquatic colloids: Concepts, definitions, and current challenges. *Limnol Oceanogr* 42:519–528.

Haitzer M, Hoss S, Traunspurger W, Steinberg C. 1998. Effects of dissolved organic matter (DOM) on the bioconcentration of organic chemicals in aquatic organisms: A review. *Chemosphere* 37:1335–1362.

Hammond DE, Giordani P, Berelson WM, Poletti R. 1999. Diagenesis of carbon and nutrients and benthic exchange in sediments of the Northern Adriatic Sea. *Mar Chem* 66:53–79.

Harkey GA, Landrum PF, Klaine SJ. 1994. Partition coefficients of hydrophobic contaminants in natural water, porewater, and elutriates obtained from dosed sediment: A comparison of methodologies. *Chemosphere* 28:583–596.

Hatcher PG, Spiker EC, Szeverenyi NM, Maciel GE. 1983. Selective preservation and origin of petroleum forming aquatic kerogen. *Nature* 305:498–501.

Hollis L, Burnison K, Playle RC. 1996. Does the age of metals-dissolved organic carbon complexes influence binding of metals to fish gills? *Aquat Toxicol* 35:253–264.

Honeyman BD, Santschi PH. 1988. Metals in aquatic systems. *Environ Sci Technol* 22:862–871.

Huettel M, Ziebis W, Forster S, Luther GW. 1998. Advective transport affecting metal and nutrient distributions and interfacial fluxes in permeable sediment. *Geochim Cosmochim Acta* 62:613–631.

Hunchak-Kariouk K, Schweitzer L, Suffet IH. 1997. Partitioning of 2,2',4,4'-tetrachlorobiphenyl by the dissolved organic matter in oxic and anoxic porewaters. *Environ Sci Technol* 31:639–645.

Jorgensen BB, Bang M, Blackburn TH. 1990. Anaerobic mineralization in marine sediments from the Baltic Sea-North Sea transition. *Mar Ecol Prog Ser* 59:39–54.

Karickhoff SW, Brown DS, Scott TA. 1979. Sorption of hydrophobic pollutants on natural sediments. *Water Res* (13) 241–248.

Koelmans AA, Radovanovic H. 1998. Predictions of trace metal distribution coefficients (K_D) for aerobic sediments. *Water Sci Tech* 37:71–81.

Krom MD, Sholkovitz ER. 1977. Nature and reactions of dissolved organic matter in the interstitial waters of marine sediments. *Geochim Cosmochim Acta* 41:1565–1573.

Krom MD, Westrich JT. 1981. Dissolved organic matter in the pore waters of recent marine sediments: A review. *Biogeochimie de la Matiere Organique a L'Interface Eau-Sediment Marin, Colloques Internationaux du C.N.R.S.* 293:105–111.

Landrum PF, Nihart SR, Eadie BJ, Herche LR. 1987. Reductions in bioavailability of organic contaminants to the amphipod *Pontoporeia hoyi* by dissolved organic matter of sediment interstitial waters. *Environ Toxicol Chem* 6:11–20.

Landrum PF, Reinhold MD, Nihart SR, Eadie BJ. 1985. Predicting the bioavailability of organic xenobiotics to *Pontoporeia hoyi* in the presence of humic and fulvic materials and natural dissolved organic matter. *Environ Toxicol Chem* 4:459–467.

Lasier PJ, Winger PV, Bogenrieder KJ. 2000. Toxicity of manganese to *Ceriodaphnia dubia* and *Hyalella azteca*. *Arch Environ Contam Toxicol* 38:298–304.

Leenheer JA. 1985. Fractionation techniques for aquatic humic substances. In: Aiken GR, McKnight DM, Wershaw RL, MacCarthy P, editors. Humic substances in soil, sediment, and water: Geochemistry, isolation, and characterization. New York NY, USA: J Wiley. p 409–429.

Lindberg SE, Harriss RC. 1974. Mercury-organic matter associations in estuarine sediments and interstitial water. *Environ Sci Technol* 8:459–462.

Lohse L, Epping EHG, Helder W, vanRaaphorst W. 1996. Oxygen pore water profiles in continental shelf sediments of the North Sea: Turbulent versus molecular diffusion. *Mar Ecol Prog Ser* 145:63–75.

Luther GW. 1995. Trace metal chemistry in pore waters. In: Allen HE, editor. Metal speciation and contamination of aquatic sediments. Chelsea MI, USA: Ann Arbor. p 65–80.

Luther GW, Brendel PJ, Lewis BL, Sundby B, Lefrancois L, Silverberg N, Nuzzio DB. 1998. Simultaneous measurement of O_2, Mn, Fe, I^-, and $S^{(-II)}$ in marine pore waters with a solid-state voltammetric microelectrode. *Limnol Oceanogr* 43:325–333.

Luther GW, Reimers CE, Nuzzio DB, Lovalvo D. 1999. In situ deployment of voltammetric, potentiometric and amperometric microelectrodes from a ROV to determine dissolved O_2, Mn, Fe, $S^{(-2)}$, and pH in porewaters. *Environ Sci Technol* 33:4352–4356.

Lyons WB, Fitzgerald WF. 1983. Trace metals speciation in nearshore anoxic and suboxic pore waters. In: Wong CS, Boyle E, Bruland KW, Burton JD, Goldberg ED, editors. Trace metals in sea water. New York NY, USA: Plenum. p 621–641.

Mahony JD, Di Toro DM, Gonzalez AM, Curto M, Dilg M, DeRosa LD, Sparrow LA. 1996. Partitioning of metals to sediment organic carbon. *Environ Toxicol Chem* 15:2187–2197.

Mantoura RFC. 1981. Organo-metallic interactions in natural waters. In: Duursma EK, Dawson R, editors. Marine organic chemistry: Evolution, composition, interactions and chemistry of organic matter in seawater. New York NY, USA: Elsevier Scientific, Elsevier Oceanography Series, 31:179–223.

Martin WR, McCorkle DC. 1993. Dissolved organic carbon concentration in marine pore waters determined by high-temperature oxidation. *Limnol Oceanogr* 38:1464–1479.

McCall PL, Tevesz MJS. 1982. Animal-sediment relations. New York NY, USA: Plenum.

McCarthy JF. 1989. Bioavailability and toxicity of metals and hydrophobic organic contaminants. In: Suffet IH, MacCarthy P, editors. Aquatic humic substances: Influence on fate and treatment of pollutants. Washington DC, USA: American Chemical Soc. p 263–277.

Meyers MB, Fossing H, Powell EN. 1987. Microdistribution of interstitial meiofauna, oxygen and sulfide gradients, and the tubes of macro-infauna. *Mar Ecol Prog Ser* 35:223–241.

Millero FJ. 1986. The thermodynamics and kinetics of the hydrogen sulfide system in natural waters. *Mar Chem* 18:121–147.

Mitra S, Dickhut RM. 1999. Three-phase modelling of polycyclic aromatic hydrocarbon association with pore-water-dissolved organic carbon. *Environ Toxicol Chem* 18:1144–1148.

Morrisey DJ, DeWitt TH, Roper DS, Williamson RB. 1999. Variation in the depth and morphology of burrows of the mud crab *Helice crassa* among different types of intertidal sediment in New Zealand. *Mar Ecol Prog Ser* 182:231–242.

Nissenbaum A, Swaine DJ. 1976. Organic matter-metal interactions in recent sediments: The role of humic substances. *Geochim Cosmochim Acta* 40:809–816.

Orem WH, Gaudette HE. 1984. Organic matter in anoxic marine pore water: Oxidation effects. *Org Geochem* 5:175–181.

Orem WH, Hatcher PG, Spiker EC, Szeverenyi NM, Macie GE. 1986. Dissolved organic matter in anoxic porewaters from Mangrove Lake, Bermuda. *Geochim Cosmochim Acta* 50:609–618.

Ozretich RJ, Smith LM, Roberts FA. 1995. Reversed-phase separation of estuarine interstitial water fractions and the consequences of C_{18} retention of organic matter. *Environ Toxicol Chem* 14:1261–1272.

Pedersen JA, Gabelich CJ, Lin C-H, Suffet IH. 1999. Aeration effects on the partitioning of a PCB to anoxic estuarine sediment pore water dissolved organic matter. *Environ Sci Technol* 33:1388–1397.

Piemontesi D, Baccini P. 1986. Chemical characteristics of dissolved organic matter in interstitial waters of lacustrine sediments and its influence on copper and zinc transport. *Environ Technol Lett* 7:577–592.

Presley BJ, Kolodny Y, Nissenbaum A, Kaplan IR. 1972. Early diagenesis in a reducing fjord, Saanich Inlet, British Columbia - II. Trace element distribution in interstitial water and sediment. *Geochim Cosmochim Acta* 36:1073–1090.

Revsbech NP, Blackburn TH, Cohen Y. 1983. Microelectrode studies of photosynthesis and O_2, H_2S, and pH profiles of a microbial mat. *Limnol Oceanogr* 28:1062–1074.

Revsbech NP, Sorensen J, Blackburn TH, Lomholt JP. 1980. Distribution of oxygen in marine sediments measured with microelectrodes. *Limnol Oceanogr* 25:403–411.

Rhoads DC, Boyer LF. 1982. The effects of marine benthos on physical properties of sediments. A successional perspective. Animal-sediment relations. London, England: Plenum. p 3–40.

Rivera-Duarte I, Flegal AR. 1997. Pore-water silver concentration gradients and benthic fluxes from contaminated sediments of San Francisco Bay, California, U.S.A. *Mar Chem* 56:15–26.

Robbins JA. 1982. Stratigraphic and dynamic effects of sediment reworking by Great Lakes zoobenthos. *Hydrobiologia* 92:611–622.

Rutherford JC, Latimer GJ, Smith RK. 1993. Bedform mobility and benthic oxygen uptake. *Water Res* 27:1545–1558.

Santschi PH, Lenhart JJ, Honeyman BD. 1997. Heterogeneous processes affecting trace contaminant distribution in estuaries: The role of natural organic matter. *Mar Chem* 58:99–125.

Shuttleworth SE, Davison W, Hamilton-Taylor J. 1999. Two-dimensional and fine structure in the concentrations of iron and manganese in sediment pore-waters. *Environ Sci Technol* 33:4169–4175.

Sigleo AC, Means JC. 1990. Organic and inorganic components in estuarine colloids: Implications for sorption and transport of pollutants. *Rev Environ Contam Toxicol* 112:123–147.

Skrabal SA, Donat JR, Burdige DJ. 2000. Pore water distributions of dissolved copper and copper-complexing ligands in estuarine and coastal marine sediments. *Geochim Cosmochim Acta* 64:1843–1857.

Socha SB, Carpenter R. 1987. Factors affecting pore water hydrocarbon concentrations in Puget Sound sediments. *Geochim Cosmochim Acta* 51:1273–1284.

Stumm E, Huang CP, Jenkins SR. 1970. Specific chemical interaction affecting the stability of dispersed systems. *Croat Chim Acta* 42:223–245.

Stumm W, Morgan JJ. 1981. Aquatic chemistry: An introduction emphasizing chemical equilibria in natural waters. New York NY, USA: J Wiley. 780 p.

Suffet IH, Jafvert CT, Kukkonen J, Servos MR, Spacie A, Williams LL, Noblet JA. 1994. Synopsis of discussion session: Influences of particulate and dissolved material on the bioavailability of organic compounds. In: Hamelink JL, Landrum PF, Bergman HL, Benson WH, editors. Bioavailability: Physical, chemical and biological interactions. Boca Raton FL, USA: CRC. p 93–108.

Sunda WG, Engel DW, Thuotte RM. 1978. Effect of chemical speciation on toxicity of cadmium to grass shrimp, *Paleomonetes pugio*: Importance of free cadmium ion. *Environ Sci Technol* 12:409–413.

Teermann IP, Jekel MR. 1999. Adsorption of humic substances onto b-FeOOH and its chemical regeneration. *Water Sci Technol* 40:199–206.

Thurman EM. 1985. Organic geochemistry of natural waters. Boston MA, USA: Martinus Nijhoff/Dr W Junk. 497 p.

Turner A, Hyde TL, Rawling MC. 1999. Transport and retention of hydrophobic organic micropollutants in estuaries: Implications of the particle concentration effect. *Estuar Coast Shelf Sci* 49:733–746.

[USEPA] U.S. Environmental Protection Agency. 1991. MINTEQA2/PRODEFA2. A geochemical assessment model for environmental systems. Atlanta GA, USA: USEPA. EPA/600/3-91/021.

[USEPA] U.S. Environmental Protection Agency. 1999. National recommended water quality criteria: Correction. Washington DC, USA: USEPA, Office of Water. EPA 822-Z-99-001.

[USEPA] U.S. Environmental Protection Agency. 2000. Methods for the derivation of site-specific equilibrium partitioning sediment guidelines (ESGs) for the protection of benthic organisms: Nonionic organics. Washington DC, USA: USEPA, Office of Science and Technology and Office of Research and Development. EPA-822-R-00-002.

Valiela I. 1984. Marine ecological processes. New York NY, USA: Springer-Verlag. 546 p.

Visscher PT, Beukema J, Vangemerden H. 1991. In situ characterization of sediments: Measurements of oxygen and sulfide profiles with a novel combined needle electrode. *Limnol Oceanogr* 36:1476–1480.

Wang FY, Chapman PM. 1999. Biological implications of sulfide in sediment: A review focusing on sediment toxicity. *Environ Toxicol Chem* 18:2526–2532.

Warnken KW, Gill GA, Griffin LL, Santschi PH. 2001. Sediment-water exchange of Mn, Fe, Ni and Zn in Galveston Bay, Texas. *Mar Chem* 73:215–231.

Wershaw RL, Burcar PJ, Goldberg MC. 1969. Interaction of pesticides with natural organic material. *Environ Sci Technol* 3:270–273.

Williamson RB, Hume TM, Molkrijnen J. 1994. A comparison of the early diagenetic environment in intertidal sands and muds of the Manukau Harbour, New Zealand. *Environ Geol* 24:254–266.

Williamson RB, Wilcock RJ, Wise BE, Pickmere SE. 1999. Effect of burrowing by the crab *Helice crassa* on chemistry of intertidal muddy sediments. *Environ Toxicol Chem* 18:2078–2086.

Yao WS, Millero FJ. 1996. Oxidation of hydrogen sulfide by hydrous Fe(III) oxides in seawater. *Mar Chem* 52:1–16.

Ziebis W, Huettel M, Forster S. 1996. Impact of biogenic sediment topography on oxygen fluxes in permeable seabeds. *Mar Ecol Prog Ser* 140:227–237.

CHAPTER 5

Porewater Chemistry: Effects of Sampling, Storage, Handling, and Toxicity Testing

William J Adams (Workgroup Leader), Robert M Burgess, Gerardo Gold-Bouchot, Lawrence Leblanc, Karsten Liber, Bruce Williamson

As a general principle, it is nearly impossible to remove a porewater sample from sediment, use it in a toxicity testing vessel with test organisms, and prevent changes in the chemistry of the natural and anthropogenic organic and inorganic constituents. The degree of change in the chemistry of various constituents that occurs in the process of porewater sampling, extraction, manipulation, storage, and testing can be significant. In spite of these suspected, known, and observed changes, we believe that with careful handling, extraction, and storage of the samples, along with rapid testing, the results of porewater toxicity tests can be meaningful. This belief is supported by several lines of evidence, including

1) reproducible toxicity tests,

2) companion solid-phase experiments,

3) toxicity identification evaluation (TIE) procedures, and

4) infaunal surveys.

In this chapter, we will offer guidance on advantages and disadvantages of different procedures used in sediment porewater sampling and on processing and storage from the viewpoint of how they affect the porewater chemistry. We will also briefly discuss sediment collection and the changes in porewater chemistry that may occur during the test procedure. Several of these issues have been discussed in other reviews (Adams 1991; American Society for Testing and Materials [ASTM] 1994; Bufflap and Allen 1995a; Burton 1998). However, the objective of this chapter is to update the state of the science. We will also conclude this review with a series of recommendations based on our expertise and experience. Finally, the chapter will end with a brief list of research needs to reduce the uncertainties surrounding porewater chemistry as it relates to toxicity testing.

Specific Considerations

The overriding factor in assessing the chemistry of pore waters and sediments is dictated by the purpose of the testing and by the questions to be answered. For example, is the investigation research oriented, routine testing, or large-scale monitoring? The type of investigation will drive several aspects of the sampling. Deciding in advance of testing what questions are to be answered will, to a large extent, influence the overall chemical and biological sampling design. In situations where the broadest types of questions are being asked (i.e., is this sediment toxic?), the sample collection, extraction, storage, and testing approach is the most difficult to design to ensure that changes in the natural chemistry of the pore water and the bioavailability and concentration of chemical contaminants are minimized. Because of the importance of this question, we encourage careful consideration of experimental design issues, including

1) contaminants of interest,

2) site selection,

3) sediment depth of interest,

4) spatial–temporal scale of sampling,

5) erosional versus depositional zones,

6) core versus grab samples, and

7) use of the data (i.e., research, monitoring, regulation).

These considerations are essential to designing an experimental approach that will reduce artifacts in the chemical form, bioavailability, and concentration of anthropogenic porewater contaminants and natural constituents, with the overall goal of producing more meaningful porewater toxicity data.

Sediment Porewater Sampling

Two broad categories of procedures exist for sampling sediment pore water: in situ and ex situ methods (Figure 5-1) (see Adams 1991, Bufflap and Allen 1995a, and Burton 1998 for more information). As the name suggests, in situ methods involve collection of pore water using samplers directly inserted into the sediment of interest and the pore water collected. Conversely, in ex situ sampling, the sediment of interest is removed from the natural setting and the pore water isolated elsewhere.

In situ porewater sampling

In situ porewater sampling is primarily performed using 1 of 3 methods: peepers (or dialysis), suction, or gel membranes/ion exchange resins. Advantages and disadvan-

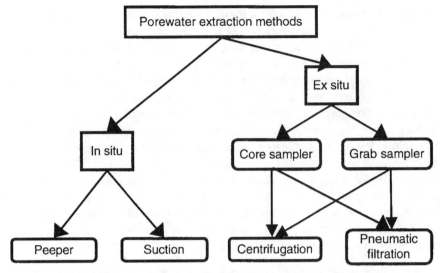

Figure 5-1 Schematic of porewater extraction methods

tages from the chemical perspective, as well as important practical considerations, are discussed below for the first 2 methods. The alternative approaches for collecting some porewater constituents and contaminants, gels and ion exchange resins, generally are used to measure chemical reaction rates or concentrations in sediments and pore waters, rather than for the actual isolation of pore water, and therefore are not of practical significance for discussion here.

Peepers

Peepers operate on the basis of passive, diffusion-controlled transport mechanisms. Sampling devices are usually constructed of Plexiglas or similar materials and generally have a filter membrane that allows diffusion of porewater constituents into the sampler. Sampling chambers can be deployed individually or arranged in a vertical configuration intended to allow for the sampling of porewater solutes at specific sediment depths. Many different designs and configurations exist, but all operate on the basis that high solute concentrations in sediment pore water will diffuse through a dialysis membrane into peeper cells, which initially contain pure water or water reconstituted to mimic the ionic composition (i.e., salinity) of the pore water being sampled. After some period of time, an equilibrium is assumed between solute concentrations in the external pore water and in the peeper cells. The use of a dialysis membrane allows for in situ filtration of pore water and therefore eliminates the need for subsequent efforts to remove fine particles from the pore water. Detailed descriptions of peeper designs and operation can be found elsewhere (e.g., Mudroch and Azcue 1995; Davison et al. 2000).

The principal advantages and disadvantages of using peepers for isolation of sediment pore water for toxicity testing are listed in Table 5-1. Despite their superior ability to accurately measure the concentration of many porewater constituents (especially inorganic compounds and metals), peepers are not the most practical method available to isolate pore water for toxicity testing. First, the sample volumes do not generally meet the requirements of most existing porewater toxicity tests. Most peeper designs have sample chambers with volumes of a few milliliters. Although larger chambers are feasible, they require very long equilibration times, especially in less porous sediments, because the large volume of uncontaminated water inside the cells would effectively dilute porewater contaminant concentrations for an extended period of time. Additionally, marine studies must consider the possible consequences of using deionized water inside the peeper cells as opposed to saline water to mimic the ionic strength of the pore water being sampled. Density-induced advection currents in the surrounding pore water may considerably slow the attainment of equilibrium (Grigg et al. 1999).

A second disadvantage is the uncertainty associated with the use of peepers for sampling hydrophobic organic compounds (HOCs) (e.g., log octanol–water partition coefficient [K_{OW}] > 4.0). For example, Carignan et al. (1985) demonstrated the loss of HOCs when pore water is sampled with peepers. In order to provide a consideration of the effect of peeper-based sampling on porewater concentrations of HOCs, we calculated whether the diffusion of sorbed organic compounds into a plastic peeper chamber would deplete the sediment adjacent to the membrane (dialysis). If the local area becomes depleted in HOCs, we would expect that the concentration of organic compounds in the peeper chamber would be lower than equilibrium porewater concentrations. This would be caused by compounds dissolved in porewater being "scavenged" as they pass through this area of the sediment. If the peeper capacity (i.e., the amount of compound in solution in the peeper chamber plus what is sorbed onto the plastic in the peeper chamber interior) is a large percentage of the sediment capacity (i.e., the total amount of chemical sorbed in a given volume of sediment that can be affected by diffusion), then the potential exists for depletion of the sediment adjacent to the dialysis membrane.

Results of this calculation are presented for 3 polycyclic aromatic hydrocarbon (PAH) compounds in Table 5-2. It can be seen that if we ignore sorption onto plastic (i.e., considering only equilibrium partitioning (EqP) from bulk sediment into the peeper chamber), then the capacity of the peeper is small relative to the local sediment, for moderately soluble and highly insoluble PAHs such as phenanthrene and benzo[a]pyrene. For HOCs of greater solubility (i.e., naphthalene), the capacity of the peeper volume begins to be more significant. This trend of decreasing peeper capacity with increasing hydrophobicity is expected because as compound hydrophobicity increases, a smaller fraction of the sorbed compound has to desorb into the porewater in order to reach equilibrium. Results also show that when an estimate of sorption onto plastic is included in the calculation, the capacity of the peeper (water plus sorption) far exceeds the capacity (by several hundred percent) of

Table 5-1 Advantages and disadvantages of several methods for isolating sediment pore water for toxicity testing

Method	Advantages	Disadvantages
Peeper	Porewater chemistry is measured without significant disturbance of the in situ equilibrium conditions.	Operates well for inorganic constituents (e.g., divalent metals, NH_3), but its utility for accurately sampling highly hydrophobic organic compounds is unknown (i.e., sorption of hydrophobic compounds onto sampler or membrane could artificially reduce porewater contaminant concentrations).
	Sample manipulation is reduced.	Extended equilibration time in the field is required (generally 15–20 d), resulting in need for 2 field trips: 1 for peeper deployment, 1 for peeper retrieval.
	Sampling influences on the oxidation state of metals are reduced.	Sample volumes are limited, generally to <10 ml. Larger peepers are limited to very porous substrates.
	Potential for loss of volatile substances, such as H_2S, and high Henry's Law constant HOCs, which occur in ex situ methods, is eliminated.	Uncontaminated water inside newly deployed peeper cells could effectively dilute porewater contaminant concentrations in low-porosity sediments.
	Use of a dialysis membrane eliminates the post-retrieval porewater filtration.	Samples must be collected from peepers immediately upon retrieval, resulting in a longer holding time for pore water outside of its natural matrix prior to toxicity testing.
	Redox conditions and pH are relatively unaltered, thereby minimizing changes in pH and oxygen-sensitive species (such as metals).	A high degree of technical competence and effort is required for proper use. Use in deeper water requires diving.
Suction	Operation is easy and low-technology.	Sorption of metals and HOCs on filter
	Functions best in highly porous sediments.	Clogging may occur in small to medium particle size sediments.
		Collection of pore waters from nontargeted depths (e.g., overlying water) may occur.
		Oxidation and degassing of pore water may occur.
Centrifugation	Extraction time is brief.	Labor intensive (e.g., sediment loading)
	Several variables (e.g., duration, speed) can be varied to optimize operation.	Lack of a generic methodology
	Large volumes of pore water are generated.	Sorption of HOCs to centrifuge tube construction material
	Operation is easy.	Lysis of cells during spinning
	Functions with fine to medium particle size sediments.	Does not function in sandy sediments
Filtration	Can be used with highly bioturbated sediments without lysis of cells.	HOCs loss on filter
		Degassing may occur.

Table 5-2 Estimation of potential depletion of 3 PAH compounds from sediments surrounding a 5 ml–volume peeper constructed from various polymers[a] and deployed for 2 weeks in Boston Harbor[b]

Compound		Surface area–normalized polymer partition coefficients[c] $(L/kg\ cm^2)$					
		HDPE	LDPE	PP	PVC	PET	PS
$K_{sa}{}^d$ - naphthalene (1 h)		2.62	3.68	2.41	0.14	0.17	0.64
$K_{sa}{}^e$ - naphthalene (24 h)		18.6	13.7	14.7	2.98	3.55	8.5
		Capacity ratio (%)[f]					
		Water plus sorption onto plastic chamber walls[h]					
	Water only[g]	HDPE	LDPE	PP	PVC	PET	PS
Naphthalene[i]	8.0	1200	840	890	270	340	600
Phenanthrene[j]	2.1	1000	1400	870	75	100	270
Benzo[a]pyrene[k]	0.5	5400	7400	4700	400	520	1500

[a] HDPE, LDPE = high density, low density polyethylene, PP = polypropylene, PVC = polyvinyl chloride, PET = polyethylene terephthalate, PS = polystyrene.

[b] Sediment PAH concentrations, Long and Morgan 1991; sediment porosity and organic carbon, Chen 1993.

[c] From a series of experiments using dissolved ^{14}C-naphthalene with polymers in pellet form (Jin 1997).

[d] K_{sa} = measurements of polymer–water distribution coefficients (K_{dp}) normalized to pellet surface area. The 1-h coefficient represents surface adsorption.

[e] 24-h measurements include absorption into the polymer as well as adsorption onto the polymer surface.

[f] Capacity ratio = peeper capacity (equilibrium PAH concentration in peeper) / sediment capacity (ng sorbed PAH) × 100. Sediment capacity (ng) = PAH concentration (ng/g) × membrane area (cm^2) × PAH diffusional distance (cm) × sediment density (g/cm^3).

[g] Water capacity (ng) = $C_s \times V_w / K_d$, where C_s = sediment concentration (ng/g), V_w = peeper volume (mls), and K_d = equilibrium sediment–water distribution coefficient (L/kg).

[h] Total capacity = water capacity + $[C_w \times K_{dp} \times m_p]$, where K_{dp} = the polymer–water distribution coefficients, based on K_{sa} above, and m_p = the mass of polymer inside the peeper chamber (g).

[i] Estimates of naphthalene sorption onto the polymers (K_{dp}) are based on the 24-h K_{sa} above × peeper surface area.

[j,k] Phenanthrene and benzo[a]pyrene polymer partition coefficients were estimated by dividing the ratio of $K_{oc}{:}K_{sa\ 1\text{-}h}$ for naphthalene into the K_{oc}s for phenanthrene and benzo[a]pyrene. K_{oc} = sediment organic carbon-normalized partition coefficient.

the sediment adjacent to the peeper. This calculation suggests that for a given deployment time (in this case, 2 weeks), depletion of the local sediment can occur, leading to peeper chamber concentrations that are lower than the porewater concentrations. However, if sorption onto the walls of the peeper chamber can be minimized, the potential exists for using peepers to sample for HOCs in pore water. Clearly, more work is needed in order to directly measure the magnitude of the sorption for various HOCs by peepers constructed of different materials.

Not considered in the above calculation is the effect of HOC sorption on porewater hydrocarbon concentrations. The type of membrane employed can have significant implications for porewater sampling. Polysulfone and polycarbonate membranes are most commonly used because of their greater stability and decreased likelihood of adsorption of HOCs. Membrane pore size could also have significant implications for measuring porewater contaminant concentrations. The majority of studies have used 0.45 or 0.2 μm, but other sizes, some as high as 70 μm (Burton 1998), have also been used. A practical disadvantage of peepers is the requirement for an extended in situ equilibration period. Peepers must be deployed and retrieved on separate occasions, generally 15 to 20 days apart, thus requiring 2 separate trips to the sampling location. This often doubles efforts and costs and can create complications related to identification of exact locations for retrieval. Occasionally, severe weather and hydrological conditions can dislodge and remove peepers, resulting in the complete loss of samples. Erosion of sediment around the peeper, especially in some marine or estuarine situations, could change the depth to which peepers are submerged in the sediment and could introduce artifacts. In addition, divers must often be used to deploy and retrieve peepers, and a high degree of technical competence is required for proper peeper preparation, deployment, handling, and subsampling. Other publications should be consulted for such details, for example, Carignan et al. 1985 and Mudroch and Azcue 1995.

Suction

In this method, porewater samples are collected in a suitable container (e.g., a syringe barrel) after the application of a vacuum (e.g., via a syringe) draws the pore water through a filter separating the liquid and particulate phases. The chief advantage of this method is that it is a simple technique for rapidly obtaining pore water in relatively porous media (Winger et al. 1998). More elaborate sampler designs have been published (Saager et al. 1990; Watson and Frickers 1990; Hursthouse et al. 1993; Bertolin et al. 1995) (Table 5-1). The sampler may be flushed with nitrogen before deployment to minimize oxidation of reduced pore water. Winger et al. (1998) found the method compared favorably to centrifugation for major cations and anions but not for metals.

A disadvantage with suction samplers is that they may be suitable only with highly permeable media and sediment particle sizes that do not clog the filter (Table 5-1). Consequently, filters with large surface areas or large pore sizes are often used (e.g., Sarda and Burton 1995; Winger et al. 1998) to circumvent blockage, but uncertainties surround the efficiency of the separation of pore water from the particulate phase because all particles are not excluded. Membrane filters can be used in sediments with a relatively low silt and organic carbon content, in order to better define filtration, but sample volume may be limited by blockage.

Hydrophobic organic compounds may be lost through adsorption onto the filter surface, a common problem with all methods that include a filtration process (see "Peepers," p 97). Heterogeneity in the permeability of the sediment will determine the sediment volume sampled because pore water will tend to flow in preferred paths (channeling). This may result in samples being collected from nontarget depths. At least 3 factors related to suction techniques require further investigation:

1) The potential of sampling overlying water instead of pore water because a permeable channel down the side of the sample tube may be created by sediment disturbance during tube insertion.

2) Degassing (e.g., CO_2, H_2S) and a concurrent change in pH may occur.

3) Oxygen may be introduced through the use of large glass sinters (e.g., air stones).

Ex situ porewater sampling

Unlike the in situ methods, ex situ methods involve an additional consideration: how to best collect the sediment sample before isolating the pore water (Figure 5-1). We believe that in sampling whole sediments, the main goal is to preserve the integrity of the sediment structure as best as possible, avoiding artifacts that might change the concentration, speciation, or bioavailability of the constituents and contaminants in the sample. This is particularly important because some porewater chemical constituents are reactive. A wide variety of specific whole-sediment sampling devices exists, not all of which will be discussed in detail here. The interested reader can consult more specific documents such as U.S. Environmental Protection Agency (USEPA 1982a, 1982b, 2001), Environment Canada (1994), and Burton (1998). Three types of samplers will be discussed: grabs, corers, and manual devices (Table 5-3).

Table 5-3 Advantages and disadvantages of several sediment sampling methods

Method	Advantage	Disadvantage
Grab	Large amount of sediment collected	Sample is frequently homogenized or altered by sampling procedure.
Corer	High-resolution depth profile	Small quantity of sediment collected
	Large box cores can be used to collect smaller subsample cores.	Large, heavy, and expensive equipment
Manual	Useful in shallow and intertidal zones	Excessive overlying water and small samples
		Not practical in deep water

Sediment sampling to obtain pore water

Grab samplers

Grab samplers are easy to use and, in general, provide a larger sample compared to other methods, thus providing sufficient pore water to perform toxicity tests and chemical analyses. Grabs can penetrate to a depth of approximately 10 to 15 cm. Their main disadvantage is that they tend to generate a bow wave during deployment, which results in a loss of any fine surficial sediments. Another possible disadvantage is that they tend to disrupt the sediment, thus destroying the vertical profile of chemical and biological constituents. Grabs that can quantitatively sample sediments have been described elsewhere (Grizzle and Stegner 1985; Murray and Murray 1987).

Corer samplers
Corer samplers are not constrained to shallow depths; in fact, basic devices can penetrate up to 2 meters and piston corers can sample up to 10 meters of sediment (Burton 1998). The main advantage of corers is that they disrupt the sample less than grabs, better preserving the vertical profile of the chemical and biological constituents of the sediment. This allows for sediments to be subsampled to specific depths. Core diameter must be considered because small diameters can create relatively large bow waves and disturb the sediment, resulting in smaller samples that may not be sufficient for all applications. The composition of core barrel material is another consideration. Some researchers prefer clear plastics, so the sample is visible and decisions can be made about what depths to subsample or to allow for the assessment of the degree of bioturbation. Care must be taken to avoid materials that will adsorb porewater constituents and contaminants from the sample.

Box corers provide a large, relatively undisturbed sample that preserves the vertical profile of the sediment sample. Horizontal sampling allows subsamples for toxicity, chemistry, benthic fauna, and other applications to be selected and collected from the same core. A disadvantage is that box corers tend to be large, heavy, and expensive, so their use in small boats can be difficult, if not impossible.

Manual devices
For shallow or intertidal sites, manual devices such as scoops can be used. They can be used to sample to a constant depth, but they must be used carefully to avoid collecting excessive surface water and losing any fine surficial sediments.

Ex situ porewater sampling techniques

Centrifugation
Centrifugation is the most commonly used ex situ method for separating pore water from sediment (Edmunds and Bath 1976; Adams et al. 1985; Adams 1991; ASTM 1994; Environment Canada 1994; Bufflap and Allen 1995a; Burton 1998) and is applicable with many porewater constituents and contaminants. For example, centrifugation has been used to study the porewater concentrations of dissolved organic carbon (DOC) (Orem and Gaudette 1984; Brownawell and Farrington 1986; Chin and Gschwend 1991; Burgess et al. 1996; Winger et al. 1998), metals (Carignan et al. 1985; Schults et al. 1992; Bufflap and Allen 1995a, 1995b; Mason et al. 1998; Winger et al. 1998), ammonia (Howes et al. 1985; Viel et al. 1991), and HOCs including PAHs, pesticides, and polychlorinated biphenyls (PCBs) (Brownawell and Farrington 1986; Schults et al. 1992; Burgess et al. 1996; Burgess and McKinney

1997; Ozretich and Schults 1998). In this method, the sediment sample is rotated at high speeds, resulting in a separation of liquid and particulate phases based on mass. After centrifugation, the sediments are compacted on the bottom of the centrifuge tube and the aqueous phase is on the sediment surface. Principal advantages of this method include relatively short (though labor-intensive) extraction times, generation of large porewater volumes (Burton 1992; ASTM 1994), ease of extraction, and published acceptance for use with several classes of contaminants (Schults et al. 1992; Bufflap and Allen 1995b; Mason et al. 1998; Ozretich and Schults 1998) (Table 5-1). Disadvantages include the potential for minor sorption of HOCs to centrifuge tubes (Schults et al. 1992; Ozretich and Schults 1998); the possibility of sample oxidation, temperature and other alterations (Adams 1991; ASTM 1994); the lack of a definitive and general methodology for applying centrifugation to all types of sediments, constituents, and contaminants (ASTM 1994); the lysis of biological cell membranes, resulting in the release of artifactual DOC and nutrients (Howes et al. 1985; Adams 1991); and equipment expense (Mason et al. 1998) (Table 5-1). A "universal" centrifugation methodology would define the optimum temperature, centrifugation force and duration, and other variables for conducting this isolation procedure. ASTM (1994) provides some specific recommendations for conducting centrifugation, but these vary on the basis of suspected contaminants. Despite the lack of a generic methodology, centrifugation is applicable to a wide variety of sediment types, including those high in organics, clays, and silt; however, because of their incompressibility, this method cannot be used with sandy sediments (ASTM 1994) (see "Pneumatic filtration," p 105).

As with all of the porewater extraction methods, there are several factors to consider when using centrifugation with sediments. Several aspects of the centrifugation method can be regulated but have not been definitively or scientifically determined for porewater toxicity. First, the centrifugation force of gravity ($\times g$) applied to a sediment during the separation can vary from relatively low values (e.g., <125 $\times g$) to extremely high forces (e.g., >10,000 $\times g$) (Adams 1991; ASTM 1994). Generally, with an increase in $\times g$, there is a decrease in the amount of sample mass that can be extracted because of the sample size limitations, but an increase in the thoroughness of separation between the particulate and interstitial phases. Second, the duration and temperature of the centrifugation can also be regulated. Third, the material from which centrifuge tubes are constructed is relevant to the porewater chemistry (e.g., metals adsorb to glass surfaces, and HOCs may partition to synthetic polymer surfaces) (Carignan et al. 1985; Schults et al. 1992; Bufflap and Allen 1995b; Ozretich and Schults 1998). Centrifuge tube materials include borosilicate glass, stainless steel, and a variety of plastics (e.g., polysulfone, Teflon). Unfortunately, as with the other variables, the scientific literature is equivocal about which construction material is most appropriate when the toxicants of concern are unknown. Along with the physical factors, there are some biological variables that should be considered when centrifugation is used. During centrifuging, low temperatures (e.g.,

4 °C) should be maintained in order to inhibit microbial alteration of the sample (e.g., degradation of DOC, generation of CO_2). To avoid contaminating the sample with biological fluids, macrobenthic organisms should be removed before centrifugation.

While a generic set of conditions does not exist for performing centrifugation with all sediments, porewater constituents, and potential contaminants, there are several recommendations that can be presented. For example, investigations particularly interested in porewater ammonia or DOC should use lower centrifugation speeds to reduce cell lysis (or use another extraction technique, such as peepers), while higher speeds may be better for investigating porewater metal concentrations (Carignan et al. 1985). In general, if metals or other ionic chemicals (including ammonia) are of interest, it is most appropriate to use a plastic centrifuge tube, while if HOCs are suspected, glass or stainless steel is recommended (Carignan et al. 1985; Schults et al. 1992; Bufflap and Allen 1995b; Ozretich and Schults 1998). Following centrifugation, both high Henry's Law constant HOCs and CO_2 may volatilize from the newly generated pore water. Consequences of this loss of volatiles can include an alteration of sample pH with resultant effects on toxicant bioavailability (e.g., ammonia). Limiting the amount of headspace in the centrifuge tube during initial loading should reduce this artifact effect. As noted above, avoiding the use of centrifuge tubes composed of highly sorptive materials is also highly recommended. Post-centrifugation filtration is not recommended because of the potential for sorption of contaminants to the filter surface.

Pneumatic filtration (squeezing)
Pneumatic filtration has been used for noncompressible sandy sediments where pore water cannot be extracted by centrifugation. In this procedure, pore water is displaced by an inert gas under pressure, passed through a filter, and collected in a receiving vessel. Elaborate, multiple extractors have been designed for high sample number processing and widely applied in the U.S., for sediments of all porosities (Carr and Chapman 1995). These authors concluded that when care was taken in the selection of materials, pneumatic filtration of sediments collected near petroleum production platforms yielded samples with toxicity similar to that of samples obtained from centrifugation and suction.

One disadvantage of pneumatic filtration is that HOCs can be lost through adsorption onto the filter surface (see "Peepers," p 97; Table 5-1). Further, in samples where the primary contaminants of concern are HOCs, Carr and Chapman (1995) recommended centrifugation as the method of choice for minimizing their loss. In contrast, pressure squeezing was the method of choice for extracting dissolved organic matter (DOM) in highly bioturbated sediments because DOM concentrations were significantly increased from cell disruption of biota during centrifugation (Martin and McCorkle 1993; Alperin et al. 1999).

We considered the possibility that pressure may increase inert gas solubility, which upon degassing in the filtrate may enhance loss of other volatile (e.g., HOC, H_2S) or saturated (e.g., CO_2) gases. However, no definitive information is available.

Storage and Handling of Sediment Samples for Porewater Extraction

Porewater extraction and testing often introduces artifacts (e.g., iron oxide precipitation) that are unavoidable. This makes it particularly difficult to devise generic guidelines for the storage and handling of sediment samples for porewater extraction. Any recommendations for procedures that minimize changes in porewater composition have to be made against a background of inevitable chemical changes in many situations. As a general principle, toxicity testing should be conducted as soon as possible after sediment collection, and storage times should be minimized. This is because some chemical changes are time dependent. However, due to scheduling and other logistical considerations, toxicity tests often cannot be performed immediately after sample collection, and sample storage and handling become important concerns.

Two options have been used:

1) storage of pore water in situ (i.e., store the sediment sample as intact as possible) or

2) extraction and storage of the pore water.

The first option attempts to leave the pore water intact within the sediment matrix and in some sort of "equilibrium" with the sediment, while maintaining it at 4 °C to slow biochemical reactions. The second option removes the sediment matrix, and the water is cooled (short-term storage) or frozen (long-term storage). The choice of the method of storage (isolated porewater samples or sediment cores or grabs) and the length of storage may depend to some extent on the suspected contaminants of interest.

Our consensus is that it is preferable to store pore water with its associated sediment, either in the form of a sediment core or a grab in a sealed container filled to the top with zero headspace, at 4 °C in the dark. This serves to maintain the integrity of the pore water in at least 2 ways:

1) Concentrations of organic compounds and metals are maintained by retaining the buffering capacity of the sediment because dissolved concentrations are often in equilibrium or in some sort of steady state with the solid phase. Artifacts such as loss of hydrophobic chemicals to sorption on container walls or volatilization into an overlying headspace are therefore minimized. This

approach also helps assure that volatile compounds like H_2S and CO_2 are not lost during storage.

2) The anoxic character of the pore water is efficiently maintained. Sediment can serve as a buffer to the diffusion of oxygen into pore waters because of the presence of reduced mineral species in the sediment. This will reduce artifacts associated with the slow oxidation of pore waters during storage, such as the precipitation of iron oxyhydroxides and manganese oxides and the coagulation of DOM.

For some studies, stirring of the sample is important to assure homogeneity. However, this should be avoided if possible to prevent oxygen from contacting the sample, or the oxic layers from mixing with the anoxic layers, thus changing the chemical characteristics and vertical profile of the sample, particularly for sediments that are relatively unbuffered to oxidation. Whether sediments should be thoroughly homogenized is dictated by the intent of the toxicity testing and use of the results. For example, Burgess and McKinney (1997) found homogenization of sediment before the collection of pore water by centrifugation resulted in significant increases in HOC concentrations. Storing intact core samples seems to be the better approach for preserving the integrity of porewater samples and their associated chemistry. Depending on the type of sediment and length of storage time, it may be necessary to store samples under nitrogen. For core samples, the isolation from the atmosphere provided by the core barrel might be sufficient for most purposes.

If the spatial integrity of the sample is an important issue, then some consideration should be given to targeting desired sediment profiles on site. Sediment samples can drain to some extent, especially those containing burrow water, because of the pressure shocks that can occur during transit (e.g., being shaken in a boat or motor vehicle).

The key principle is that sediment storage time should be kept to a minimum, preferably less than 2 weeks. Although anaerobic metabolism is slowed considerably at 4 °C, it will still occur at this temperature. Long-term storage of sediment cores will result in death of resident benthic organisms. In strongly bioturbated samples, this could result in releases of DOM and changes in concentrations of decomposition products such as ammonia and sulfide. Similarly, sediments enriched in decomposing organic material may accumulate these decomposition products over longer storage times.

Storage and Handling of Porewater Samples

If it is deemed necessary to isolate pore water from sediment for storage purposes, a number of potential artifacts should be considered, especially if oxidation is allowed to occur:

1) Dissolved concentrations of HOCs, such as many PCBs and PAHs and other compounds with log K_{OW}s > 4.0, will decrease because of sorption on the sides of the storage container.

2) Oxidation of pore waters that were anoxic in situ will lead to precipitation of hydrous iron oxides over very short time scales (hours) and iron and manganese oxides over longer timescales (several days) if the sample contains significant concentrations of dissolved iron and manganese. These precipitates provide surfaces for adsorption of metal ions such as Cu, Cd and Zn, ammonium ions, colloidal organic carbon (COC) and DOC, and organic contaminants (e.g., PCBs and PAHs). This oxidation–precipitation reaction also occurs during the test (see next section, "Chemical Changes to Pore Water in the Toxicity Testing Vessel," p 109).

3) As pore water becomes oxidized, partitioning of HOCs to DOC may be enhanced (fresh water) or suppressed (salt water), changing the bioavailability of any remaining dissolved HOCs (Hunchak-Kariouk et al. 1997; Pedersen et al. 1999).

4) Oxidation of pore waters will also lead to the oxidation of dissolved reduced species such as hydrogen sulfide, which will also occur during the test.

5) Oxidation of pore waters can lead to stimulation of aerobic bacterial processes, causing metabolism of DOC, production of metabolites such as CO_2, and utilization of nutrients such as NH_3.

To minimize these artifacts, we recommend the following: Store pore water in cleaned vessels that have low adsorptive capacity for trace metals and HOC under N_2 or with no headspace. Samples should be stored at low temperatures (4 °C) in the dark to minimize bacterial activity and photooxidation. Storage times ideally should be kept to less than 24 hours, as in standard water analysis procedures. However, longer times (several days) may be suitable. Filtration should be avoided because of potential loss of HOCs and metals onto filter surfaces. Also recommended is routine monitoring of dissolved oxygen (DO), NH_3, and H_2S concentrations before storing and before testing to evaluate alterations of porewater chemistry during storage. This can be done by using rapid methods of analysis, such as electrodes (e.g., DO and various ion-specific probes) and/or colorimetric analyses. The intent is to uncover gross changes due to storage that may impact porewater toxicity.

The freezing of water samples is a complex issue. Freezing preserves the integrity of a water sample against bacterial decay and, in this way, would provide an excellent means for storing large numbers of samples for porewater toxicity testing. It has been used extensively (Carr 1998) in sampling programs that require long delays between extraction and testing. Only a few studies have examined the effects of freezing on toxicity. Carr and Chapman (1995) found the toxicity of pore water from petroleum platforms to be unaltered by freezing. Also, Carr et al. 2001 found that freezing and thawing produced no difference in the concentration of the toxic organics (PAH, organochlorines, organotins), but several of the metals were reduced by ~50%. There was no difference in the toxicity between frozen or thawed and unfrozen samples, however. No systematic evaluation has been conducted to evaluate changes in chemistry and toxicity over time for both organic substances and metals.

For surface waters, as a rule, freezing is not recommended when the integrity of the sample is an issue with respect to its complexation and adsorbent properties because it can lead to coagulation of DOC (Giesy and Briese 1978), precipitation of dissolved species such as silica and metals, and alteration of major ion chemistry (Adams et al. 1980; Hilton et al. 1997; Tallberg et al. 1997). In addition, because of the expansion of water upon freezing and thawing, there is a high probability of loss of volatile organic compounds into the headspace. The importance of these effects when testing porewater toxicity is highly dependent on the chemistry of the porewater sample. The chemical reactions induced by freezing will also occur during the test procedures (e.g., metal oxide precipitation, adsorption of toxicants onto precipitates, degassing, and volatilization), although probably to different extents. Such effects (which occur with or without freezing) may be unimportant when high concentrations of contaminants are involved, or when the contaminants do not adsorb on any precipitated phases. However, until we have a much better understanding of the chemical and toxicological changes that occur in porewater samples as they are taken through the test procedure, it will be difficult to make recommendations about the utility of freezing for long-term storage.

Cooling of pore water below in situ temperatures can also lead to artifacts. Cooler temperatures increase the capacity of the water for DO, thereby increasing oxidation. This effect is thought to be acceptable because of the benefit of reduced bacterial activity, which otherwise can produce major changes in porewater chemistry, including increased CO_2, decreased pH, and decreased nutrient and DOC concentrations upon prolonged storage.

Chemical Changes to Pore Water in the Toxicity Testing Vessel

When pore water is introduced to the test vessel, a cascade of chemical changes may occur (Figure 5-2). The major change is the oxidation of the anoxic or suboxic porewater sample because toxicity tests are frequently conducted in open containers, which promotes the exchange of gases between the test solution (i.e., pore water) and the atmosphere. Furthermore, several new surfaces are available for porewater constituents and contaminants to interact with, including the testing chamber, toxicity testing organisms, and their food and waste. What follows is a discussion of the specific chemical changes that may be expected to occur under toxicity testing conditions.

Noncontaminants
- DO concentrations increase
- pH increases as a result of CO_2 degassing
- Aeration causes positive increase in oxidation-reduction potential

---

Metals
- Precipitation from solution on Fe and Mn oxides and sulfide
- Sequestration by DOC
- Formation of oxidized species of metals (e.g., Cr(VI), As(V))

Hydrophobic organic compounds (HOCs)
- Precipitation of DOC on Fe and Mn oxides
- Alteration of HOC–DOC partitioning caused by oxidation
- Sorption to testing vessel surfaces

Hydrophilic organic compounds
- Alteration of partitioning to DOC caused by oxidation
- *Unknown effects*

Ammonia
- Changes in pH caused by degassing may alter toxicity
- Sorption of NH_4^+ to testing vessel surfaces

Sulfur
Loss of H_2S by oxidation and volatilization

Figure 5-2 Review of chemical changes to the pore water occurring in toxicity testing chambers

Noncontaminant chemistry changes: DO, pH, redox

Briefly, once toxicity testing has commenced, the DO concentrations in the testing chamber will be increased to avoid stressing the test organisms. Saturated gases like CO_2 will exchange with the atmosphere, which may result in an increase in pH. However, if during the toxicity test there is an accumulation of food and fecal material in the chamber and insufficient aeration, an increase in biological oxygen demand may occur and cause an increase in CO_2 and a reduction in pH. However, in general, toxicity tests are performed with acceptable aeration to avoid stressing the test organisms. Finally, aeration will result in a positive increase in the oxidation–reduction potential of the solution.

Metals

Iron

The natural outcome of aeration of toxicity test solutions is the oxidation of dissolved iron (Fe^{2+}) with the rapid formation of hydrous iron oxide colloids and the slow formation of hydrous ferric oxide (HFO) precipitates. These colloids and precipitates adsorb trace metals and reduce their bioavailability (see Chapter 4). The importance of this to porewater toxicity testing depends on the relative concentrations of HFO, toxic metals, and competing ions (e.g., other trace metals, Mn^{2+}, Ca^{2+}, Mg^{2+}). The effect is well characterized and can be predicted at a useful level of accuracy (Dzombak and Morel 1990). Dissolved iron concentrations vary widely with depth, organic enrichment, bioturbation, and texture (see Chapter 4). As a rule, shallow sampling may avoid the higher Fe^{2+} concentrations and minimize the effect of precipitation. In addition, DOC can also affect trace metal adsorption through complexation of dissolved metal, changes in concentration by adsorption onto HFO, and alteration of the adsorption process (Santschi et al. 1997). Consequently, accurate prediction is not possible without considering the interaction of DOC. In addition, the nature of DOC may be changed by the oxidation process (see "Hydrophobic organic compounds and dissolved organic carbon," p 112).

Manganese

Mn^{2+}, like Fe^{2+}, is common in pore water from surface sediments but is more slowly oxidized (see Chapter 4). Hydrous manganese oxide precipitates formed in the presence of oxygen also adsorb trace metals (Murray 1975; Tessier 1992). At the start of the toxicity tests, following aeration of the test solution, any Fe^{2+} is precipitated as FeOOH, while Mn^{2+} stays in solution and MnO_2 will play only a minor role in any contaminant removal. Over longer test periods (e.g., 4 days), Mn^{2+} oxidation and MnO_2 precipitation may be significant for removing some trace contaminants (e.g., metals and HOCs). The oxidation–precipitation reaction may proceed to a significant extent at high pH and FeOOH concentrations (see Chapter 4). Therefore, in some special circumstances, manganese oxidation–precipitation may change trace contaminant bioavailability over the time course of the test. This is distinct from Fe^{2+} oxidation–precipitation, which has the largest effect before or at the start of the test. However, while Mn^{2+} oxidation and MnO_2 precipitation are well understood for natural systems, there is presently little information on its effects in porewater toxicity tests.

Aluminum

Elevated dissolved aluminum concentrations have been measured in pore waters collected from sediments in the Clark Fork River, Montana, USA (Brumbaugh et al. 1994). In theory, these elevated concentrations should not exist because aluminum does not undergo reductive solubilization (e.g., as Fe^{2+} and Mn^{2+}), and the maximum solubility of aluminum (as $Al(OH)_3$, an amphoteric metal) is at pH 6.4, which is well below the pH measured in Clark Fork River sediments. At the same time, concentrations of aluminum above 1 mg/L could be toxic to some organisms. This requires investigation before we can predict its importance and behavior in porewater toxicity tests.

Sulfide

In sediments and pore waters, sulfides play a dominant role in controlling concentrations of dissolved metals. In most sediments, sulfur and iron are in greater concentrations than trace metals. Sulfur interactions with iron result in the formation of (black) iron sulfides in anaerobic sediments. Under these same conditions, iron is replaced by divalent metals that have a greater affinity for sulfide than does iron. Metal sulfides are highly insoluble, which reduces their bioavailability. In oxic zones or locations where oxic and anoxic waters are mixed, metal concentrations in pore waters are controlled by the oxidation of metal sulfides and the liberation of soluble metals. Under these conditions, metal bioavailability is controlled by sorption to iron and manganese oxides. In porewater toxicity tests, the oxidation of the pore water results in the dissolution of iron sulfides and the formation of iron and manganese oxides that precipitate and reduce metal concentrations because of complexation reactions between divalent metals and iron and manganese oxides. The metal precipitation usually occurs well before the test is carried out.

Chromium

Chromium exists in both the +3 and +6 oxidation states, of which the +6 state is the most toxic (Richard and Bourg 1991). The dominant form in pore waters is expected to be the +3 state. Chromate (VI) is the thermodynamically stable state in natural oxygenated waters, so there is the potential for a change in toxicity of dissolved chromium upon aeration in pore water. However, Cr (III) is the usual form found in waters at low to moderate concentrations, probably because of complexation with DOC, kinetic factors, and the presence of natural reducing agents (DOC, H_2O_2) (Richard and Bourg 1991). Likewise, in reducing sediments, chrome +3 is the usual form detected. On this basis, it is predicted that oxidation of chrome +3 to +6 should not be a concern in porewater tests. However, the oxidation process may need to be considered when testing pore waters with high Cr levels to determine whether this condition (predominance of Cr (III)) is still applicable.

Arsenic

Arsenic exists in 2 different oxidation states in natural waters: +3 (arsenite) and +5 (arsenate). The As (III) form is marginally more toxic (2× to 3×) (USEPA 1984). While the As (III) may be the stable form in anoxic waters, and As (V) the stable form in oxygenated waters, the evidence is equivocal, with both forms being found in both anoxic and oxic pore waters (Riedel et al. 1987; Riedel 1993; Soto et al.

1994). While the toxicological consequences of the conversion of As +3 to As +5 during porewater toxicity testing is unknown, it is not expected to have a major impact on toxicity.

Hydrophobic organic compounds and dissolved organic carbon

Unlike the metals discussed above, which are affected directly by changes in porewater oxidation, HOCs are less significantly impacted by redox status. However, changes in the porewater chemistry of HOCs and DOC do occur as the result of porewater oxidation. Oxidation of pore waters upon initiation of a porewater toxicity test is expected to produce a number of changes to bulk DOC and dissolved HOCs of differing hydrophobicity. As noted in Chapter 4, the concentration of DOC in pore water generally increases with sediment depth. This increase is presumed to be due to condensation and polymerization of DOC under anaerobic conditions. The decrease in concentration at the water–sediment interface has been attributed to bacterial and chemical oxidation of DOC (Krom and Sholkovitz 1977; Orem and Gaudette 1984; Brownawell and Farrington 1986; Orem et al. 1986). Coagulation of DOC onto iron oxyhydroxide precipitates will reduce the concentration of DOC in isolated pore waters (Orem and Gaudette 1984). In addition, a limited number of studies show modifications in the degree of partitioning between HOCs and DOC (i.e., K_{DOC}) after aeration of anoxic pore waters. Partitioning of a tetrachlorobiphenyl isomer into anoxic porewater DOC has been shown to be altered upon aeration, causing an increase in porewater–DOC partitioning in fresh water (Hunchak-Kariouk et al. 1997) and a decrease in estuarine water (Pedersen et al. 1999). Pedersen et al. (1999) invoke differences in the nature of the conformation of hydrophilic and hydrophobic portions of DOC molecules between salt and fresh water to explain the changes in the direction of HOC–DOC partitioning upon aeration in saltwater and freshwater systems. While the contradictory nature of these results may confuse the understanding of the underlying mechanisms governing this partitioning, it serves as a warning of the potential for alteration of HOC–DOC partitioning when anoxic pore waters become oxidized. The time scale over which the above processes occur is on the order of minutes to hours and therefore important to consider for toxicity tests.

There are several other potential ways in which the dissolved concentrations of HOCs can be reduced during porewater toxicity testing. Because of the hydrophobic nature of HOCs, their concentrations will be reduced by sorption onto any surfaces for which they have an affinity. Generally, this loss is a function of the organic carbon content of the surface; however, the entropic effect will make almost any surface a viable site (Chiou et al. 1979). Loss through volatilization can also be important for HOCs with high Henry's Law constants, such as naphthalene. Further, any alteration of the ionic strength of the pore water will alter HOC solubility, which subsequently will affect dissolved concentrations. During toxicity testing, ionic strength is most frequently altered in testing solutions by the addition of hypersaline brine (or dehydrated sea salts) to increase salinity or the addition of deionized water to reduce salinity. Chemically, the addition of ions to a solution will reduce the ability of HOCs to remain in solution (Schwarzenbach et al. 1993). The magnitude of the change in solubility can be calculated if the Setschenow constant and solubility (in pure fresh water) of the chemical are known and if the salinity of the solution is available. Figure 5-3A provides a general relationship between log K_{OW} and the expected reduction in solubility for a range of HOCs (e.g., PAHs, PCBs, toluene) when salinity is adjusted from 0 ppt to 30 ppt. The relatively soluble organic compound phenol is included in this figure to represent effects in the low K_{OW} range. Under these conditions and for chemicals with log K_{OW}s of 1.0 to 5.5, a decrease in solubility of approximately 10% to 40% can be anticipated. Finally, temperature will also affect the solubility of HOCs, generally increasing solubility with increasing temperature (Schwarzenbach et al. 1993). Based on a relatively small dataset, Figure 5-3B demonstrates the reduction in solubility of several HOCs (e.g., PAHs and chlorinated solvents) as aqueous temperature is changed from approximately 30 °C to 4 °C. For the low K_{OW} compounds used in this figure, the solubility actually decreases with increasing temperature because of the competing heats of reaction.

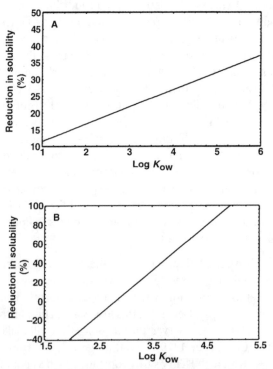

Figure 5-3 General relationship between **A**) the log K_{OW} of several organic compounds and reduction in solubility when adjusting solution salinity from 0 ppt to 30 ppt and **B**) the log K_{OW} of several organic compounds and reduction in solubility resulting from a decrease in temperature from approximately 30 °C to 4 °C. (Data from Schwarzenbach et al. 1993.)

In attributing porewater toxicity to HOCs, it is important to consider the saturation concentrations of the compounds of interest. Specifically, it is important to consider whether it is physically possible, under even ideal conditions, to attain concentrations of single compounds or mixtures of compounds sufficient to cause toxicity. Figure 5-4A presents a general treatment of chemical aqueous solubility (represented by log K_{OW}) to the actual exposure quantity for a variety of testing volumes. Figure 5-4B provides a specific example for fluoranthene. If the maximum possible exposure concentration in a given testing volume is small relative to known LC50 (lethal concentration to 50% of the test population) values, there may be insufficient HOC available to cause adverse effects. In other words, for very insoluble HOCs, frequently too little chemical is bioavailable to result in any sort of toxic effect.

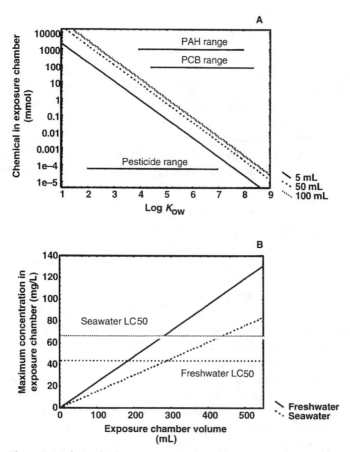

Figure 5-4 Relationship between organic chemical concentrations in toxicity test chambers and chamber volume: **A)** expressed as a function of log K_{OW} for a selection of contaminant classes and **B)** specifically for the PAH fluoranthene in both freshwater and seawater systems. Solubility data was collected from Schwarzenbach et al. (1993), and LC50 values are cited from Spehar et al. (1999).

Hydrophilic Organic Compounds

Dissolved concentrations of hydrophilic organic compounds, such as phenolics, cationic and anionic surfactant compounds, and estrogenic pharmaceutical chemicals, may attain concentrations sufficient to cause toxicity in the small volumes associated with some porewater toxicity tests. The degree of complexation of hydrophilic organic compounds with sediments is an area of current research, which indicates they may persist in sediments and therefore may partition into pore waters along with their products of biologically induced degradation (John et al. 2000). Therefore, like the HOCs, considerations such as reduced bioavailability of dissolved compounds due to complexation with DOC, and artifacts associated with changing the nature and/or degree of this partitioning, may be important.

Ammonia

Although ammonia is not as sensitive to oxidation and surface reactivity or as likely to undergo precipitation during toxicity tests as are some of the porewater constituents that have been discussed earlier, ammonia chemistry is worth considering further. Ammonia is formed during breakdown of organic matter, usually under anaerobic conditions. It can be oxidized by nitrifying bacteria, first to nitrite, then to nitrate under aerobic conditions (Russo 1985). Nitrite (NO_2^-) is known to be toxic to aquatic organisms, but it is rapidly oxidized to nitrate (NO_3^-), which is relatively nontoxic. Nitrification is considered unlikely for samples held for short periods of time (i.e., up to 96 hours).

More likely to be an issue during toxicity tests are changes in ammonia toxicity that result from changes in pH as the test proceeds. Ammonia speciation between NH_3 and NH_4^+ is regulated by pH (Miller et al. 1990), with unionized NH_3 as the toxic form. Consequently, pH alterations that occur during some porewater toxicological exposures need to be considered and minimized. There are many reactions that can change the pH of an unbuffered solution, for example, CO_2 volatilization or iron hydrous oxide precipitation. Therefore, it may be difficult to maintain pH at values measured in the newly isolated pore water. The importance of pH changes on NH_3 can be evaluated by calculating free NH_3, checking its likely toxicity against known water quality criteria, and calculating the magnitude of any change in concentration that results from change in pH (USEPA 1985, 1998). Some researchers have expressed concerns about adsorptive loss of NH_4^+ onto surfaces (apparatus and precipitates) with the resulting change in NH_3 concentrations, but we have insufficient information to evaluate this. Another concern that has been expressed is the lysing of algal and bacterial cells and the release of NH_3. However, this is not thought to be significant because algal cells do not store NH_3. A valid concern, however, is the release of NH_3 from the decomposition of algal blooms. Where this occurs in the environment, for example, it can be large enough to kill fish in overlying water. A further concern expressed is the loss of NH_3 through volatilization.

However, this is insignificant at the pH of most natural pore waters and will be significant only at pH > 9.5 (Stratton 1968). A secondary concern related to ammonia speciation is the potential for NH_4^+ to adsorb to toxicity testing surfaces (and particles), particularly at pH < 7 where NH_4^+ is the dominant form. Rosenfeld (1979) reported that NH_4^+ associated with the particulate phase in marine sediment was 1× to 2× greater than the concentration in the pore waters. There was evidence that the NH_4^+ was associated with both inorganic and organic components of the sediments (e.g., clays and organic carbon) (Rosenfeld 1979). How this relates to the potential for NH_4^+ to adsorb to toxicity testing materials is unclear but suggests there may be some level of NH_4^+ loss. In conclusion, due to the high solubility of ammonia, we generally assume there is relatively little concern with artifacts like sorption occurring during toxicity testing; however, the potential does exist. We recommend that porewater ammonia concentrations be monitored during sampling and toxicity testing so that any loss can be documented.

Hydrogen sulfide

H_2S is extremely toxic to test animals and will react with trace metals to form insoluble sulfides (see "Metals," p 110) (Wang and Chapman 1999). It is oxidized by DO, and oxidation is enhanced by iron oxides. H_2S is also easily volatilized in solution, especially at low pH. Oxidation usually results in sulfur precipitates. The sulfur cycle in sediments is well understood and plays a major role in the control of metals through the formation of insoluble metal sulfides. The importance of dissolved sulfide and precipitated sulfur in porewater testing has not been extensively studied.

General Recommendations and Research Needs

General recommendations

In assembling these recommendations, 3 principles are emphasized:

1) The specific purpose of each porewater investigation should dictate the method and procedures used to collect pore water for the objectives of conducting toxicity tests.

2) The collection of pore water by necessity introduces artificiality to the sample that does not exist in the field.

3) Nevertheless, the objective of the extraction procedure should be to preserve the chemical and toxicological characteristics of the pore water as best as possible.

Given these general principles, several specific recommendations are provided:

1) Several methods exist for collecting pore water, and each has advantages and disadvantages (Tables 5-3, 5-4).

2) Peepers may provide better information on actual in situ chemistry, at least for inorganic contaminants and major constituents, but may be unsuitable for collecting HOCs.

3) Ex situ sampling using centrifugation appears to be most widely applicable for routine porewater toxicity testing and is most suitable for collecting HOCs.

4) Core or grab sampling can be used to collect sediment for porewater isolation, but disturbance and manipulation of the sediment and sediment profile should be minimized before the porewater extraction. Samples should be stored at 4 °C in the dark under nitrogen (or another inert gas) and/or zero headspace.

5) Porewater isolation can be performed in either the field or the laboratory, but the latter is generally recommended because the pore water is maintained during transport and storage in as close to its original state as possible.

6) When pore water is isolated in the field, samples should be stored with the container completely filled without any headspace, kept cool (4 °C), and toxicity testing initiated as soon after porewater isolation as possible.

7) When pore water is isolated in the laboratory, sediment samples should be transported in sealed containers either with no headspace or with nitrogen or other inert gas atmosphere and should be kept cool (4 °C) until pore water is isolated (generally under a nitrogen atmosphere), and toxicity tests should be initiated immediately after isolation. For static renewal tests, pore water should not be isolated all at once, but in small batches as needed.

8) Easily measured and toxicologically important chemical characteristics, including salinity, pH, conductivity, DO, NH_3, H_2S, and Eh, should be recorded after collection and storage.

9) Further, these parameters are expected to change during sample manipulation and toxicity testing.

10) To improve accuracy, a combination of both in situ and ex situ methods to assess field and laboratory conditions should be considered.

11) Minimize handling and storage of isolated pore water prior to testing, and initiate toxicity testing as quickly as possible after porewater isolation.

Table 5-4 Comparison of different methods of isolating pore water for toxicity testing (assuming the use of optimal experimental procedures)

Consideration	In situ methods		Ex situ methods	
	Peeper	Suction	Centrifugation	Pneumatic filtration
Potential for changing chemical equilibrium conditions during sample isolation	Inorganics: Low Nonpolar organics: High	Inorganics: Intermediate Nonpolar organics: High	Inorganics: Low Nonpolar organics: Low	Inorganics : Low to intermediate Nonpolar organics: High
Potential for changing chemical equilibrium conditions during sample holding (prior to test start)	High	High	Low	Low
Potential for sample oxidation	Low	Intermediate	Intermediate to high	High
Potential for loss of volatile compounds (e.g., H_2S, CO_2, volatile organics)	Low	High	Intermediate	High
Filtration	Automatic	Some degree of mechanical screening required	Not necessary	Some degree of mechanical screening or filtering required
Length of porewater holding time prior to testing	Long	Long	Short	Short
Appropriateness for sampling hydrophilic compounds	Good	Good	Good	Good
Appropriateness for sampling hydrophobic compounds	Questionable	Questionable	Reasonable to good	Questionable
Appropriateness for use with ammonia-rich sediments	High	Questionable	High	Questionable
Appropriateness for sampling metals	Very good	Reasonable	Good	Reasonable
Required level of overall sampling and isolation effort	High	Low to intermediate	Intermediate	Intermediate
Volume of pore water isolated	Small	Small to intermediate	High	High
Availability of methodological information	High	Low	High	High
Ability to sample pore water at specific sediment depths	High	Low to intermediate	High (if isolated from core sections)	Intermediate
Applicability in a wide range of sediment particle sizes	All	Permeable sediments only	Compactable sediments	All sediment types

Research Needs

We conclude this chapter with a series of research needs that will advance the science of porewater chemistry in the special context of the toxicity testing of pore waters. The most fundamental questions revolve around the limitations of the porewater toxicity test imposed by the inevitable chemical changes that occur during porewater testing procedures. These include oxidation of dissolved Fe^{2+}, Mn^{2+}, HS^-, and DOC; precipitation of iron and manganese oxides and sulfur; adsorption of metals, DOC, NH_4^+, and HOCs; and degassing or volatilization of CO_2, H_2S, and volatile hydrocarbons. The following research needs are identified:

1) Investigate development of a standard procedure allowing for the oxidation–precipitation of iron and manganese oxides and sulfides, along with alterations to DOC prior to the initiation of toxicity testing. Pore waters are frequently anoxic, and oxidation–precipitation reactions are commonly observed. These changes are inevitable in many porewater toxicity tests, so it may be more effective to let these reactions occur in a prescribed manner during the test procedure. The impact of iron and manganese oxide precipitation on toxicity test results needs to be investigated for both metals and HOCs.

2) Develop a better understanding of the concentrations and distributions of prominent colloid species in porewater tests, including DOC and iron and manganese oxides. Specifically, a quantitative understanding of the interactions of colloids with surfaces and with dissolved contaminants is needed.

3) Conduct simple aqueous phase experiments to scope the extent of degassing and volatilization of CO_2, H_2S, and NH_3 with sampling, extraction, and storage procedures to provide guidance for method standardization and for assessing the importance of these effects.

4) Test the validity of extraction methods in simple systems (i.e., spiked solutions and sediments) to evaluate volatilization and sorption of HOCs, metals, and NH_4^+ to the apparatus.

5) Perform studies to measure the effects of sample holding time and holding method on the fundamental constituents in both pore water and whole sediments (e.g., DO, redox, CO_2, DOC, H_2S, NH_3, Fe, and Mn).

References

Adams DD. 1991. Sampling sediment pore water. In: Mudroch A, MacKnight SD, editors. CRC handbook of techniques for aquatic sediments sampling. Boca Raton FL, USA: CRC. p 171–202.
Adams DD, Darby DA, Young RJ. 1980. Selected analytical techniques for characterizing the metal chemistry and geology of fine-grained sediments and interstitial water. In: Baker RA, editor.

Contaminants and sediments, Volume 2. Analysis, chemistry and biology. Ann Arbor MI, USA: Ann Arbor Science. p 373–392.

Adams WJ, Kimerle RA, Mosher RG. 1985. Aquatic safety assessment of chemicals sorbed to sediments. In: Cardwell RD, Purdy R, Bahner RC, editors. Aquatic toxicology and hazard assessment, Seventh symposium. Philadelphia PA, USA: American Society for Testing and Materials. ASTM STP 854. p 429–453.

Alperin MJ, Martens CS, Albert DB, Suayah IB, Benninger LK, Blair NE, Jahnke RA. 1999. Benthic fluxes and porewater concentration profiles of dissolved organic carbon in sediments from the North Carolina continental slope. *Geochim Cosmochim Acta* 63:427–448.

[ASTM] American Society for Testing and Materials. 1994. Standard guide for collection, storage, characterization, and manipulation of sediments for toxicological testing. Designation E 1391-94. Volume 11.04, ASTM annual book of standards. Philadelphia PA, USA: ASTM.

Bertolin A, Rudello D, Ugo P. 1995. A new device for in-situ pore-water sampling. *Mar Chem* 49:233–239.

Brownawell BJ, Farrington JW. 1986. Biogeochemistry of PCBs in interstitial waters of a coastal marine sediment. *Geochim Cosmochim Acta* 50:167–169.

Brumbaugh WG, Ingersoll CG, Kemble NE, May TW, Zajicek JL. 1994. Chemical characterization of sediments and pore water from the upper Clark Fork River and Milltown Reservoir, Montana. *Environ Toxicol Chem* 13:1971–1983.

Bufflap SE, Allen HE. 1995a. Sediment pore water collection methods for trace metal analysis: A review. *Water Res* 29:165–177.

Bufflap SE, Allen HE. 1995b. Comparison of pore water sampling techniques for trace metals. *Water Res* 29:2051–2054.

Burgess RM, McKinney RA. 1997. Effects of sediment homogenization on interstitial water PCB geochemistry. *Arch Environ Contamin Toxicol* 33:125–129.

Burgess RM, McKinney RA, Brown WA. 1996. Enrichment of marine sediment colloids with polychlorinated biphenyls: Trends resulting from PCB solubility and chlorination. *Environ Sci Technol* 30:2556–2566.

Burton Jr GA. 1992. Sediment collection and processing: Factors affecting realism. In: Burton Jr GA, editor. Sediment toxicity assessment. Boca Raton FL, USA: Lewis. p 37–66.

Burton Jr GA. 1998. Assessing aquatic ecosystems using pore waters and sediment chemistry. Ottawa ON, Canada: Natural Resources Canada, CANMET. Technical Evaluation, Aquatic Effects Technology Evaluation Program Project 3.2.2a.

Carignan R, Rapin F, Tessier A. 1985. Sediment pore water sampling for metal analysis: A comparison of techniques. *Geochim Cosmochim Acta* 49:2494–2497.

Carr RS. 1998. Marine and estuarine porewater toxicity testing. In: Wells PG, Lee K, Blaise C, editors. Microscale testing in aquatic toxicology: Advances, techniques and practices. Boca Raton FL, USA: CRC. p 523–538.

Carr RS, Chapman DC. 1995. Comparison of methods for conducting marine and estuarine sediment porewater toxicity tests: Extraction, storage, and handling techniques. *Arch Environ Contam Toxicol* 28:69–77.

Carr RS, Nipper M, Biedenbach JM, Hooten RL, Miller K, Saepoff S. 2001. Sediment toxicity identification evaluation (TIE) studies at marine sites suspected of ordnance contamination. *Arch Environ Contam Toxicol* 41:298–307.

Chen H-W. 1993. Fluxes of organic pollutants from the sediments in Boston Harbor [master's thesis]. Cambridge MA, USA: Dept Civil and Environmental Engineering, Massachusetts Institute of Technology.

Chin Y-P, Gschwend PM. 1991. The abundance, distribution, and configuration of porewater organic colloids in recent sediments. *Geochim Cosmochim Acta* 55:1309–1317.

Chiou CT, Peters LJ, Freed VH. 1979. A physical concept of soil-water equilibria for nonionic compounds. *Science* 206:831–832.

Davison W, Fones G, Harper M, Teasdale P, Zhang H. 2000. Dialysis, DET and DGT: In situ diffusional techniques for studying water, sediments and soils. In: Buffle J, Horvai G, editors. In situ monitoring of aquatic systems: Chemical analysis and speciation. Chichester, UK: J Wiley. p 495–569.

Dzombak DA, Morel FMM. 1990. Surface complexation modeling; Hydrous ferric oxide. New York NY, USA: J Wiley.

Edmunds WM, Bath AH. 1976. Centrifuge extraction and chemical analysis of interstitial waters. *Environ Sci Technol* 10:467.

Environment Canada. 1994. Guidance document on collection and preparation of sediment for physicochemical characterization and biological testing. Ottawa ON, Canada: Environment Canada, Technology Development Directorate. EPS 1/RM/29.

Giesy Jr JP, Briese LA. 1978. Particulate formation due to freezing in humic waters. *Water Resour Res* 14:542–544.

Grigg NJ, Webster IT, Ford PW. 1999. Pore-water convection induced by peeper emplacement in saline sediment. *Limnol Oceanogr* 44:425–430.

Grizzle RE, Stegner WE. 1985. A new quantitative grab for sampling benthos. *Hydrobiol* 126:91–95.

Hilton J, Nolan L, Geelhoed-Bonouvrie P, Comans RNJ. 1997. The effect of different treatment processes on estimates of radionuclide distribution coefficients in freshwater sediments. *Water Res* 31:49–54.

Howes BL, Dacey JWH, Wakeham SG. 1985. Effects of sampling technique on measurements of porewater constituents in salt marsh sediments. *Limnol Oceanogr* 30:221–227.

Hunchak-Kariouk K, Schweitzer L, Suffet IH. 1997. Partitioning of 2,2',4,4'-tetrachlorobiphenyl by the dissolved organic matter in oxic and anoxic porewaters. *Environ Sci Technol* 31:639–645.

Hursthouse AS, Iqbal PP, Denman R. 1993. Sampling interstitial waters from intertidal sediments: An inexpensive device to overcome an expensive problem? *Analyst* 118:1461–1462.

Jin B. 1997. Investigation of the sorption characteristics of volatile organic compounds (VOCs) by waste polymers [master's thesis]. Stony Brook NY, USA: State Univ New York, Marine Sciences Research Center.

John DM, House WA, White GF. 2000. Environmental fate of nonylphenol ethoxylates: Differential adsorption of homologs to components of river sediments. *Environ Toxicol Chem* 19:293–300.

Krom MD, Sholkovitz ER. 1977. Nature and reactions of dissolved organic matter in the interstitial waters of marine sediments. *Geochim Cosmochim Acta* 41:1565–1573.

Long ER, Morgan LG. 1991. The potential for biological effects of sediment-sorbed contaminants tested in the National Status and Trends Program. Seattle WA, USA: National Oceanic and Atmospheric Administration. NOAA Technical Memorandum NOS OMA 52.

Martin WR, McCorkle DC. 1993. Dissolved organic carbon concentration in marine pore waters determined by high-temperature oxidation. *Limnol Oceanogr* 38:1464–1479.

Mason R, Bloom N, Cappellino S, Gill G, Benoit J, Dobbs C. 1998. Investigation of porewater sampling methods for mercury and methylmercury. *Environ Sci Technol* 32:4031–4040.

Miller DC, Poucher S, Cardin JA, Hansen D. 1990. The acute and chronic toxicity of ammonia to marine fish and a mysid. *Arch Environ Contam Toxicol* 19:40–48.

Mudroch A, Azcue J. 1995. Manual of aquatic sediment sampling. Boca Raton FL, USA: Lewis. 219 p.

Murray JW. 1975. Interaction of metal ions at the manganese dioxide-solution interface. *Geochim Cosmochim Acta* 39:505–519.

Murray WG, Murray J. 1987. A device for obtaining representative samples from the water sediment interface. *Mar Geol* 76:313–317.

Orem WH, Gaudette HE. 1984. Organic matter in anoxic marine pore water: Oxidation effects. *Org Geochem* 5:175–181.

Orem WH, Hatcher PG, Spiker EC, Szeverenyi NM, Macie GE. 1986. Dissolved organic matter in anoxic porewaters from Mangrove Lake, Bermuda. *Geochim Cosmochim Acta* 50:609–618.

Ozretich RJ, Schults DW. 1998. A comparison of interstitial water isolation methods demonstrates centrifugation with aspiration yields reduced losses of organic constituents. *Chemosphere* 36:603–615.

Pedersen JA, Gabelich CJ, Lin C-H, Suffet IH. 1999. Aeration effects on the partitioning of a PCB to anoxic estuarine sediment pore water dissolved organic matter. *Environ Sci Technol* 33:1388–1397.

Richard FC, Bourg AC. 1991. Aqueous geochemistry of chromium: A review. *Water Res* 25:807–816.

Riedel GF. 1993. The annual cycle of arsenic in a temperate estuary. *Estuaries* 16:533–540.

Riedel GF, Sanders JG, Osman RW. 1987. Biogeochemical control on the flux of trace elements from estuarine sediments: Water column oxygen concentrations and benthic infauna. *Estuar Coast Shelf Sci* 44:23–38.

Rosenfeld JK. 1979. Ammonium adsorption in nearshore anoxic sediments. *Limnol Oceanogr* 24:356–364.

Russo RC. 1985. Ammonia, nitrite, and nitrate. In: Rand GM, Petrocelli SR, editors. Fundamentals of aquatic toxicology. Washington DC, USA: Hemisphere. p 455–471.

Saager PM, Sweerts J-P, Ellermeijer HJ. 1990. A simple pore-water sampler for coarse, sandy sediments of low porosity. *Limnol Oceanogr* 35:747–751.

Santschi PH, Lenhart JJ, Honeyman BD. 1997. Heterogeneous processes affecting trace contaminant distribution in estuaries: The role of natural organic matter. *Mar Chem* 58:99–125.

Sarda N, Burton Jr GA. 1995. Ammonia variation in sediments: Spatial, temporal and method-related effects. *Environ Toxicol Chem* 14:1499–1506.

Schults DW, Ferraro SP, Smith LM, Roberts FA, Poindexter CK. 1992. A comparison of methods for collecting interstitial water for trace organic compounds and metals analyses. *Water Res* 26:989–995.

Schwarzenbach RP, Gschwend PM, Imboden DM. 1993. Environmental organic chemistry. New York NY, USA: J Wiley. 681 p.

Soto EG, Rodriguez EA, Rodriguez DP, Mahia PL, Lorenzo SM. 1994. Extraction and speciation of inorganic arsenic in marine sediments. *Sci Total Environ* 141:87–91.

Spehar RL, Poucher S, Brooke LT, Hansen DJ, Champlin D, Cox DA. 1999. Comparative toxicity of fluoranthene to freshwater and saltwater species under fluorescent and ultraviolet light. *Arch Environ Contamin Toxicol* 37:496–502.

Stratton FE. 1968. Ammonia losses from streams. *J San Eng Div, Proc Am Soc Civil Eng* 94:1085–1092.

Tallberg P, Hartikainen H, Kairesalo T. 1997. Why is soluble silicon in interstitial and lake water samples immobilized by freezing? *Water Res* 31:130–134.

Tessier A. 1992. Sorption of trace elements on natural particles in oxic environments. *Environ Part* 1:425–453.

[USEPA] U.S. Environmental Protection Agency. 1982a. Sampling protocols for collecting surface water, bed sediment, bivalves, and fish for priority pollutant analysis. Washington DC, USA: Monitoring and Data Support Division, Office of Water Regulations and Standards. Final draft report.

[USEPA] U.S. Environmental Protection Agency. 1982b. Handbook for sampling and sample preparation of water and wastewater. Cincinnati OH, USA: Environment Monitoring and Support Laboratory. EPA-600/4-82-029.

[USEPA] U.S. Environmental Protection Agency. 1984. Ambient water quality criteria for arsenic. Washington DC, USA: Office of Water. EPA-440/5-85-033.

[USEPA] U.S. Environmental Protection Agency. 1985. Ambient water quality criteria for ammonia. Washington DC, USA: Office of Water. EPA-440/5-85-001.

[USEPA] U.S. Environmental Protection Agency. 1998. 1998 update of ambient water quality criteria for ammonia. Washington DC, USA: Office of Water. EPA-822-R-98-008.

[USEPA] U.S. Environmental Protection Agency. 2001. Methods for collection, storage and manipulation of sediments for chemical and toxicological analyses: Technical manual. Washington DC, USA: USEPA, Office of Science and Technology, Office of Water. EPA-823-B-01-002.

Viel M, Barbanti A, Langone L, Buffoni G, Paltrinieri D, Rosso G. 1991. Nutrient profiles in the pore water of a deltaic lagoon: Methodological considerations and evaluation of benthic fluxes. *Estuar Coast Shelf Sci* 33:361–382.

Wang FY, Chapman PM. 1999. Biological implications of sulfide in sediment: A review focusing on sediment toxicity. *Environ Toxicol Chem* 18:2526–2532.

Watson PG, Frickers TE. 1990. A multilevel, in situ pore-water sampler for use in intertidal sediments and laboratory microcosms. *Limnol Oceanogr* 35:1381–1389.

Winger PV, Lasier PJ, Jackson BP. 1998. The influence of extraction procedure on ion concentrations in sediment pore water. *Arch Environ Contam Toxicol* 35:8–13.

CHAPTER 6

Porewater Toxicity Testing: An Overview

Kenneth G Doe, G Allen Burton Jr, Kay T Ho

S ediments act as sinks for contaminants, where they may build up
to toxic levels. Sediments containing toxic levels of contaminants
pose a risk to aquatic life, human health, and wildlife. There is an overwhelming
amount of evidence that demonstrates chemicals in sediments are responsible for
toxicological (Williams et al. 1986; Chapman 1988; Ankley et al. 1989; Giesy and
Hoke 1989, 1990; Swartz et al. 1989) and ecological effects (Swartz et al. 1982, 1994;
Anderson et al. 1987; Bailey, Day et al. 1995; Hartwell et al. 1997; Hatakeyama and
Yokoyama 1997).

Pore water (interstitial water) is a major route of exposure to contaminants for many
benthic organisms. Contaminants in pore water can be transported to ground water
or to overlying water by a variety of processes, thus exposing hyporheic and water
column organisms. Porewater assessments are challenging because sediment
gradients and microenvironments that control the physicochemical characteristics
are disrupted during the porewater collection process, thus bioavailability may be
significantly altered. However, if proper in situ and ex situ collection methods are
used, then reliable and accurate conclusions can be obtained. The optimal methods
used depend on the questions being asked and the site-specific conditions.

An accurate assessment of the importance of porewater contamination in an
ecosystem requires an understanding of

1) the in situ bioavailability of contaminants,

2) the exposure of indigenous organisms,

3) the predictive capability of surrogate species effects to the ecosystem, and

4) the predictive capability of laboratory results to field conditions.

Historically, assessment of sediment quality has been carried out by using 3 types of
characterization techniques: analysis for chemical contaminants, benthic community structure, and toxicity. A weight-of-evidence approach, using a combination of
these techniques (the Sediment Quality Triad [SQT] is an example; see Chapters 8
and 9) is now generally accepted as the superior approach to sediment assessments

Porewater Toxicity Testing: Biological, Chemical, and Ecological Considerations. R. Scott Carr and Marion Nipper, editors.
© 2003 Society of Environmental Toxicology and Chemistry (SETAC). ISBN 1-880611-65-1

(Burton 1998). Toxicity testing may be carried out on whole sediment using benthic organisms, and this has been the approach for assessing the suitability of dredged material for different disposal options in the regulatory arena in the U.S. (U.S. Environmental Protection Agency/U.S. Army Corps of Engineers [USEPA/USACE] 1991). However, measurement of total levels of contaminants in whole sediments or in the pore water does not necessarily relate the observed bioavailability of the contaminants. In the mid 1980s, the USEPA proposed an approach for developing sediment quality guidelines (SQGs) based on equilibrium partitioning (EqP) of contaminants in pore water. EqP theory predicts that pore water is the controlling exposure medium in the toxicity of sediments to organisms (USEPA 1993). The USEPA Equilibrium Partitioning Approach suggests that a major route of exposure is pore water, and it accounts for acute toxicity effects when sediment concentrations of nonpolar organics and metals are normalized to the primary sorption sites (total organic carbon [DOC] and acid volatile sulfide [AVS], respectively). For this reason, TOC and AVS measurements are useful in sediment and porewater toxicity assessments. This relationship has been shown to work in many lentic and estuarine environments with a few chemicals. However, its applicability to flowing water systems and more dynamic sediment environments has not been well tested. In addition, the influence of sediment sampling, manipulation, and other confounding factors (discussed later in this chapter) is likely, yet ill defined. These uncertainties provide compelling reasons to assess porewater toxicity directly.

Both lethal and sublethal endpoints have been used in porewater tests with a variety of test organisms (Carr 1998). Well-developed toxicity identification evaluation (TIE) procedures exist for liquid samples, and these can be applied to porewater samples, thus allowing the ability to identify which class or specific chemical is responsible for toxicity in contaminated sediments (Ho et al. 1997). Whole sediment TIEs are under development but are not as advanced as those for pore waters. Knowing the contaminants responsible for toxicity can help improve regulatory control, source identification, and/or remediation techniques. For these reasons, porewater tests are a useful complement to whole sediment toxicity tests; they have been included in batteries of tests to assess the quality of contaminated sediments (e.g., Long et al. 1990; Burton 1998; Porebski et al. 1999; Carr et al. 2000) and are used in Canada as part of a suite of regulatory tests to assess the suitability of dredged material for different disposal options (Porebski and Osborne 1998; see Chapters 10 and 12).

Porewater Testing: Methodological Aspects

A number of methodological factors can alter porewater chemistry (Chapter 5) and toxicity. The following discussions highlight these factors and reflect the state of the science on the available procedures.

Sample collection

Most sediment sampling procedures promote significant chemical alterations, for example, changes in redox potential, temperature, and oxidation. These changes can affect the equilibrium of contaminants contained in the sediment and, therefore, the porewater characteristics, compared to the state in the field. Although these changes are expected to affect porewater toxicity, they are not unique to porewater tests because these same alterations in the equilibrium of contaminants can affect results of solid-phase sediment toxicity tests (see Burton 1992 for a review; Environment Canada 1994).

Another critical sampling consideration is matching the sediment and/or porewater sampling depth with the exposures of indigenous biota. The primary route of exposure varies by organism type and age, ranging from overlying water, surficial sediments, and sediment and/or pore waters at depths from <1 cm to >10 cm (fresh water) and to 1 m in marine sediments. Contaminants enter sediments via ground-water and surfacewater inputs. Most contaminants enter sediments associated with settling particulates, which through time are slowly buried by newer layers of particulates. These processes, in combination with hydrological and biological perturbations and biogeochemical cycling processes, dramatically affect contaminant distribution with depth. Contaminant gradients can vary by orders of magnitude over ranges of millimeters to centimeters.

Pore waters can be collected by a wide variety of methods. Most often, pore waters are extracted in the laboratory from whole sediments collected by cores, dredges, or grabs (American Society for Testing and Materials [ASTM] 1994). Pore waters can also be collected directly, in situ, by suction or passive diffusion into peepers (ASTM 1994). This helps reduce potential artifacts related to the sediment collection and porewater extraction process (discussed later in this chapter). Large amounts of sediment can be quickly collected by cores, dredges, or grabs, but samples must be quickly cooled and stored for only a short time because, reportedly, ammonia can increase significantly during storage (Carr 1998). However, losses of ammonia have also been observed with storage (Sarda and Burton 1995) (see "Sample storage," p 130). Many studies have documented the effects of various porewater collection techniques on porewater chemistry. The chemical alterations that occur have been attributed to various factors such as temperature shifts, gradient disruption, sorption and/or desorption, oxidation and precipitation, complexation, and pressure factors (e.g., reviews in Burton 1991, 1998; ASTM 1994; USEPA 2000). Use of in situ peepers helps maintain the in situ conditions of the pore water, thereby reducing sampling artifacts. For instance, ammonia concentrations in pore water increase with the degree of sediment disruption that occurs in the assessment process, with peepers having the lowest concentrations and grab samples the highest (Sarda and Burton 1995).

In situ versus laboratory testing

There are benefits and limitations associated with both in situ and laboratory porewater toxicity testing (Burton, Hickey et al. 1996; Burton and Rowland 2000; Chappie and Burton 2000). Each approach provides unique information that may not be provided by the other, and as such, provides additional weight of evidence in environmental assessments (Table 6-1). However, there are logistical limitations to the use of peepers, such as the difficulty of deployment in deep or high energy systems, the need for long equilibration periods (chamber and sediment dependent), and the collection of only small volumes (Burton 1998; USEPA 2000).

Table 6-1 Advantages and limitations of in situ chamber (chemistry and toxicity) evaluations

Advantages	Limitations
Measure site-specific, dynamic exposures and/or effects Not a "snapshot," grab exposure	Require equilibration period after deployment
Provide integrated exposures to potentially important factors such as suspended solids, sunlight, flow, suspended organic matter, biogeochemical gradients in sediments. Many of these issues are more important in surface exposures than in pore waters.	Test organisms may require on-site acclimation.
Reduce potential artifacts associated with sampling, processing, and storage	Chamber effects may be significant and must be minimized.
Can provide real-time or rapid information for decision-making	Nonstandardized and less historical use
Can separate stressor sources: low flow, high flow, sediment interface, surficial sediments, pore water, suspended solids, flow, oxygen, photo-induced toxicity, ammonia, metals, nonpolar organics, and predators. Many of these issues are more important in surface exposures than in pore waters.	Deployment and retrieval of chambers may be difficult and impractical in deep or high-flow waters. Potential for loss of test chambers due to human, animal, or environmental disturbances would cause loss of information.
Successfully conducted with many species and life stages, in many environments, for various regulatory applications, by numerous investigators	Less control over treatment variables, such as overlying water quality, temperature, DO, suspended solids. Difficult to establish threshold effect levels. Unforeseen changes in uncontrollable environmental variables may make comparison to references difficult.

In situ testing here refers to both exposure of chamber-confined organisms in the field (water, sediment, and/or pore water) and chemical collection of pore water via passive porewater collection. These methods have been widely reported in the literature over the past 3 decades in marine and freshwater studies in the UK, North

America, Italy, Portugal, and New Zealand (e.g., reviews by Burton, Hickey et al. 1996; Chappie and Burton 2000; Baird and Burton 2001). Most of these studies, however, have been focused on bioaccumulation and toxicity in water column exposures with a number of porewater peeper studies that assess porewater chemistry using microanalytical methods.

In situ exposures also may suffer from significant artifacts if done incorrectly (Burton, Ingersoll et al. 1996, see Table 6-2). The mesh or pore size of the chamber openings controls equilibration rates, flow impedance, and entrance of suspended solids and indigenous organisms. If these are not properly designed to answer the study questions, then in situ exposure chamber responses may bear little resemblance to the exposures of indigenous organisms. However, if in situ exposures are correctly done, they can provide a more realistic exposure than the laboratory environment, although they allow less control of treatment variables (such as for establishing threshold levels). More recently, there have been porewater studies that used larger chambers with larger mesh designs (e.g., Fisher 1991; Sarda and Burton 1995). Peepers are appropriate for organism exposure in oxic porewater environments such as are found in large grain sediments (Fisher 1991). Test chambers must be allowed to equilibrate with surrounding pore waters. Too large or too small mesh sizes can create artifactual results due to flow, solids, or biofouling. If, however, the proper controls and references are used to account for artifacts (such as manipulation effects) with supporting porewater physicochemistry, then the results will provide greater realism with respect to dynamic exposure and stressor interactions than is possible with laboratory exposures.

Table 6-2 In situ chamber deployment cautions

Factor	Potential effects
Mesh size of deployment chambers can have dramatic effect on exposure, influencing the listed potential effects	Equilibration and water renewal rates Flow Suspended solids Predator or indigenous organism presence Biofouling
Biological stress	Chamber may increase or decrease organism exposure to stress because it is confined and cannot escape. This can be an advantage (allowing delineation of stressor sources) or disadvantage (organism unduly stressed because it is confined and cannot retreat to refugia). However, stress due to confinement can also occur in laboratory tests and therefore is not a unique feature of in situ experiments.
Controls and references	Appropriate references and controls are essential to account for deployment stress and ambient, background site characteristics that may confound study objectives (e.g., presence of natural toxicants).

Laboratory tests, on the other hand, have the advantage of being performed under controlled conditions and therefore suffer less from the influence of environmental variables. Laboratory exposures conducted with pore water collected in situ join the main advantages of in situ and laboratory test methods, that is, exposure to a realistic sample obtained with the least possible amount of manipulation and exposure under controlled conditions, therefore avoiding a number of confounding factors.

Porewater extraction methods

Pore water has been extracted from sediment samples by centrifugation, squeezing (using pneumatic pressure), vacuum, and in situ methods (using suction and peepers) (for a discussion on the latter, see "In situ versus laboratory testing," p 128). In a recent review on the subject, Carr (1998) stated that both centrifugation and squeezing were commonly used to extract pore waters for toxicity testing. Both methods have advantages over other methods in some situations. Centrifugation minimizes contact with surfaces, compared with the other methods, and works well with clay sediments. However, centrifugation does not work well with sandy sediments, as compared with squeezing (e.g., see citations in reviews by ASTM 1994; Burton 1998). Squeezing can be performed in the field. Vacuum extraction can be used in the field, but it is very tedious with clay samples and can result in a large amount of particulate material in the pore water. Carr (1998) found similar toxicities between samples that were squeezed and centrifuged. However, chemical differences in pore water obtained with different methods have been reported (see citations in ASTM 1994).

Sample storage

Prolonged porewater sample storage has resulted in loss of volatile components, precipitation of unstable metals (e.g., iron and manganese, with co-precipitation of other metals), change in sample pH, or degradation of other contaminants (e.g., ASTM 1994; Burton 1998). Prolonged storage can result in increased or decreased ammonia levels (Sarda and Burton 1995; Carr 1998). It is not surprising, given the diversity of sediment types and contaminant characteristics, that there is no consensus on appropriate sediment or porewater storage times. Given the varying nature of contaminant bioavailability, storage effects cannot be predicted a priori. Options for porewater storage and their advantages and disadvantages are discussed in Chapter 7.

Sample preparation and manipulation

There are several (necessary) sample preparation and manipulation techniques that can affect the integrity of porewater samples and the results of toxicity tests conducted on them (see also "Sample collection," p 127). Essentially all porewater collection methods may result in colloidal to fine grained particulate matter that may affect toxicity and variability (Carr 1998). Before testing, all porewater samples (even those obtained by centrifugation) should be centrifuged at a high speed to remove fine particulate matter. Double centrifugation at high speed (e.g., 10,000 ×g) has been recommended (Ho et al. 1997). Filtration is not recommended because it could lead to loss of some contaminants (ASTM 1994). The importance of this issue has been largely overlooked and has dramatic implications for the relationship of effects to exposure. By simply varying the speed of centrifugation, or by repeating centrifugation on porewater supernatants, toxicity can change significantly (Ankley, Schubauer-Berigan et al. 1992). It is not advisable to compare toxicity or chemistry between different porewater collection methods because of these particulate and colloidal differences.

Low dissolved oxygen (DO) (and supersaturation) is often toxic to aquatic organisms. In addition, physical aeration will strip off volatile contaminants (e.g., some petroleum hydrocarbons, unionized ammonia, and sulfide), oxidize constituents (e.g., sulfide, iron, and manganese), and alter pH. Thus, aeration may change the chemistry of the porewater sample and could result in erroneous toxicity results. However, meiofauna and macrofauna require oxygen and often create oxidized microenvironments in anoxic sediments (e.g., Burton 1998). This suggests that some degree of sediment and/or porewater oxidation is acceptable. The degree of change that occurs in sediment chemistry and bioavailability because of aeration and subsequent oxidation is a rate phenomenon and depends on the rate of oxygen introduction to a unit area of sediment and the physicochemical nature of the sediment. Currently, we cannot predict these changes or the significance of these changes on our assessment methods. It seems logical, however, that current methods of toxicity testing should attempt to minimize the degree of oxidation based on in situ conditions and organism life histories.

Salinity of samples from differing sampling locations within an estuary may differ significantly. This could affect both the bioavailability of the contaminants and the viability of the test organisms. Most test methods recommend adjustment of the salinity of the porewater samples to the physiological limits of the test organisms to control the effect of this variable on the test results (e.g., Environment Canada 1992; Carr 1998). Again, care should be taken to ensure that laboratory conditions mimic those that organisms are being exposed to in situ, unless regulatory or programmatic requirements dictate a standard method parameter.

Freshwater versus estuarine and marine samples

The issues surrounding toxicity testing of freshwater porewater samples have been reviewed by Burton (1998). Metals are a common contributor to toxicity of pore waters from freshwater sediments (refer to "Toxicity Identification Evaluations," p 133). Because metals are unstable in freshwater pore waters (Burton 1998), extreme care must be taken to avoid sampling, extraction, and storage artifacts. Metals have not yet been identified as acute toxicants in estuarine and marine pore waters; however, they have been suspected as the primary toxicants at some sites (Carr et al. 1996, 2003). There are several possible explanations for this: the effects of salinity on the bioavailability and acute toxicity of metals, the effects of differing concentrations of AVS on the bioavailability and toxicity of metals (discussed further in "Role of sulfide," below), or possibly the fact that not enough estuarine and marine porewater TIE studies have been conducted to date. It is our opinion that the issue of freshwater versus estuarine and marine samples will have major implications on how samples for porewater testing are collected, extracted, and stored, and that the recommended methods may be different for the 2 types of media.

The Role of Confounding Factors in Porewater Testing

Oxidation and metal precipitation

Precipitation of unstable metals (e.g., iron and manganese, with co-precipitation of other metals) could result from changes in the oxidation state of the porewater sample. Multitudes of studies have documented the importance and complexity of these reactions in controlling metal fate in aquatic environments (e.g., review by Burton 1991). The extensive dredged materials research program by the USACE in the 1970s showed that iron and manganese oxyhydroxides are rapidly formed when anoxic sediments are resuspended and quickly scavenge trace metals out of the water column (e.g., review by Burton 1991). Others have shown that when sediments are oxidized, sulfides are lost and metals are released, thereby increasing toxicity (e.g., review by Burton 1998).

Role of ammonia

In both freshwater and marine pore waters, ammonia has been characterized as a cause of toxicity at many sites (e.g., Gupta and Karuppiah 1996a; Sprang et al. 1996; Sprang and Janssen 1997). While the presence of ammonia has always been noted in sediments, the widespread and prominent role it plays in sediment toxicity has been underestimated until recent years. Metals and synthetic organics are frequently implicated as the cause of toxicity when ammonia co-occurs and may be contributing to toxicity. This can lead to incorrect management decisions regarding causality, remediation, and enforcement actions.

Role of sulfide

In contrast to what is known about ammonia toxicity and the role it plays in the toxicity of contaminated sediments, the role that sulfide plays in sediment toxicity is virtually unknown (refer to "Toxicity Identification Evaluations," p 133; Wang and Chapman 1999). Sulfide is abundant in marine and freshwater sediments and is more toxic than ammonia (USEPA 1986). At pH 8 (typical of marine waters), about 8% of sulfide is in the unionized or toxic form. At pH 7 (representative of many freshwater environments), about 91% of sulfide is in the unionized or toxic form. Sulfide should be viewed both as a toxicant and for the role it plays in regulating the toxicity of metals (Di Toro et al. 1992). Sulfides are often lost because of extraction and sample preparation techniques (refer to "Sample preparation and manipulation," p 130). However, there is a risk of sulfide toxicity even in porewater toxicity tests (Sims and Moore 1995). As is the case for ammonia, incorrectly identifying organic or metallic contaminants as the cause of observed sediment toxicity when sulfide was the real cause could lead to unnecessary and costly remediation actions, and to inaccurate interpretation of databases used to develop SQGs.

Applicability and bioavailability of porewater tests to organisms with different life histories

Porewater methods may reasonably approximate field exposures for water-soluble compounds such as ammonia and for organisms that have pore water as their major route of exposure. On the other hand, where high log octanol–water partition coefficient (K_{OW}) compounds are suspected of causing toxicity, porewater testing may severely underestimate the exposure to contaminants of infaunal benthic organisms because the contaminants may sorb onto the test containers or may remain bound to sediment particles. Porewater testing may underestimate exposure by failing to account for exposure via sediment ingestion or direct contact with particles (Hare et al. 2000; Lee et al. 2000). Porewater testing may also overestimate exposure if organisms are normally exposed to overlying waters (via downwelling). More realistic exposures would include pore water diluted with overlying water.

Toxicity Identification Evaluations

The ability to identify which class or specific chemical is responsible for toxicity in contaminated sediments is the objective of porewater TIE methods. These methods are useful in a variety of contexts. Once a toxicant is identified, it can be linked to a discharger and steps can be taken to prevent further discharge of the toxicant. The identification of specific classes of compounds may also be helpful in designing effective sediment remediation schemes or reasonable options for disposal of sediments (Ankley, Schubauer-Berigan, Hoke et al. 1992). In addition, identification of major causes of toxicity in sediments may guide programs such as environmental sediment guidelines and pesticide registration. Finally, knowledge of the causes of toxicity that drive ecological changes such as community structure would be useful in performing ecological risk assessments.

For all of the above reasons, researchers at the USEPA's Ecology Divisions in Duluth, Minnesota and Narragansett, Rhode Island, USA, developed a set of toxicity-based

guidelines for identifying toxic compounds in complex effluents (Mount and
Anderson-Carnahan 1988, 1989; Norberg-King, Durhan et al. 1991; Norberg-King,
Mount et al. 1991; Durhan et al. 1992; Mount et al. 1993; Burgess et al. 1996). These
TIE procedures use toxicity-based fractionation schemes to characterize (Phase I),
identify (Phase II), and confirm (Phase III) contaminants responsible for observed
toxicity. TIE methods have been widely and successfully used for identifying toxi-
cants in effluents (Burkhard and Ankley 1989; Amato et al. 1992; Ankley and
Burkhard 1992; Burkhard and Jenson 1993; Schubauer-Berigan et al. 1993; Jop and
Askew 1994; Wells et al. 1994; Bailey, Miller et al. 1995; Burgess et al. 1995; Jin et al.
1999a, 1999b), fresh (Norberg-King, Durhan et al. 1991; Steidl-Pulley et al. 1998;
Riveles and Gersberg 1999) and marine waters (Hunt et al. 1999; Rumbold and
Snedaker 1999), ballast waters (Ertan-Unal et al. 1998), and wastewater treatment
plants (Adamsson et al. 1998).

Porewater TIEs

More recently, effluent methods have been adapted for pore waters (Ankley,
Schubauer-Berigan et al. 1992), and porewater TIEs have been performed on
freshwater pore waters to identify ammonia (Ankley et al. 1990; Wenholz and
Crunkilton 1995; Gupta and Karuppiah 1996a; Karuppiah and Gupta 1996; Sprang
et al. 1996; Sprang and Janssen 1997), organic chemicals (Schubauer-Berigan and
Ankley 1991; Gupta and Karuppiah 1996a; Karuppiah and Gupta 1996), and metals
(Schubauer-Berigan et al. 1993; Wenholz and Crunkilton 1995; Gupta and
Karuppiah 1996a, 1996b; Boucher and Watzin 1999) as toxicants. In marine pore
waters, ammonia and organic chemicals (Kuhn et al. 1995; Ho et al. 1997) have been
characterized as causes of toxicity by the application of TIE procedures.

From these relatively limited results, we can propose some hypothesis about causes
of toxicity in sediments. First, the causes of toxicity are fairly widespread, that is,
there is no one predominant cause of toxicity such as polycyclic aromatic hydrocarbons (PAHs), and metals, organics, and ammonia all play a role in about equal
amounts in causing toxicity. Second, in about half of the sediments, more than one
toxicant (or class of toxicants) was identified or characterized (Figure 6-1). Further,
if you divide the sediments into marine or freshwater sediments, TIEs performed on
freshwater pore waters indicate a variety of toxicants in fairly equal proportions as
the cause of toxicity (Figure 6-2), while TIEs performed on marine pore waters have
identified only ammonia and organics as toxicants (Figure 6-3). Note that Carr et al.
(1996) in a study of gas production platforms in the Gulf of Mexico found metals in
pore waters at concentrations at or above toxic levels but did not perform a TIE. This
study indicates that metals may also be a cause of toxicity in marine waters in some
cases. This result supports the finding of AVS research that metals may not play as
large a role as previously assumed in the acute toxicity of marine sediments (Hansen
et al. 1996). Finally, ammonia has often been implicated as a primary toxicant (Kuhn
et al. 1995; Ho et al. 1997). While the presence of ammonia has always been noted

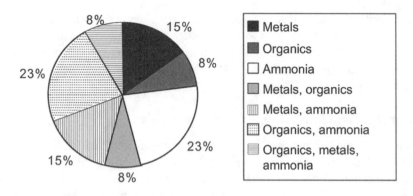

Figure 6-1 Causes of acute toxicity in pore waters evaluated by TIEs: freshwater and marine

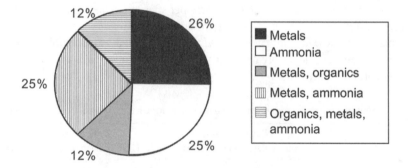

Figure 6-2 Causes of acute toxicity in freshwater pore waters

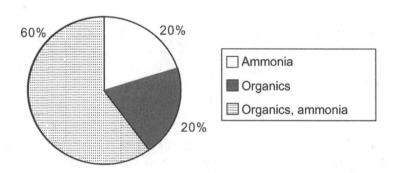

Figure 6-3 Causes of acute toxicity in marine pore waters

in sediments, the widespread and prominent role it plays in sediment toxicity has been unknown until revealed by these porewater TIEs. However, it is necessary to keep in mind that a very small number of porewater TIEs have been performed and that they measure only acute toxicity. Therefore, these trends may change as larger numbers of porewater TIEs are performed.

Advantages and limitations of porewater TIEs

Some limitations of porewater TIEs include changes in metal toxicity due to oxidation during aeration of pore waters, underexposure to high log K_{OW} compounds that may sorb to test containers, overexposure of organisms that normally are not exposed to 100% pore water, elimination of other routes of exposure, such as sediment ingestion, to the organism, and difficulty in conducting chronic toxicity testing. These limitations may be responsible for the differences observed between whole sediment toxicity tests and porewater toxicity tests (Table 6-3). The whole sediment tests presented in Table 6-3 were static with aeration using 20 g of sediment, 60 ml overlying water, and 10 organisms per species (*Ampelisca abdita* and *Americamysis bahia*). Porewater tests were conducted in 10 ml of water with 5 organisms per species (*A. abdita* and *A. bahia*) in separate exposure chambers. In both types of tests, *A. bahia* were fed daily with newly hatched *Artemia*, and *A. abdita* were not fed during the test.

Table 6-3 Comparison of porewater and whole sediment testing, in 96-h tests with mysids and amphipods[a]

Site	Whole sediment (100% unless otherwise noted) % survival		Pore water (100% unless otherwise noted) % survival	
	A. bahia	*A. abdita*	*A. bahia*	*A. abdita*
Westport MA	100[b] (0) *n* = 3	100[b] (0) *n* = 3	0[c] (0) *n* = 3	93[c] (11) *n* = 3
New York Harbor NY	90 (14) *n* = 2	27 (25) *n* = 3	0 (0) *n* = 3	10 (14) *n* = 3
Fox Point RI	40 (35) *n* = 3	87 (15) *n* = 3	0 (0) *n* = 3	27 (31) *n* = 3

[a] Standard deviation (SD) in parentheses.
[b] Tested at 75% whole sediment.
[c] Tested at 50% pore water.

A major advantage of porewater TIEs is that the methods for porewater TIEs exist (Ankley, Schubauer-Berigan et al. 1992) and are not under development like whole sediment TIE methods. These porewater methods have given us useful information on the causes of toxicity, such as highlighting the widespread role of ammonia in sediment toxicity. Currently, research suggests that some stressors may be identified in situ (Burton, Hickey et al. 1996; Burton 1999; Burton and Moore 1999; Nordstrom et al. 1999; Burton and Rowland 2000; Baird and Burton 2001). These in situ exposures separate acute toxicity into the following categories: overlying water (low versus high flow), sediment (surficial versus 0 to 2 cm), pore water, suspended solids and flow, photoinduced toxicity, and metals, nonpolar organics, or ammonia in pore waters. In situ porewater TIEs may alleviate some of the concerns over sorption of high log K_{OW} compounds and oxidation of metals (Burton, Hickey et al. 1996; Nordstrom et al. 1999); however, all porewater testing eliminates potentially important routes of exposure such as sediment ingestion and direct contact with sediment particles. In addition, in situ testing is limited to low-energy, shallow water systems and has been compared to traditional TIE procedures only to a limited extent.

References

Adamsson M, Dave G, Forsberg L, Guterstam B. 1998. Toxicity identification evaluation of ammonia, nitrite and heavy metals at the Stensund Wastewater Aquaculture plant, Sweden. *Water Sci Technol* 38:151–157.

Amato JR, Mount DI, Durhan EJ, Lukasewycz MT, Ankley GT, Robert ED. 1992. An example of the identification of diazinon as a primary toxicant in an effluent. *Environ Toxicol Chem* 11:209–216.

Anderson J, Birge W, Gentile J, Lake J, Rodgers JJ. 1987. Biological effects, bioaccumulation and ecotoxicology of sediment associated chemicals. In: Dickson KL, Maki AW, Brungs WA, editors. Fate and effects of sediment-bound chemicals in aquatic systems. New York NY, USA: Pergamon. p 267–296.

Ankley GT, Burkhard LP. 1992. Identification of surfactants as toxicants in a primary effluent. *Environ Toxicol Chem* 11:1235–1248.

Ankley GT, Hoke RA, Giesy J, Winger PV. 1989. Evaluation of the toxicity of marine sediments and dredge spoils with the Microtox bioassay. *Chemosphere* 18:2069–2075.

Ankley GT, Katko A, Arthur J. 1990. Identification of ammonia as an important sediment-associated toxicant in the lower Fox River and Green Bay, Wisconsin. *Environ Toxicol Chem* 9:312–322.

Ankley GT, Schubauer-Berigan MK, Dierkes JR, Lukasewycz MT. 1992. Sediment toxicity identification evaluation: Phase I (characterization), phase II (identification) and phase III (confirmation) modifications of effluent procedures. Duluth MN, USA: U.S. Environmental Protection Agency, Environmental Research Laboratory. Technical report 08-91.

Ankley GT, Schubauer-Berigan MK, Hoke RA. 1992. Use of toxicity identification evaluation techniques to identify dredged material disposal options: A proposed approach. *Environ Manag* 16:1–6.

[ASTM] American Society for Testing and Materials. 1994. Guide for collection, storage, characterization and manipulation of sediments for toxicological testing. Designation E1391-94. Volume 11.04, Annual book of ASTM standards. Philadelphia PA, USA: ASTM. 1786 p.

Bailey HC, Miller JL, Miller MJ, Dhaliwal BS. 1995. Application of toxicity identification procedures to the echinoderm fertilization assay to identify toxicity in a municipal effluent. *Environ Toxicol Chem* 14:2181–2186.

Bailey RC, Day KE, Norris RH, Reynoldson TB. 1995. Macroinvertebrate community structure and sediment bioassay results from nearshore areas of North American Great Lakes. *J Great Lakes Res* 21:42–52.

Baird D, Burton Jr GA. 2001. Ecological variability: Separating natural from anthropogenic causes of ecosystem impairment. Pensacola FL, USA: Society of Environmental Toxicology and Chemistry (SETAC).

Boucher AM, Watzin MC. 1999. Toxicity identification evaluation of metal-contaminated sediments using an artificial pore water containing dissolved organic carbon. *Environ Toxicol Chem* 18:509–518.

Burgess RM, Ho KT, Morrison GE, Chapman G, Denton DL. 1996. Marine toxicity identification evaluation (TIE) procedures manual: Phase I guidance document. Washington DC, USA: U.S. Environmental Protection Agency, Office of Research and Development. EPA 600/R-96/054.

Burgess RM, Ho KT, Tagliabue MD, Kuhn A, Comeleo R, Comeleo P, Modica G, Morrison GE. 1995. Toxicity characterization of an industrial and a municipal effluent discharging to the marine environment. *Mar Pollut Bull* 30:524–535.

Burkhard LP, Ankley GT. 1989. Identifying toxicants: NETAC's toxicity-based approach. *Environ Sci Technol* 23:1438–1443.

Burkhard LP, Jenson JJ. 1993. Identification of ammonia, chlorine, and diazinon as toxicants in a municipal effluent. *Arch Environ Contam Toxicol* 25:506–515.

Burton Jr GA. 1991. Assessing the toxicity of freshwater sediments. *Environ Toxicol Chem* 10:1585–1627.

Burton Jr GA, editor. 1992. Sediment toxicity assessment. Boca Raton FL, USA: CRC. 457 p.

Burton Jr GA. 1998. Assessing aquatic ecosystems using pore waters and sediment chemistry. Ottawa, ON, Canada: Natural Resources Canada, Aquatic Effects Technology Evaluation Program. K1A 0G1. Contract No NRCan 97-0083, AETE Project 322a. Final report, December 31, 1998.

Burton Jr GA. 1999. Realistic assessments of ecotoxicity using traditional and novel approaches. *J Aquat Ecosyst Health Manag* 2:1–8.

Burton Jr GA, Hickey C, DeWitt T, Morrisey D, Roper D, Nipper M. 1996. In situ toxicity testing: Teasing out the environmental stressors. *SETAC NEWS* 16(5):20–22.

Burton Jr GA, Ingersoll C, Burnett L, Henery M, Klaine S, Landrum P, Ross P, Tuchman M. 1996. A comparison of sediment toxicity test methods at three Great Lakes Areas of Concern. *J Great Lakes Res* 22:495–511.

Burton Jr GA, Moore L. 1999. An assessment of storm water runoff effects in Wolf Creek, Dayton OH, USA. Dayton OH, USA: City of Dayton. Final report.

Burton Jr GA, Rowland C. 2000. Assessment of in situ toxicity at the Eastern Woolen Mill Superfund site. Portland ME, USA: Final report to Harding Lawson Assoc.

Carr RS. 1998. Marine and estuarine pore water toxicity testing. In: Wells PG, Lee K, Blaise C, editors. Microscale testing in aquatic toxicology: Advances, techniques, and practices. Boca Raton FL, USA: CRC. p 523–538.

Carr RS, Chapman DC, Presley BJ, Biedenbach JM, Robertson L, Boothe P, Kilada R, Wade T, Montagna P. 1996. Sediment porewater toxicity assessment studies in the vicinity of offshore oil and gas production platforms in the Gulf of Mexico. *Can J Fish Aquat Sci* 53:2618–28.

Carr RS, Montagna PA, Biedenbach JM, Kalke R, Kennicutt MC, Hooten R, Cripe G. 2000. Impact of storm-water outfalls on sediment quality in Corpus Christi Bay, Texas, USA. *Environ Toxicol Chem* 19:561–574.

Carr RS, Nipper M, Plumlee G. 2003. Survey of marine contamination from mining-related activities on Marinduque Island, Philippines: Porewater toxicity and chemistry. *Aquat Ecosyst Health Manag* (in press).

Chapman PM. 1988. Marine sediment toxicity tests. In: Lichtenberg JL, Winter JA, Weber CI, Fradkin L, editors. Chemical and biological characterization of municipal sludges, sediments, dredge spoils and drilling muds. Philadelphia PA, USA: American Society for Testing and Materials. STP 976. p 391-402.

Chappie DJ, Burton Jr GA. 2000. Applications of aquatic and sediment toxicity testing in situ. *Soil Sed Contam* 9:219–245.

Di Toro DM, Mahony JD, Hansen DJ, Scott KJ, Carlson AR, Ankley GT. 1992. Acid volatile sulfide predicts the acute toxicity of cadmium and nickel in sediments. *Environ Sci Technol* 26:96–101.

Durhan EJ, Norberg-King TJ, Burkhard LP. 1992. Methods for aquatic toxicity identification evaluations: Phase II toxicity identification procedures for samples exhibiting acute and chronic toxicity. Duluth MN, USA: U.S. Environmental Protection Agency, Environmental Research Laboratory. EPA/600/6-92/080.

Environment Canada. 1992. Biological test method: Fertilisation assay using echinoids (sea urchins and sand dollars). Ottawa ON, Canada: Environment Canada, Environmental Protection. Report EPS 1/RM/27.

Environment Canada. 1994. Guidance document on collection and preparation of sediments for physicochemical characterization and biological testing. Ottawa ON, Canada: Environment Canada. EPS 1/RM/29.

Ertan-Unal M, Gelderloos AB, Hughes JS. 1998. A toxicity reduction evaluation for an oily waste treatment plant exhibiting episodic effluent toxicity. *Sci Total Environ* 218:141–152.

Fisher R. 1991. Sediment interstitial water toxicity evaluations using *Daphnia magna* [MS thesis]. Dayton OH, USA: Wright State Univ.

Giesy JP, Hoke RA. 1989. Freshwater sediment toxicity bioassessment: Rationale for species selection and test design. *J Great Lakes Res* 15:539–569.

Giesy JP, Hoke RA. 1990. Freshwater sediment quality criteria: Toxicity bioassessment. In: Baudo R, Giesy J, Muntau H, editors. Sediments: Chemistry and toxicity of in-place pollutants. Ann Arbor MI, USA: Lewis. p 265–348.

Gupta G, Karuppiah M. 1996a. Toxicity study of a Chesapeake Bay tributary: Wicomico River. *Chemosphere* 32:1193–1215.

Gupta G, Karuppiah M. 1996b. Toxicity identification of Pocomoke River pore water. *Chemosphere* 33:939–960.

Hansen DJ, Berry WJ, Mahony JD, Boothman WS, Di Toro DM, Robson DL, Ankley GT, Ma D, Yan Q, Pesch CE. 1996. Predicting the toxicity of metal-contaminated field sediments using interstitial concentration of metals and acid-volatile sulfide normalizations. *Environ Toxicol Chem* 15:2080–2094.

Hare L, Tessier A, Warren L. 2000. Cadmium accumulation by invertebrates living at the sediment-water interface. *Environ Toxicol Chem* 20:880–889.

Hartwell SI, Dawson CE, Durell EQ, Alden RW, Adolphson PC, Wright D, Coelho GM, Magee JA, Ailstock S, Norman M. 1997. Correlation of measures of ambient toxicity and fish community diversity in Chesapeake Bay, USA, tributaries: Urbanizing watersheds. *Environ Toxicol Chem* 16:2556–2567.

Hatakeyama S, Yokoyama N. 1997. Correlation between overall pesticide effects monitored by shrimp mortality test and change in macrobenthic fauna in a river. *Ecotox Environ Saf* 36:148–161.

Ho KT, McKinney RA, Kuhn A, Pelletier MC, Burgess RM. 1997. Identification of acute toxicants in New Bedford Harbor sediments. *Environ Toxicol Chem* 16:551–558.

Hunt JW, Anderson BS, Phillips BM, Tjeerderma RS, Puckett HM, deVlaming V. 1999. Patterns of aquatic toxicity in an agriculturally dominated coastal watershed in California. *Agric Ecosyst Environ* 75:75–91.

Jin H, Yang X, Yin D, Yu H. 1999a. A case study on identifying the toxicant in effluent discharged from a chemical plant. *Mar Pollut Bull* 39:122–125.

Jin H, Yang X, Yu H, Yin D. 1999b. Identification of ammonia and volatile phenols as primary toxicants in a coal gasification effluent. *Bull Environ Contam Toxicol* 63:399–406.

Jop KM, Askew AM. 1994. Toxicity identification evaluation using a short-term chronic test with *Ceriodaphnia dubia*. *Bull Environ Contam Toxicol* 53:91–97.

Karuppiah M, Gupta G. 1996. Impact of point and nonpoint source pollution on pore waters of two Chesapeake Bay tributaries. *Ecotox Environ Saf* 35:81–85.

Kuhn A, Ho KT, Pelletier M, Burgess RM. 1995. Phase I toxicity identification evaluation (TIE) for New York/New Jersey Harbor estuary sediments. Narragansett RI, USA: U.S. Environmental Protection Agency, Atlantic Ecology Division. Internal report to USEPA Region II.

Lee B-G, Griscom SB, Lee J-S, Choi HJ, Koh C-H, Luoma SN, Fisher NS. 2000. Influences of dietary uptake and reactive sulfides on metal bioavailability from aquatic sediments. *Science* 287:282–284.

Long, ER, Buchman MF, Bay SM, Breteler RJ, Carr RS, Chapman PM, Hose JE, Lissner AL, Scott J, Wolfe, DA. 1990. Comparative evaluation of five toxicity tests with sediments from San Francisco Bay and Tomales Bay, California. *Environ Toxicol Chem* 9:1193–1214.

Mount DI, Anderson-Carnahan L. 1988. Methods for aquatic toxicity identification evaluations: Phase I toxicity characterization procedures. Duluth MN, USA: U.S. Environmental Protection Agency. EPA/600-3-88/034.

Mount DI, Anderson-Carnahan L. 1989. Methods for aquatic toxicity identification evaluations: Phase II toxicity identification procedures. Duluth MN, USA: U.S. Environmental Protection Agency. EPA/600-3-88/035.

Mount DI, Norberg-King T, Ankley G, Burkhard LP, Durhan EJ, Schubauer-Berigan MK, Lukasewycz M. 1993. Methods for aquatic toxicity identification evaluations: Phase III toxicity confirmation procedures for samples exhibiting acute and chronic toxicity. Duluth MN, USA: U.S. Environmental Protection Agency. EPA/600/R-92/081.

Norberg-King T, Durhan E, Ankley GT, Robert E. 1991. Application of toxicity identification and evaluation procedures to the ambient waters of the Coulsa Basin Drain, California. *Environ Toxicol Chem* 10:881–900.

Norberg-King, T, Mount DI, Amato JR, Jensen DA, Thompson JA. 1991. Toxicity identification evaluation: Characterization of chronically toxic effluents, phase I. Duluth MN, USA: U.S. Environmental Protection Agency. EPA-600/6-91/005.

Nordstrom JF, Burton Jr GA, Greenberg MS, Moore LA, Rowland CD. 1999. A novel stressor identification method for pore water and sediment. *Abstr Ann Meeting Soc Environ Toxicol Chem*, Philadelphia PA, USA. No. PWA197.

Porebski LM, Doe KG, Zajdlik BA, Lee D, Pocklington P, Osborne JM. 1999. Evaluating the techniques for a tiered testing approach to dredged sediment assessment: A study over a metal concentration gradient. *Environ Toxicol Chem* 18:2600–2610.

Porebski LM, Osborne JM. 1998. The application of a tiered testing approach to the management of dredged sediments for disposal at sea in Canada. *Chem Ecol* 14:197–214.

Riveles K, Gersberg RM. 1999. Toxicity identification evaluation of wet and dry weather runoff from the Tijuana River. *Bull Environ Contam Toxicol* 63:625–632.

Rumbold D, Snedaker S. 1999. Sea-surface microlayer toxicity off the Florida Keys. *Mar Environ Res* 47:457–472.

Sarda N, Burton Jr GA. 1995. Ammonia variation in sediments: Spatial, temporal, and method-related effects. *Environ Toxicol Chem* 14:1499–1506.

Schubauer-Berigan MK, Amato JR, Ankley GT, Baker SE, Burkhard LP, Dierkes JR, Jenson JJ, Lukasewycz MT, Norberg-King TJ. 1993. The behavior and identification of toxic metals in complex mixtures: Examples from effluent and sediment pore water toxicity identification evaluations. *Arch Environ Contam Toxicol* 24:298–306.

Schubauer-Berigan MK, Ankley GT. 1991. The contribution of ammonia, metals and nonpolar organic compounds to the toxicity of sediment interstitial water from an Illinois River tributary. *Environ Toxicol Chem* 10:925–939.

Sims JG, Moore DW. 1995. Risk of pore water hydrogen sulfide toxicity in dredged material bioassays. Vicksburg MS, USA: U.S. Army Corps of Engineers Waterways Experiment Station. Miscellaneous paper D-95-4.

Sprang PAV, Janssen CR. 1997. Identification and confirmation of ammonia toxicity in contaminated sediments using a modified toxicity identification evaluation approach. *Environ Toxicol Chem* 16:2501–2507.

Sprang PAV, Janssen CR, Sabayasachi M, Benijts F, Persoone G. 1996. Assessment of ammonia toxicity in contaminated sediments of the Upper Scheldt (Belgium): The development and application of toxicity identification evaluation procedures. *Chemosphere* 33:1967–1974.

Steidl-Pulley T, Nimmo DWR, Tessari JD. 1998. Characterization of toxic conditions above Wilson's Creek National Battlefield Park, Missouri. *J Am Water Res Assoc* 34:1087–1098.

Swartz RC, Cole FA, Lamberson JO, Ferraro SP, Schults DW, DeBen WA, Lee HI, Ozretich RJ. 1994. Sediment toxicity, contamination and amphipod abundance at a DDT and dieldrin contaminated site in San Francisco Bay. *Environ Toxicol Chem* 13:949–962.

Swartz RC, DeBen WA, Sercu KA, Lamberson JO. 1982. Sediment toxicity and the distribution of amphipods in Commencement Bay, Washington, USA. *Mar Pollut Bull* 13:359–364.

Swartz RC, Kemp PF, Schultzs DW, Ditsworth GR, Ozretich RJ. 1989. Acute toxicity of sediments from Eagle Harbor, Washington, to the infaunal amphipod *Rhepoxynius abronius*. *Environ Toxicol Chem* 8:215–222.

[USEPA] U.S. Environmental Protection Agency. 1986. Quality criteria for water. Cincinnati OH, USA: USEPA. EPA-550/5-86-001.

[USEPA] U.S. Environmental Protection Agency. 1993. Technical basis for deriving sediment quality criteria for nonionic organic contaminants for the protection of benthic organisms by using equilibrium partitioning. Washington DC, USA: USEPA, Office of Water. EPA-822-R-93-011.

[USEPA] U.S. Environmental Protection Agency. 2000. Methods for measuring the toxicity and bioaccumulation of sediment-associated contaminants with freshwater invertebrates. Washington DC, USA: USEPA, Office of Research and Development, Office of Water. EPA/600/R-99/064.

[USEPA/USACE] U.S. Environmental Protection Agency/U.S. Army Corps of Engineers. 1991. Evaluation of dredged material proposed for ocean disposal. Washington DC, USA: USEPA. EPA-503/8-91/001.

Wang F, Chapman PM. 1999. Biological implications of sulfide in sediment: A review focusing on sediment toxicity. *Environ Toxicol Chem* 18:2526–2532.

Wells MJM, Rossano AJJ, Roberts EC. 1994. Textile wastewater effluent toxicity identification evaluation. *Arch Environ Contam Toxicol* 27:555–560.

Wenholz M, Crunkilton R. 1995. Use of toxicity identification evaluation procedures in the assessment of sediment pore water toxicity from an urban stormwater retention pond in Madison, Wisconsin. *Bull Environ Contam Toxicol* 54:676–682.

Williams LG, Chapman PM, Ginn TC. 1986. A comparative evaluation of marine sediment toxicity using bacterial luminescence, oyster embryo and amphipod sediment bioassays. *Mar Environ Res* 19:225–249.

CHAPTER 7

Issues and Recommendations for Porewater Toxicity Testing: Methodological Uncertainties, Confounding Factors, and Toxicity Identification Evaluation Procedures

Marion Nipper (Workgroup Leader), G Allen Burton Jr, Duane C Chapman, Ken G Doe, Mick Hamer, Kay T Ho

During the last decade, porewater toxicity tests have gained popularity as tools for assessing the presence and biological effects of bioavailable contaminants in aquatic sediments. However, there are numerous gaps in our knowledge of the biogeochemical processes that occur in sediments and pore waters and of the artifacts introduced during removal of pore water from sediments, which may affect the accuracy and precision of the methods used for porewater sampling, toxicity testing, and toxicity identification evaluation (TIE) procedures.

Some of the methodological uncertainties associated with porewater testing include appropriate sediment sampling and porewater extraction and storage procedures. In addition to these uncertainties, several features inherent to sediments, such as salinity for marine and estuarine pore waters, hardness for freshwater samples, dissolved oxygen (DO), pH, and numerous others, can affect the results and act as confounding factors in porewater toxicity tests. Some chemicals, such as ammonia and sulfides, can be natural features in some sediments but anthropogenically altered in others. All these factors must be considered when pore water is sampled, toxicity tests are conducted, and results are interpreted.

Several species and endpoints are currently used in porewater toxicity tests throughout the world (Carr 1998; see Chapter 12), but the use of indigenous species is often advocated. The adequacy of different available types of porewater toxicity tests, encompassing a variety of species from different habitats (e.g., planktonic versus benthic), life stages, endpoints, and in situ and laboratory tests, and the use of indigenous versus non-indigenous species for both marine and freshwater environments are addressed in this chapter.

The use of pore water allows the performance of TIE procedures, whereas TIE methods are not yet fully developed for solid-phase sediment tests. Therefore, porewater tests present a major advantage when there is a need to identify the types and/or the sources of contaminants in a particular area.

The state of the scientific knowledge regarding the aspects above was discussed in the Society of Environmental Toxicology and Chemistry (SETAC) Technical Workshop on Porewater Toxicity Testing and will be presented in this chapter, concluding with a series of recommendations, identified information gaps, and research needs.

Methodological Uncertainties

Chapter 6 contains a review of methods used for porewater toxicity testing and discusses some of the issues associated with them, and Chapter 5 presents a thorough review of the chemical changes of concern in sediment sampling and in porewater extraction and storage, including sample oxidation, sorption of organics, volatilization, pH alterations, and precipitation. The toxicological implications of these chemical changes are discussed in this section and listed in Table 7-1.

Sediment collection

An issue regarding sediment sampling for porewater extraction concerns the depth of sediment to be sampled. The traditional methods collect sediments or pore waters over depths of several centimeters and homogenize any concentration gradients that likely exist. This makes relationships between subsequent chemical analyses and biological responses tenuous at best. The sampling depth should match the expected exposure of organisms in a real world situation. The decision on sampling depth should be done on a case-by-case basis. For instance, if porewater toxicity is to be assessed for dredging purposes, the whole depth to be dredged should be sampled for porewater extraction and toxicity analyses. On the other hand, if a field survey intends to assess exposure of benthic biota, only the depth that is inhabited by benthic organisms, which varies according to the kind of sediment and environment, should be sampled. Local dynamics, severity of storm events and consequent perturbation, and other similar factors should be considered for study design (Figure 7-1).

Porewater extraction

While all methods of porewater extraction may in some way affect the chemical properties (see Chapter 5) and consequently the toxicity of the sample (Table 7-1), certain methods may minimize these changes. In situ porewater extraction was identified as the method that causes the least disruption of natural sediment conditions (see Chapter 6) and therefore would be the method of choice for the extraction of pore water to be used in toxicity testing. Ammonia concentrations

Project design question: *Is there a need for porewater chemistry or toxicity information?*

Yes, if: Need to verify effects based on sediment quality guidelines (SQGs) (equilibrium partitioning [EqP])
Need to build a weight-of-evidence–based conclusion
Advantageous to assess the more bioavailable fraction of sediments (i.e., pore water)
Advantageous to use standardized water toxicity assay methods
Advantageous to apply TIE methods
Upwelling, downwelling, or dynamic porewater conditions exist (that do not allow for accurate assessment of sediment quality or risk based on chemistry or whole sediment toxicity testing)

Objective of survey:

Monitoring
Are peepers feasible?
Yes, if:
Shallow — accessible for manual deployment
Minimal porewater volumes needed
Expertise available
Sediment depth of concern can be matched with peeper exposure
Equilibration time can be met

Yes

Oxic sediment
In situ organism exposure
Options: Toxicity testing 1 to 4 d
Bioaccumulation 1 to 4 d
Mesh size: dialysis to 150 μm depending on questions and sediment contaminants
NOTE: Certain plastics and small mesh membranes may sorb organics; larger mesh membranes allow initial entry of colloids and particulates. Consider effects of these factors versus project questions and objectives.

Anoxic sediment
Equilibrate 2 to 14 d, depending on chamber size and mesh pore size, sediment
Pre-purge chambers with nitrogen prior to deployment

Important measures for most porewater sites:
Porewater flow, total suspended solids, DO, pH, temperature, NH_3, sulfide, salinity, total organic carbon (TOC), alkalinity, hardness, conductivity, + chemicals of concern
NOTE: Attempt to match sampling depth with biota exposure depths. Minimize oxidation.
Minimize processing manipulation and time until analyses and toxicity testing.

Dredging
Use grab or core samplers, collect sediment throughout the whole depth to be dredged, homogenize and extract pore water using laboratory method of choice, and conduct laboratory porewater toxicity tests.

No: *Are corers feasible?*

Yes, if: Sediment unconsolidated, mainly clay or silt
Appropriate equipment, logistical support, and time available
NOTE: Use least destructive coring method possible. Match sampling depth with biological effect zone.
Use in situ core suction if available.

No: Use grab sampler.
NOTE: Use least destructive grab method available, e.g., Ekman > Ponar > Van Veen

Figure 7-1 Optimizing sediment and porewater sampling design

Table 7-1 Toxicological effects of chemical changes induced by porewater extraction

Extraction method	Methodological and chemical factors	Effect on toxicity or on testing condition
Peeper	1) Sample manipulation is reduced. 2) There is no significant disturbance of the in situ equilibrium conditions.	Realistic assessment of toxicity is provided.
	3) Potential for loss of volatile substances which occurs in ex situ methods is eliminated.	
	4) When dialysis membrane is used, need for post-retrieval porewater filtration is eliminated, therefore minimizing the loss of some contaminants.	
	5) Redox and pH conditions are relatively unaltered, therefore minimizing changes in pH and oxygen-sensitive species (such as metals).	
	Sampling influences on the oxidation state of metals are reduced.	Assessment of toxicity is realistic when metals are the contaminants of concern.
	Highly hydrophobic organic compounds (HOCs) are sorbed onto the sampler.	Actual toxicity is underestimated.
	Uncontaminated water inside newly deployed "peeper" could dilute porewater contaminant concentrations in low porosity sediments.	Actual toxicity is underestimated if equilibrium is not reached.
	Peepers generally produce less sample volume than grab sampling.	May limit tests to methods using small organisms
Suction	Metals and HOCs are sorbed on filtering medium.	Actual toxicity is underestimated.
	Can be used with highly porous sediments.	Toxicity assessments in sandy areas, such as in the vicinity of coral reefs, are feasible.
	Collection of pore waters from nontargeted depths as well as overlying water may occur.	Actual toxicity can be under- or overestimated, depending on quality of overlying water and pore water at different depths.
	Oxidation and degassing of pore water may occur, requiring aeration.	Aeration can introduce artifacts on toxicity (e.g., loss of volatiles, oxidation of toxicants), causing underestimation of toxicity.

Table 7-1 *(cont'd.)*

Extraction method	Methodological and chemical factors	Effect on toxicity or on testing condition
Centrifugation	Several variables (e.g., duration, speed) can be varied to optimize operation.	Lack of a generic methodology makes comparison of toxicity data from different surveys unreliable.
	HOCs are sorbed to centrifuge tube, depending on construction material.	Actual toxicity is underestimated.
	Method does not function with sandy sediments.	Toxicity assessments are prevented in several critical areas that are typically sandy, such as the vicinity of coral reefs.
	Lysis of cells can occur during spinning, releasing dissolved organic carbon (DOC).	Released DOC can bind contaminants, causing underestimation of actual toxicity.
Pressurization	Method can be used with highly bioturbated sediments without lysis of cells.	DOC release from broken cells is prevented and consequently allows realistic toxicity assessment.
	HOCs may be lost on filter.	Actual toxicity is underestimated.
	Degassing may occur, requiring aeration.	Aeration can introduce artifacts on toxicity test results, loss of volatiles, oxidation of toxicants, causing underestimation of toxicity.

have been reported to increase with a variety of laboratory porewater extraction methods (Sarda and Burton 1995), which would increase porewater toxicity to highly ammonia-sensitive organisms. This increase in ammonia concentration is likely to be related to enhanced microbial activity and could therefore enhance microbial degradation. This could introduce an additional artifact that results in an increase or reduction in toxicity, depending on the degradation products generated in the pore water (Nipper et al. 2002). Other chemical changes caused by porewater extraction are pointed out in Chapter 5. Most of them are expected to reduce porewater toxicity relative to natural conditions (Table 7-1).

Although in situ porewater extraction is considered the preferred method, the practicality of toxicity testing must also be considered, in addition to toxicological concerns over changing chemistry. Peeper methods often do not yield enough porewater for toxicity testing and/or TIE manipulations. Often, large-scale centrifugation or squeezing is the only practical method of porewater extraction. If surveys are conducted in shallow areas or if sediment cores can be collected, vacuum extraction of pore water (Winger and Lasier 1991) can be conducted on site. This would cause the least disruption and possibly allow the collection of large enough volumes of pore water (Winger et al. 1998) for use in toxicity tests and TIE procedures. When porewater extraction by vacuum is performed in subtidal areas, care must be taken to insert the filtering medium in the sediment properly in order to avoid contamination of the porewater sample with overlying water. Although in situ porewater extraction is recommended as the preferred method, for the sake of consistency and comparability, only one porewater extraction method should be used throughout a single survey, no matter which method is chosen. Chemical differences in pore water obtained with different methods have been identified (Chapter 6), and further testing is needed to determine the toxicological significance of these differences in a range of sediments, with varying contaminants and assay methods.

The chemicals of concern are rarely known before a sediment survey is conducted, and therefore the selection of porewater extraction methods based on contamination characteristics can seldom be established a priori. However, when organics are known to be the main contaminants of concern, centrifugation would be the method of choice because it minimizes losses that can occur through sorption (Carr 1998), but the type of sediment must be considered because sandy sediments do not yield sufficient pore water when centrifuged.

Porewater storage

Storage time and conditions can also introduce a series of artifacts that can considerably affect the toxicity of pore waters. Porewater samples should be extracted and tested as soon as possible. However, reality dictates that testing cannot always be performed immediately, and therefore, we need to balance our concern for minimizing changes with practicalities. No consensus on appropriate methods to store pore waters exists; one school of thought is that sediments should be stored at 4 ± 2 °C and pore water should be extracted within 24 hours of toxicity testing (Environment Canada 1994), while the other school is that pore water can be extracted immediately after sediment collection and frozen after particle removal without changing toxicity (Carr and Chapman 1995). In all cases, the interpretation of results should reflect any storage conditions.

Porewater storage is clearly an area that needs more research. Carr and Chapman (1995) performed a study wherein they analyzed a variety of marine samples for porewater toxicity when extracted fresh versus frozen, before and after removal of particles, and when extracted at different storage times from sediment kept at 4 °C. In that study, differences in toxicity from freezing were not observed. Certain chemical changes are known to occur during the freezing and thawing of pore water (see Chapter 5), which might logically be predicted to alter the toxicity of at least some pore waters, depending on the toxic constituents and other water quality factors. While the results of Carr and Chapman (1995) were conclusive for the samples in that study, similar research is needed with a broader range of sample types for a better understanding of the effects of freezing on porewater storage.

Despite the chemical changes that occur with porewater extraction and storage, the applicability of porewater toxicity tests for sediment contamination assessments has been demonstrated through their concurrence with whole sediment test results and field surveys (see Chapters 3 and 9). In addition, the use of porewater tests can be justified on the basis of their increased sensitivity, compared to whole sediment tests (Carr and Chapman 1992; Carr, Long et al. 1996; Carr et al. 2000).

Confounding Factors

Definition and relevance

A confounding factor is a substance or physicochemical parameter that can mask or interfere with the results and/or interpretation of a porewater toxicity test. The classification of a substance as a confounding factor is dynamic, and it may change depending upon the objective of the assessment.

Confounding factors are inherent to uncontaminated sediments, but almost any of them may be changed by anthropogenic activity such that they become a pollution problem. A substance may be a confounding factor in some situations and a contaminant of concern in others. An example is ammonia, which is naturally occurring (Chambers et al. 1992) and, as such, may cause toxicity in pore waters from uncontaminated sediments. In other cases, ammonia may be directly introduced into the environment or increase as a result of, for example, effluents and wastewater discharges (Monda et al. 1995; Tegner et al. 1995), organic or nutrient enrichment (Ankley et al. 1990; Heip 1995; Frazier et al. 1996), or reduced bioturbation resulting from the removal of benthic organisms by anthropogenic chemical or physical disturbances (Pearson and Rosenberg 1976; Rhoads et al. 1978; Flint and Kalke 1983; Gaston 1985). Reduced bioturbation, however, can also be caused by the temporary removal of benthic organisms by non-anthropogenic factors such as extreme variations in freshwater inflow to estuaries due to high rainfall. The presence of elevated concentrations of ammonia should not be automatically considered a confounding factor without taking into account the objectives of the study and the potential sources of ammonia. In most cases involving anthropogenically induced ammonia input or increase, it should be considered a contaminant and not a confounding factor. Similarly, a number of other porewater characteristics can be anthropogenically induced or natural. The interpretation of test results therefore needs to take study objectives and these factors into account on a case-by-case basis.

The most commonly recognized confounding factors are ammonia, sulfide, pH, salinity (for marine and estuarine water), conductivity, hardness and alkalinity (for fresh water), DO, and total and dissolved organic carbon (TOC and DOC), but there are many other porewater constituents or physicochemical parameters that may be important, such as

1) physicochemical parameters — particulates, ion ratios, oxidation status, temperature;

2) inorganic constituents — nitrate, nitrite, sulfate, phosphate, silicate, CO_2, bicarbonate, manganese, iron, aluminum, potassium; and

3) organic constituents — methane, lignins.

Some of the above are unique or more relevant to freshwater or saltwater situations, and therefore measurements should be site specific. Ammonia, DO, pH, and sulfide are relevant for all kinds of environments and should always be measured. Salinity should also be measured in marine samples, and hardness, alkalinity, and conductivity in freshwater samples. If organic pollution is suspected, the organic carbon content should be determined, and in phytotoxicity tests the nutrient status of the pore water should be analyzed. These should be considered minimum requirements, and other measurements may be necessary to produce useful data in porewater toxicity tests.

Effects of confounding factors on porewater toxicity test results

One way in which a confounding factor may affect the test result is presence at a concentration that is directly toxic to the test organism (Ankley et al. 1990; Sims and Moore 1995; Sprang et al. 1996; Tay et al. 1998; Wang and Chapman 1999). It is therefore important that the tolerance levels of the test organism be determined (e.g., Knezovich et al. 1996). Alternatively, confounding factors may affect the bioavailability of contaminants or change them to a toxic form. For example, pH, temperature, and salinity affect the toxicity of ammonia (Bower and Bidwell 1978;

USEPA 1985, 1986; Miller et al. 1990); pH and hardness affect the toxicity of many metals by transforming them into more or less toxic species (Mount 1966; Calamari et al. 1980; Ajmal and Khan 1984; Bradley and Sprague 1985; U.S. Environmental Protection Agency [USEPA] 1986; Paulauskis and Winner 1988; Stouthart et al. 1996), and DOC affects bioavailability of organics (Caron 1989; Ingersoll 1995) and some metals (Wiener and Giesy 1979; Playle et al. 1993).

The interactions of confounding factors can be complex. For example, manganese is toxic in its reduced form (Lasier et al. 2000), but when oxidized, it precipitates and sorbs other potentially toxic metals (Stumm and Morgan 1996). Chapter 5 addresses the chemical processes involved.

Manipulations of pore waters for mitigation of confounding factors

Organisms will tolerate only a given range for confounding factors, outside which toxicity testing is impractical. For toxicity tests that have standard methods (American Society for Testing and Materials [ASTM], USEPA, Organization for Economic Cooperation and Development [OECD], Environment Canada), guidance is provided for adjustment of factors such as temperature, DO, and hardness or salinity. This is necessary to optimize or standardize test conditions but may affect test results. For example, aeration may result in loss of volatiles or it may alter pH, resulting in changes in the bioavailability or toxicity of contaminants. It may also cause oxidation of dissolved iron and manganese compounds, resulting in their precipitation and subsequent scavenging of metals and organic contaminants of concern (Orem and Gaudette 1984; Adams 1991).

In porewater testing, there can be many confounding factors to consider other than those that are addressed in standardized methods. Identification of confounding factors requires knowledge of the physicochemical status of the pore water, the concentration of potential confounding factors, and how these may be changed through manipulations. It also requires knowledge of the tolerance range of the test organisms.

It is necessary to understand the impact of confounding factors, and in order to achieve the test objectives, it might be appropriate to manipulate a porewater sample to adjust or remove confounding factors. However, this should be done with caution because adjustments may well alter other parameters that will influence sample toxicity. For example, *Ulva lactuca*, the sea lettuce, effectively removes ammonia; however, it also removes organic contaminants (see "Toxicity Identification Evaluation Procedures," p 156; Ho et al. 1999), and therefore, such sample manipulations require careful forethought.

Some confounding factors may change during sample storage, and therefore, it is important that factors such as pH, DO, ammonia, and sulfides are measured at appropriate times. To allow accurate interpretation of test results, it is necessary to measure these factors at the beginning of exposure, and they should be monitored throughout and/or at the end of tests with relatively long exposure periods. It is also useful to measure these variables as soon as possible after collection because knowledge of how they may have changed during any subsequent storage can aid later interpretation of test data.

The understanding of confounding factors is vital to the interpretation of porewater toxicity test results. A consequence of not recognizing their role may be an unwise regulatory or management decision based on incorrect or false toxicity test results, or incorrect identification of the cause of sample toxicity.

Toxicity Testing

Selection of organisms, life stages, and endpoints for toxicity testing

In porewater toxicity testing, it usually is necessary to work with small sample volumes. This limits the number of species that can be used to those that can be tested with a limited amount of water. However, as in whole sediment or surfacewater toxicity testing, there is a need for a battery of toxicity test organisms and endpoints from which to choose (see Chapter 12). Different endpoints and organisms are required to answer different questions, and using a variety of endpoints and organisms adds to the robustness of sediment quality assessments (see Chapters 3 and 9). Organisms vary in their sensitivity to different contaminants and in their susceptibility to different confounding factors. The researcher should consider these factors when selecting a test organism. For example, fish tend to be more sensitive to ammonia than *Ceriodaphnia dubia* (USEPA 1985) and benthic invertebrates (Kohn et al. 1994; Monda et al. 1995). Therefore, a larval fish would not be a good choice of test organism if the researcher wishes to minimize the effects of ammonia toxicity. Understanding the sensitivity of different organisms allows the design of experiments that better address the objectives of a particular study.

One advantage of porewater toxicity testing is that it is amenable to testing with organisms that dwell in aqueous media. Many standardized aqueous medium toxicity tests are immediately transferable to porewater testing. These tend to be as sensitive as longer-term whole sediment toxicity tests (see Chapter 2). In the limited number of surveys directly comparing the sensitivity of nonbenthic organisms and porewater-dwelling organisms in porewater toxicity tests, there was a high degree of agreement (Carr, Chapman, Presley et al. 1996). Therefore, the use of nonbenthic organisms in porewater toxicity testing as surrogate organisms and as standardized toxicity test organisms is considered appropriate. Aquatic systems should not be compartmentalized because surface waters, sediments, and ground water are continuously interacting physically, chemically, and biologically. When contaminants enter an aquatic ecosystem, they move from water to sediment and possibly back to water, thereby exposing multiple communities. Water column organisms are frequently in contact with surficial sediment and even sometimes consume it. Water column, benthic, and epibenthic species have been shown to be of comparable sensitivity to toxicants, based on evaluations of the extensive water quality criteria database (Di Toro et al. 1991).

Some examples of nonbenthic animals that have been commonly used in saltwater sediment porewater toxicity tests are echinoderm gametes and larvae (Burgess et al. 1993; Carr, Chapman, Howard, Biedenbach 1996; Carr, Chapman, Presley et al. 1996; Carr, Long et al. 1996; Nipper et al. 1998; Carr et al. 2000), shellfish larvae (ASTM 1994), and larval fish (Carr and Chapman 1992). Freshwater nonbenthic organisms that have been used in porewater toxicity tests include cladocerans (Giesy et al. 1990; Ankley et al. 1991), insects (Giesy et al. 1990; Harkey et al. 1994), and fish (Ankley et al. 1991; Kemble et al. 1994). Zoospores of marine macroalgae (Hooten and Carr 1998) and the freshwater alga *Selenastrum capricornutum* (Ankley et al. 1990) have been used to assess phytotoxicity of pore waters. Infaunal animals may also be used to test the toxicity of pore water, if it can be shown that the absence of sediment particles does not unduly stress the organism. Examples of infaunal animals that have been used in porewater toxicity tests are the polychaete *Dinophilus gyrociliatus* (Carr et al. 1989; Nipper and Carr 2003), harpacticoid copepods (Carr, Chapman, Presley et al. 1996), fresh and saltwater amphipods (Landrum et al. 1987; Ankley et al. 1991; Ho et al. 1997), chironomids (Giesy et al. 1990; Sibley et al. 1997), mayflies (Giesy et al. 1990), and oligochaetes (Ankley et al. 1991). The Microtox and Mutatox assays, which incorporate luminescent bacteria, have been used to assess the toxicity of fresh and saline pore waters (Johnson and Long 1998). Ciliates, which are resident in sediment pore water and have a moderately well-developed toxicological database (Persoone and Dive 1978; Lynn and Gilron 1992), have a rapid life cycle, and are small enough to use with the small volumes required, were recommended for use in sediment porewater toxicity testing (Nipper 2000) but to our knowledge have not yet been used for this purpose.

In some limited cases, a species indigenous to the area of the sample collection might be required. These cases would include in situ tests in which there is danger of introduction of an invasive species and circumstances in which the question to be answered concerns the response of that particular organism. However, we believe that in most situations it is not necessary to use organisms indigenous to the area of sample collection. Particularly when a nonbenthic organism is being used in the test, there is little to be gained from a scientific standpoint by requiring the use of an organism native to the area. In this case, the organism is already a surrogate species, and using an indigenous animal does not make it less so. Use of standardized toxicity tests across wide geographic areas allows for easier ranking of sediment quality and provides a stronger regulatory foundation. However, there may be practical (e.g., availability of test organisms), political, and social reasons to use a locally occurring species.

Developing new porewater toxicity test methods may be required for a variety of reasons, such as the need for an indigenous test species, a different endpoint, or a species that is less sensitive to confounding factors. When a new species is used in porewater toxicity testing, information on species sensitivity to confounding factors should be determined and test methods should be adequately developed. When investigating the suitability of a new species for porewater toxicity testing, researchers should bear in mind the difficulty in obtaining large volumes of pore water. If large organisms are used in small test volumes, it can result not only in oxygen depletion but also in depletion of low solubility contaminants, which may occur before critical body residues are reached.

Most porewater toxicity tests currently in practice are acute tests. Although some of these tests are very sensitive, they do not address the effects of long-term or life-cycle exposures. The 7-day tests with freshwater cladocerans (Adams et al. 1990; Lasier et al. 2000) and with the saltwater polychaete *D. gyrociliatus* (Carr et al. 1989; Carr and Chapman 1992; Carr, Chapman, Presley et al. 1996; Nipper and Carr 2003) are perhaps the only laboratory life-cycle porewater chronic tests that have been performed. Porewater toxicity tests are not well suited to exposures of a longer term than the short-term chronic tests, except in the case of in situ testing or possibly through the use of sequential field collection of porewater samples. This is because of the changes in porewater chemistry with holding time. There is a need to develop a broader range of short-term chronic tests with pore water.

Acclimation of test organisms

Aquatic toxicity test organisms are sensitive to a wide range of toxicants. Therefore, by their very nature, they are sensitive to environmental stressors that include natural factors such as temperature, pH, alkalinity, hardness, and salinity. The conditions in which organisms reside prior to testing may differ markedly from those that occur during toxicity testing. For example, the in situ temperature where indigenous test organisms are collected for toxicity testing may differ from standard laboratory test temperatures. The reverse situation arises where laboratory-cultured test organisms may be exposed in situ at a different temperature than that to which they are accustomed. In addition, test organisms purchased from aquatic suppliers may be raised under conditions that differ from reference sample conditions or laboratory test conditions. If these differences between pretest and test conditions vary markedly, then they may introduce an unacceptable level of stress that confounds interpretations of sample toxicity. This dictates that general water quality factors (such as temperature, pH, alkalinity, hardness, salinity) be considered and appropriate acclimation conducted immediately prior to toxicity testing. Because species differ in their sensitivity to these stressors, generalized statements as to the degree of acclimation necessary cannot be made. Standardized test methods (e.g., ASTM, USEPA, OECD, Environment Canada) contain some guidance on acclimation, and anecdotal information exists in the literature.

Use of controls and references

Perhaps two of the most critical aspects of an environmental assessment of potential stressor impacts are proper controls and references and documentation of acceptable biological response. Without good performance of toxicity test organisms in the controls and references, adverse effects at the test site cannot be determined. The definitions of reference and control samples differ widely. For purposes of this chapter, a "control sample" is one that is not a test treatment and has conditions similar to those that the test organisms were exposed to during their pretest life cycle. For example, for laboratory-cultured organisms, this would be a sample of water similar to what they were cultured in, and for field-collected organisms, it would be water (or sediment or pore water) from the site where they were collected or reconstituted water similar to water from the collection site. Additional positive and/or negative controls may be necessary to determine whether other sample manipulation issues had an effect (e.g., a solvent control for studies in which an organic chemical was dissolved in a solvent and spiked into the test sample; use of reference toxicant tests to assess the sensitivity of the test organisms).

"Reference sample" refers to samples collected from a field site similar in characteristics to the test samples and of acceptable quality. The definitions of "similar" characteristics and "acceptable" quality will be study specific. The selection of the appropriate reference site may be influenced by several factors, such as program needs, presence or absence of keystone species, meteorological conditions, and logistical constraints. For pore waters, similar characteristics usually will refer to whole sediments with similar salinity, grain size distributions, and organic carbon and/or acid volatile sulfide (AVS) concentrations. However, "similar" characteristics such as organic matter type, hardness or alkalinity, pH, and/or mineralogy may be of importance. Ideally, reference sites designated as being of acceptable quality will be based on the best achievable conditions (e.g., meeting beneficial uses, good fishery and benthic invertebrate communities) and from a nearby location. In human-dominated watersheds (e.g., urban and agricultural), it may be inappropriate to use a pristine reference site from a natural, undisturbed watershed. This does not suggest, however, that nearby sites that are contaminated with toxic levels of chemicals should be used. High sediment ammonia has been discounted by some as a "nuisance" or natural constituent and ignored in classifications of sediment toxicity or contamination. In most human-dominated watersheds, elevated sediment ammonia can be attributed to pollutant loadings from point and nonpoint sources. These loadings add nitrogen compounds directly to sediments (which are converted to ammonia) or other chemical toxicants that disrupt biological control mechanisms. Because of loadings from uncontrolled point or nonpoint sources, there may not be any nearby reference site that is of acceptable quality. In this situation, an ecoregion reference site should be used. While this may pose a logistical challenge, it is an essential component for determinations of hazard and risk.

Toxicity Identification Evaluation Procedures

Laboratory TIE procedures

Currently, only TIE methods for pore waters are routinely used to identify stressors in sediments. Although pore water is the only medium for which TIE procedures have successfully been performed to date, it is envisioned that when whole sediment methods are developed and validated, porewater and whole sediment methods will be used together as weight-of-evidence procedures to ensure that the correct stressors are identified. In addition, even when whole sediment TIE methods exist, porewater methods would be used if the test organism of choice is compatible with a water-only matrix, for example, sea urchin sperm cell and embryological development tests.

In general, TIE methods used for effluents and pore waters are similar. The exception may be the closed chamber manipulation used to maintain pH in the graduated pH method. In pore waters, the chemical oxygen demand of the sample, combined with the biological oxygen demand of the organisms may decrease DO concentrations below those required by the test organisms. This is of particular concern for marine TIE manipulations in which the test organisms are large (e.g., *Americamysis bahia, Ampelisca abdita*), as opposed to smaller freshwater test organisms (e.g., *C. dubia*).

In porewater TIE procedures, as in all TIE methods, appropriate assessment of sample toxicity is critical to the correct identification of stressors. Incorrect assessment of porewater toxicity, resulting from methodological artifacts created during sampling or handling of the pore waters, will likely result in the incorrect identification of a stressor. The same uncertainties and confounding factors that affect porewater toxicity testing will affect porewater TIE procedures (see " Methodological Uncertainties," p 144, and "Confounding Factors," p 149). These include changes in the bioavailability of porewater toxicants during collection, extraction, or storage, as well as exposure of organisms to porewater concentrations that are unlike field exposures.

If the initial objective of the testing procedure requires removal of ammonia, hydrogen sulfide, or other toxicants in order to assess signals from remaining toxicants, and the researchers plan to use TIE procedures, one should proceed with caution. Researchers should be aware that TIE manipulations are designed to work as a group and not singly. Very few of the manipulations are selective for only one class of toxicants, and use of a single manipulation can lead to false results. *U. lactuca*, the sea lettuce, can be used effectively to remove ammonia, but it also removes organic contaminants (Ho et al. 1999). If the *U. lactuca* procedure is used to remove ammonia outside of normal TIE protocols, then sequential use of the C_{18} column manipulation (which removes organic contaminants) and *U. lactuca* is recommended. For example, the sample may be passed through a C_{18} column; then an aliquot of the sample is removed for testing, with the remaining sample subjected to *U. lactuca* exposure and then tested for toxicity. One may be able to discern the true ammonia signal by comparing the toxicity signals after each manipulation. Alternatively, the researcher may perform both manipulations in parallel and then determine the difference between the signals; however, the latter may be less successful, depending upon the toxicity of the sample.

Recommendations

As a result of our workgroup's discussion during the Porewater Toxicity Testing Workshop, we present the following recommendations to researchers applying porewater toxicity tests:

1) Sampled sediment depth should match expected exposure of organisms and should be based on the objective of the study (e.g. dredging, assessment of potential impact on benthic biota) and/or on the dynamics of the system to be surveyed (e.g., prone to storm events that disturb sediment to extensive depth, coarse grain size, and high currents).

2) The use of in situ peepers to collect pore water is recommended whenever it is feasible. The second-best option would be sampling pore water in situ by vacuum. If in situ sampling is not possible, whole sediment samples should be collected by cores, dredges, or grabs and stored at 4 °C for the minimum possible amount of time, and pore water should be extracted as close as practical to the time of the toxicity test.

3) Either centrifugation or pressurization is recommended for ex situ porewater extraction because both result in relatively large amounts of pore water. The method of choice is dictated by logistics and sample characteristics. Centrifugation is recommended if organics are the contaminants of concern.

4) Porewater filtration is not recommended. Pore water should be double-centrifuged if centrifugation is the selected extraction method or centrifuged after other methods of extraction.

5) The same porewater extraction method should be used throughout a survey, regardless of which method is selected.

6) Most test methods recommend adjustment of the salinity of the porewater samples to within 1 or 2 parts per thousand of organism requirements to control the effect of this variable on the test results.

7) Ammonia and sulfides should be measured in porewater samples submitted for toxicity testing, and the tolerance of the test organisms to unionized ammonia and sulfide should be established for comparison to the levels measured in the samples.

8) In addition to ammonia and sulfides, the following porewater quality and confounding factors need to be measured: DO, pH, salinity (marine and estuarine samples), hardness, alkalinity, and conductivity (freshwater samples); nutrients, in samples used for phytotoxicity tests; TOC and DOC if contamination by organics is suspected.

9) The use of indigenous organisms in laboratory tests is seldom necessary, unless there is a particular species of concern in the study area.

10) Some species may be invasive and should not be used for in situ testing unless they are resident species.

11) The use of water column organisms for porewater toxicity tests is considered scientifically appropriate.

12) Appropriate reference sites should be selected from a location near the study site, or at least in the same ecoregion, and should have similar sediment (and therefore porewater) characteristics.

13) If confounding factors are manipulated (e.g., TIE), caution should be exercised to avoid removal of contaminants of concern. As a general rule, TIE manipulations should not be performed separately, but as a suite.

Research Needs

During the discussions of our workgroup at the Porewater Toxicity Testing Workshop, we identified a variety of research needs to improve the understanding of the effects of methodological variables, confounding factors, and TIE manipulations on the results of porewater toxicity tests. Research is needed in the following areas:

1) Acceptable sediment storage time prior to porewater extraction

2) The effect of porewater extraction and storage methods on microbial processes such as degradation of contaminants and ammonia production

3) The effect of freezing and other storage methods on chemistry and toxicity of contaminants contained in pore water

4) The development of improved (larger-volume) in situ porewater collection devices

5) Different ways of introducing oxygen into the samples while minimizing changes to the pore water and contaminants contained in it

6) The effect of the oxidation of metals and organics on porewater toxicity

7) The effect of the sorption of contaminants to test and storage vessels, and the relation of sorption to container type and surface area to water volume

8) The tolerance of test organisms to a variety of confounding factors, for example, ammonia, sulfides, pH

9) The relationship between porewater tests and field conditions to help assess the utility of porewater testing and TIE procedures

10) Appropriate sample volumes for porewater toxicity tests to maintain exposure conditions (e.g., hydrophobic contaminants, DO, metabolic waste)

11) The development of a broader range of short-term chronic tests in small volumes for use with porewater tests.

References

Adams DD. 1991. Sediment pore water sampling. In: Mudroch A, MacKnight SD, editors. Handbook of techniques for aquatic sediments sampling. Boca Raton FL, USA: CRC. p 171–202.

Adams JR, Newsted JL, Ankley GT, Giesy JP, Hoke RA. 1990. Toxicity of sediments from Western Lake Erie and the Maumee River at Toledo, Ohio, 1987: Implications for current dredged material disposal practices. *J Great Lakes Res* 16:457–470.

Ajmal M, Khan AU. 1984. Effect of water hardness on the toxicity of cadmium to micro-organisms. *Water Res* 10:1487–1491.

Ankley GT, Katko A, Arthur J. 1990. Identification of ammonia as an important sediment-associated toxicant in the lower Fox River and Green Bay, Wisconsin. *Environ Toxicol Chem* 9:312–322.

Ankley GT, Schubauer-Berigan MK, Dierkes JR. 1991. Predicting the toxicity of bulk sediments to aquatic organisms with aqueous test fractions: Pore water versus elutriate. *Environ Toxicol Chem* 10:1359–1366.

[ASTM] American Society for Testing and Materials. 1994. Standard guide for conducting static acute toxicity tests starting with embryos of four species of saltwater bivalve mollusks. Designation E724-94. Volume 11.04, ASTM annual book of standards. Philadelphia PA, USA: ASTM. 1786 p.

Bower CE, Bidwell JP. 1978. Ionization of ammonia in seawater: Effects of temperature, pH, and salinity. *J Fish Res Board Can* 35:1012–1016.

Bradley RW, Sprague JB. 1985. The influence of pH, water hardness, and alkalinity on the acute lethality of zinc to rainbow trout (*Salmo gairdneri*). *Can J Fish Aquat Sci* 42:731–736.

Burgess RM, Schweitzer KA, McKinney RA, Phelps DK. 1993. Contaminated marine sediments: Water column and interstitial toxic effects. *Environ Toxicol Chem* 12:127–138.

Calamari D, Marchetti R, Vailati G. 1980. Influence of water hardness on cadmium toxicity to *Salmo gairdneri* Rich. *Water Res* 14:1421–1426.

Caron G. 1989. Modeling the environmental distribution of nonpolar organic compounds: The influence of dissolved organic carbon in overlying and interstitial water. *Chemosphere* 19:1473–1482.

Carr RS. 1998. Marine and estuarine porewater toxicity testing. In: Wells PG, Lee K, Blaise C, editors. Microscale testing in aquatic toxicology: Advances, techniques, and practice. Boca Raton FL, USA: CRC. p 523–538.

Carr RS, Chapman DC. 1992. Comparison of solid-phase and pore-water approaches for assessing the quality of marine and estuarine sediments. *Chem Ecol* 7:19–30.

Carr RS, Chapman DC. 1995. Comparison of methods for conducting marine and estuarine sediment porewater toxicity tests: Extraction, storage, and handling techniques. *Arch Environ Contam Toxicol* 28:69–77.

Carr RS, Chapman DC, Howard CL, Biedenbach JM. 1996. Sediment quality triad assessment survey of the Galveston Bay, Texas system. *Ecotoxicology* 5:341–64.

Carr RS, Chapman DC, Presley BJ, Biedenbach JM, Robertson L, Boothe P, Kilada R, Wade T, Montagna P. 1996. Sediment porewater toxicity assessment studies in the vicinity of offshore oil and gas production platforms in the Gulf of Mexico. *Can J Fish Aquat Sci* 53:2618–28.

Carr RS, Long ER, Windom HL, Chapman DC, Thursby G, Sloane GM, Wolfe DA. 1996. Sediment quality assessment studies of Tampa Bay, Florida. *Environ Toxicol Chem* 15:1218–31.

Carr RS, Montagna PA, Biedenbach JM, Kalke R, Kennicutt MC, Hooten R, Cripe G. 2000. Impact of storm-water outfalls on sediment quality in Corpus Christi Bay, Texas. *Environ Toxicol Chem* 19:561–74.

Carr RS, Williams JW, Fragata CTB. 1989. Development and evaluation of a novel marine sediment pore water toxicity test with the polychaete *Dinophilus gyrociliatus*. *Environ Toxicol Chem* 8:533–43.

Chambers RM, Harvey JW, Odum WE. 1992. Ammonium and phosphate dynamics in a Virginia salt marsh. *Estuaries* 15:349–359.

Di Toro DM, Zarba CS, Hansen DJ, Berry WJ, Swartz RC, Cowan CE, Pavlou SP, Allen HE, Thomas NA, Paquin PR. 1991. Technical basis for establishing sediment quality criteria for nonionic organic chemicals using equilibrium partitioning. *Environ Toxicol Chem* 10:1541–1583.

Environment Canada. 1994. Guidance document on collection and preparation of sediments for physicochemical characterization and biological testing. Ottawa ON, Canada: EPS 1/RM/29.

Flint RW, Kalke RD. 1983. Environmental disturbance and estuarine benthos functioning. *Bull Environ Contam Toxicol* 31:501–511.

Frazier BE, Naimo TJ, Sandheinrich MB. 1996. Temporal and vertical distribution of total ammonia nitrogen and un-ionized ammonia nitrogen in sediment pore water from the Upper Mississippi River. *Environ Toxicol Chem* 15:92–99.

Gaston GR. 1985. Effects of hypoxia on macrobenthos of the inner shelf off Cameron, Louisiana. *Estuar Coast Shelf Sci* 20:603–613.

Giesy JP, Rosiu CJ, Graney RL. 1990. Benthic invertebrate bioassays with toxic sediment and pore water. *Environ Toxicol Chem* 9:233–248.

Harkey GA, Landrum PF, Klaine SJ. 1994. Comparison of whole-sediment, elutriate and pore-water exposures for use in assessing sediment-associated organic contaminants in bioassays. *Environ Toxicol Chem* 13:1315–1329.

Heip C. 1995. Eutrophication and zoobenthos dynamics. *Ophelia* 41:113–136.

Ho KT, Kuhn A, Pelletier MC, Burgess RM, Helmstetter A. 1999. Use of *Ulva lactuca* to distinguish pH-dependent toxicants in marine waters and sediments. *Environ Toxicol Chem* 18:207–12.

Ho KT, McKinney RA, Kuhn A, Pelletier MC, Burgess RM. 1997. Identification of acute toxicants in New Bedford harbor sediments. *Environ Toxicol Chem* 16:551–558.

Hooten RL, Carr RS. 1998. Development and application of a marine sediment pore-water toxicity test using *Ulva fasciata* zoospores. *Environ Toxicol Chem* 17:932–40.

Ingersoll CG. 1995. Sediment tests. In: Rand GM, editor. Fundamentals of aquatic toxicology: Effects, environmental fate, and risk assessment. Washington DC, USA: Taylor & Francis. p 231–255.

Johnson BT, Long ER. 1998. Rapid toxicity assessment of sediments from estuarine ecosystems: A new tandem in vitro testing approach. *Environ Toxicol Chem* 17:1099–1106.

Kemble NE, Brumbaugh WG, Brunson EL, Dwyer FJ, Ingersoll CG, Monda DP, Woodward DF. 1994. Toxicity of sediments from the upper Clark Fork River, Montana, to aquatic invertebrates and fish in laboratory exposures. *Environ Toxicol Chem* 13:1985–1997.

Knezovich JP, Steichen DJ, Jelinski JA, Anderson SL. 1996. Sulfide tolerance of four marine species used to evaluate sediment and porewater toxicity. *Bull Environ Contam Toxicol* 57:450–457.

Kohn NP, Word JQ, Niyogi DK, Ross LT, Dillon T, Moore DW. 1994. Acute toxicity of ammonia to four species of marine amphipod. *Mar Environ Res* 38:1–15.

Landrum PF, Nihart SR, Eadie BJ, Herche LR. 1987. Reduction in bioavailability of organic contaminants to the amphipod *Pontoporeia hoyi* by dissolved organic matter of sediment interstitial water. *Environ Toxicol Chem* 6:11–20.

Lasier PJ, Winger PV, Bogenrieder KJ. 2000. Toxicity of manganese to *Ceriodaphnia dubia* and *Hyalella azteca*. *Arch Environ Contam Toxicol* 38:298–304.

Lynn DH, Gilron GL. 1992. A brief review of approaches using ciliated protists to assess aquatic ecosystem health. *J Aquat Ecosyst Health* 1:263–270.

Miller DC, Poucher S, Cardin JA, Hansen D. 1990. The acute and chronic toxicity of ammonia to marine fish and a mysid. *Arch Environ Contam Toxicol* 19:40–48.

Monda DP, Galat DL, Finger SE, Kaiser MS. 1995. Acute toxicity of ammonia (NH_3-N) in sewage effluent to *Chironomus riparius*. II. Using a generalized linear model. *Arch Environ Contam Toxicol* 28:385–390.

Mount DI. 1966. The effect of total hardness and pH on the acute toxicity of zinc to fish. *Air Water Pollut Int J* 10:49–56.

Nipper M. 2000. Current approaches and future directions for contaminant-related impact assessments in coastal environments: Brazilian perspective. *Aquat Ecosyst Health Manag* 3:433–447.

Nipper M, Carr RS. 2003. Recent advances in the use of meiofaunal polychaetes for ecotoxicological assessments. *Hydrobiologia* (in press).

Nipper M, Carr RS, Biedenbach JM, Hooten RL, Miller K. 2002. Toxicological and chemical assessment of ordnance compounds in marine sediments and pore waters. *Mar Pollut Bull* 44:789–806.

Nipper MG, Roper DS, Williams EK, Martin ML, VanDam LF, Mills GN. 1998. Sediment toxicity and benthic communities in mildly contaminated mudflats. *Environ Toxicol Chem* 17:502–510.

Orem WH, Gaudette HE. 1984. Organic matter in anoxic pore water: Oxidation effects. *Org Geochem* 5:175–181.

Paulauskis JD, Winner RW. 1988. Effects of water hardness and humic acid on zinc toxicity to *Daphnia magna* Straus. *Aquat Toxicol* 12:273–290.

Pearson TH, Rosenberg R. 1976. A comparative study of the effects on the marine environment of wastes from cellulose industries in Scotland and Sweden. *Ambio* 5:77–79.

Persoone G, Dive D. 1978. Toxicity tests on ciliates, a short review. *Ecotox Environ Saf* 2:105–114.

Playle RC, Dixon DG, Burnison K. 1993. Copper and cadmium binding to fish gills: Modification by dissolved organic carbon and synthetic ligands. *Can J Fish Aquat Sci* 50:2667–2677.

Rhoads DC, McCall PL, Yigst JY. 1978. Disturbance and production on the estuarine seafloor. *Am Sci* 66:577–578.

Sarda N, Burton Jr GA. 1995. Ammonia variation in sediments: Spatial, temporal and method-related effects. *Environ Toxicol Chem* 14:1499–1506.

Sibley PK, Legler J, Dixon DG, Barton DR. 1997. Environmental health assessment of the benthic habitat adjacent to a pulp mill discharge. I. Acute and chronic toxicity of sediments to benthic macroinvertebrates. *Arch Environ Contam Toxicol* 32:274–284.

Sims JG, Moore DW. 1995. Risk of porewater hydrogen sulfide toxicity in dredged material bioassays. Vicksburg MS, USA: U.S. Army of Engineers Waterways Experiment Station. Miscellaneous Paper D-95-4.

Sprang PAV, Janssen CR, Sabayasachi M, Benjits F, Persoone G. 1996. Assessment of ammonia toxicity in contaminated sediments of the Upper Scheldt (Belgium): The development and application of toxicity identification evaluation procedures. *Chemosphere* 33:1967–1974.

Stouthart XJHX, Haans JLM, Lock RAC, Bonga SEW. 1996. Effects of water pH on copper toxicity to early life stages of the common carp (*Cyprinus carpio*). *Environ Toxicol Chem* 15:376–83.

Stumm W, Morgan JJ. 1996. Aquatic chemistry: Chemical equilibria and rates in natural waters. 3rd ed. New York NY, USA: J Wiley. 1022 p.

Tay K-L, Doe KG, MacDonald AJ, Lee K. 1998. The influence of particle size, ammonia, and sulfide on toxicity of dredged materials for ocean disposal. In: Wells PG, Lee K, Blaise C, editors. Microscale testing in aquatic toxicology: Advances, techniques, and practice. Boca Raton FL, USA: CRC. p 559–574.

Tegner MJ, Dayton PK, Edwards PB, Riser KL, Chadwick DB, Dean TA, Deysher L. 1995. Effects of a large sewage spill on a kelp forest community: Catastrophe of disturbance? *Mar Environ Res* 40:181–224.

[USEPA] U.S. Environmental Protection Agency. 1985. Ambient water quality criteria for ammonia: 1984. Washington DC, USA: USEPA. EPA 440/5-85-001.

[USEPA] U.S. Environmental Protection Agency. 1986. Quality criteria for water. Washington DC, USA: USEPA. EPA 440/5-86-001.

Wang F, Chapman PM. 1999. Biological implications of sulfide in sediment: A review focusing on sediment toxicity. *Environ Toxicol Chem* 18:2526–2532.

Wiener JG, Giesy JP. 1979. Concentrations of Cd, Cu, Mn, Pb, and Zn in fishes in a highly organic softwater pond. *J Fish Res Board Can* 36:270–279.

Winger PV, Lasier PJ. 1991. A vacuum-operated pore-water extractor for estuarine and freshwater sediments. *Arch Environ Contam Toxicol* 21:321–324.

Winger PV, Lasier PJ, Jackson BP. 1998. The influence of extraction procedure on iron concentrations in sediment pore water. *Arch Environ Contam Toxicol* 35:8–13.

Chapter 8

Porewater Toxicity Tests: Value as a Component of Sediment Quality Triad Assessments

Edward R Long, R Scott Carr, Paul A Montagna

T he purpose of this chapter is to describe how toxicity tests of sediment pore waters can be used in Sediment Quality Triad (SQT) surveys. Data are reviewed and summarized from surveys of U.S. estuaries to illustrate some of these applications. The data presented were compiled from the National Oceanic and Atmospheric Administration's (NOAA) National Status and Trends (NS&T) Program and the U.S. Environmental Protection Agency's (USEPA) Environmental Monitoring and Assessment Program (EMAP). Spatial patterns in toxicity identified with porewater tests are contrasted with those from amphipod survival tests in several areas. The spatial scales of toxicity in the porewater tests are compared among major estuarine regions of the U.S. The frequency of toxicity and distribution of results are compared among tests with data compiled from surveys conducted nationwide. Relationships among observations of toxicity, concentrations of toxicants in sediments, and measures of benthic community composition are described with data from selected survey areas. Applications of the data from these tests in SQT assessments are described with the aid of principal component analyses (PCAs) to identify both degraded and undegraded (reference) conditions.

Frequency and Severity of Toxic Responses

Amphipod survival tests are the most commonly used acute toxicity tests in North America. NOAA, USEPA, and their collaborators have performed them on numerous samples in large-scale and regional environmental assessments of sediment quality. In these programs, samples were collected at locations determined by a random process within numerous estuaries and bays along each of the 3 U.S. coastlines (Paul et al. 1992; Long, Robertson et al. 1996). Sampling was not focused upon known toxic hotspots or toxicant sources. Therefore, the data from both programs represent conditions throughout major regions of each survey area, including locations near and far from toxicant sources.

Porewater Toxicity Testing: Biological, Chemical, and Ecological Considerations. R. Scott Carr and Marion Nipper, editors.
© 2003 Society of Environmental Toxicology and Chemistry (SETAC). ISBN 1-880611-65-1

Based upon their review of results from 22 estuarine survey areas, Long, Robertson et al. (1996) estimated that the spatial extent of toxicity in amphipod survival tests represented about 11% of the surficial area, whereas in the sea urchin fertilization test with porewater, it was 43%. When they examined concordance between the 2 tests, the spatial extent estimate decreased to about 4% of the combined survey area, suggesting that the 2 types of tests rarely were in agreement as to which samples were classified as toxic, therefore producing complementary, not redundant, information. Porewater tests will become more important complements to the amphipod tests if, among other attributes, it can be demonstrated that they do not duplicate results from the more commonly used amphipod tests.

Data compiled in Table 8-1 were generated by NOAA in Hudson-Raritan Estuary, Newark Bay, Long Island Sound, Biscayne Bay, Boston Harbor, northern Puget Sound, 6 estuaries of South Carolina and Georgia, Galveston Bay, Sabine Lake, 4 western Florida bays, and Tampa Bay. Data were generated by the USEPA for the Louisianan Province in the Gulf of Mexico and the Virginian Province along the Atlantic coast. For these databases, the amphipod tests were the most commonly employed procedure for evaluating the toxicological properties of saltwater sediments.

Table 8-1 Comparison of the incidence of toxicity among 2 solid-phase amphipod survival tests and 4 porewater tests performed with sea urchins[a]

Test type	No. of samples	Percent significant[b]	Percent highly significant[c]
Amphipod survival			
Ampelisca abdita	2060	23.2	12.3
Rhepoxynius abronius	1668	63.4	40.1
Urchin fertilization[d]			
Arbacia punctulata	640	39.8	38.0
Strongylocentrotus purpuratus	354	65.5	64.7
Urchin embryo development[d]			
A. punctulata	500	63.4	63.2
S. purpuratus	463	62.0	60.5

[a] Data were assembled from numerous studies conducted nationwide in U.S. estuaries and marine bays.
[b] Sample means different from control means, $p \leq 0.05$.
[c] Sample means different from control means at $p \leq 0.05$, and differences exceed minimum significant differences (Carr and Biedenbach 1999).
[d] Toxicity tests run with 100% pore waters.

The percentages of samples classified as having either significant toxicity (i.e., survival less than controls with $p \leq 0.05$) or highly significant toxicity (i.e., survival less than controls with $p \leq 0.05$ and mean survival $\leq 80\%$ of controls) in the *Ampelisca abdita* tests were considerably lower than in the other tests (Table 8-1). Whereas about 12% of samples had highly significant toxicity in tests with *A. abdita*, about 40% had highly significant toxicity in tests with *Rhepoxynius abronius*, and 38%

to 65% were toxic in the 4 kinds of sea urchin tests with 100% porewater concentrations. Samples were classified as highly toxic in the *Arbacia punctulata* tests of fertilization success about as frequently (38%) as in the amphipod tests performed with *R. abronius* (40%). A considerably higher incidence of toxicity occurred in samples tested for embryo development with either *A. punctulata* or *Strongylocentrotus purpuratus* or for fertilization success with *S. purpuratus*.

The frequency distributions of results from tests with 2 amphipod survival tests (*A. abdita* and *R. abronius*), 2 tests of sea urchin fertilization (*A. punctulata* and *S. purpuratus*), and 2 tests of sea urchin embryological development (*A. punctulata* and *S. purpuratus*) are compared in Table 8-2. Data from NOAA and EMAP studies along the Atlantic and Gulf of Mexico coasts were assembled from results of tests done with *A. abdita* and compared to those from studies performed by NOAA and the states of California and Washington along the Pacific coast with *R. abronius*. Data generated by the U.S. Geological Survey (USGS) for NOAA in surveys of Atlantic coast and Gulf of Mexico estuaries (*A. punctulata*) are compared to those generated by NOAA, USGS, and the states of California and Washington (*S. purpuratus*).

Table 8-2 Frequency distribution of results (% survival or % normal) of 6 toxicity tests performed with marine sediments and pore waters[a]

Test type	No. of samples	Percent of samples within each range of control-adjusted results				
		0–20	20–40	40–60	60–80	≥80
Amphipod survival (*A. abdita*)	3186	2.4	1.7	2.1	6.4	87.4
Amphipod survival (*R. abronius*)	1668	2.8	4.1	9.2	25.4	58.3
Urchin fertilization (*A. punctulata*)[b]	640	15.2	4.2	5.6	8.4	65.9
Urchin fertilization (*S. purpuratus*)[b]	354	35.3	10.5	9.0	10.7	32.5
Urchin development (*A. punctulata*)[b]	500	53.8	2.4	3.2	4.4	35.4
Urchin development (*S. purpuratus*)[b]	463	51.2	4.8	4.8	4.3	35.0

[a] Data assembled from numerous studies conducted nationwide in U.S. estuaries and marine bays.

[b] Toxicity tests performed with 100% pore waters.

These data indicate that the distributions of results for both amphipod species were somewhat different (Table 8-2). There were more samples with low survival (<80% of controls) in the tests with *R. abronius* (42%) than in the tests with *A. abdita* (13%). Survival ranged from 60% to 80% of negative controls in 25% of the samples tested with *R. abronius*, whereas only 6% of samples had equivalent results in tests with *A. abdita*. In tests with *R. abronius*, 9% indicated survival from 40% to 60%, compared to 2% of the samples tested with *A. abdita*. In both tests, however, there was a steady and gradual decrease in the proportion of samples in each classification as survival decreased. Similar frequency distributions were not apparent in the tests performed with the sea urchins. In the tests of both fertilization success and embryological development, there appeared to be bimodal distributions in the results. Most samples indicated percent fertilization and percent normal development either >80% or from 0% to 20%. The distribution of the fertilization success data differed between the 2 species, whereas they were very similar for results of embryological development.

Mean effects range median (ERM) quotients were calculated for numerous studies conducted in U.S. estuaries and bays, as the means of chemical concentrations in sediment samples normalized to (divided by) the respective ERM values for 25 substances (Long et al. 1998). These quotients were compared to toxicity data from the same studies. The amphipod data were compiled from NOAA and EMAP surveys, and the urchin data were compiled from NOAA surveys done in conjunction with USGS and the state of California. The responses of the amphipod survival tests (*A. abdita, R. abronius* combined) and urchin fertilization tests (*A. punctulata* and *S. purpuratus* combined) to 4 and 5 ranges, respectively, in mean ERM quotients are compared in Table 8-3. The incidence of highly significant toxicity in the amphipod tests was lowest (9%) in samples with mean ERM quotients < 0.1 and peaked at 76% in samples with quotients > 1.5. Average control-adjusted survival in these samples decreased from 93% to 41% in the 4 chemical categories. In contrast, for the sea urchin fertilization test, the incidence of toxicity was much higher (21%) in samples with lower chemical concentrations (mean ERM quotients < 0.01) and increased quickly to 64% with quotients ranging from 0.011 to 0.9. The incidence of toxicity in the urchin tests was variable thereafter as the chemical concentrations increased and did not show a consistent increasing trend. However, average percent fertilization was highest (86%) in the least contaminated samples and steadily decreased with increasing chemical concentrations. These data, therefore, demonstrated that whereas amphipod survival was very high in the cleanest samples and diminished incrementally with increasing concentrations of 25 substances, sea urchin fertilization did not indicate such a pattern, suggesting that the tests were responding to other (unmeasured) etiological factors.

Collectively, these data (Tables 8-1, 8-2, and 8-3) suggest that the amphipod and urchin tests respond differently to the sediment samples. Both the frequency of toxicity and the frequency distributions of results differed between the 2 kinds of tests. The data also suggest that toxic conditions rarely occurred in tests of survival performed with *A. abdita*. Toxic conditions occurred more frequently in tests of survival with *R. abronius* and fertilization success with *A. punctulata*. Toxicity was most frequent when tests were run for *S. purpuratus* fertilization and embryological development in both sea urchin species. Therefore, the porewater tests added considerably to the sensitivity of the toxicity tests, that is, the frequency of toxic responses increased when these tests were performed in conjunction with the amphipod survival tests. However, as demonstrated in previous analyses (Long, Robertson et al. 1996), the 2 kinds of tests rarely show duplicative results. It appears that this lack of agreement may be attributable to the different relationships between toxicological responses in the 2 kinds of tests and ranges in chemical

Table 8-3 Comparison of results of sea urchin fertilization tests and amphipod survival tests to 5 ranges in mean ERM quotients[a]

Ranges in mean ERM quotients	Percent highly toxic[b]		Average control-adjusted response[b] (%)	
	Amphipod survival	Urchin fertilization	Amphipod survival	Urchin fertilization
<0.01	nd[c]	21	nd[c]	86
0.011–0.10	9	64	93	74
0.11–0.5	21	50	86	60
0.51–1.5	49	49	70	58
>1.5	76	67	41	36

[a] Data assembled from numerous studies conducted nationwide in U.S. estuaries and marine bays.
[b] Amphipod data, $n = 1513$; urchin data, $n = 824$.
[c] nd = Not determined.

concentrations in bulk sediments. Therefore, as recommended previously (Ingersoll et al. 1997), these data illustrate the need to use a battery of multiple tests in comprehensive assessments because the different tests appear to respond to different contaminants and, therefore, rarely duplicate or parallel each other in responses to samples.

Spatial Patterns in Toxicity

Spatial (geographic) patterns in toxicity have been reported for many U.S. bays and estuaries (Carr, Long et al. 1996; Long, Robertson et al. 1996; Turgeon et al. 1998; Long 2000). The degree (severity) and magnitude (size) of toxicity in estuaries and marine bays are functions of the characteristics of the sources of toxicants, the physical distribution and sedimentation of the toxicants in the receiving system, and the bioavailability of the toxicants. Spatial patterns can be determined with any of the toxicity tests done as a part of a triad survey or by composing an index based upon results from all the tests. The types of spatial patterns in toxicity observed in the many large-scale surveys performed in U.S. estuaries are outlined in the discussion chapter from the SQT workgroup (Chapter 9).

One of the strengths of using a weight-of-evidence approach such as the SQT is the opportunity to identify sites that are toxic in different tests or in multiple tests. If different tests gave the same results, there would be no point in running more than 1 test. However, the notion that an increasing number of samples inevitably would be classified as toxic by chance alone if enough different tests were run (O'Connor et al. 1998) is not supported by evaluations of actual data. As noted above, the porewater toxicity tests often identify more samples as toxic than do solid-phase tests done with amphipods. Therefore, they have proven useful in classification of grossly polluted sites in which all tests show toxicity, slightly contaminated sites in which only significant results are obtained with the most sensitive tests, and uncontaminated sites in which toxicity is not apparent in any tests.

In the following 3 examples, the results of porewater toxicity tests are compared with those from another test performed on a portion of the same samples. Agreement between the 2 tests as to which samples were toxic or nontoxic was apparent at some locations in each example. However, as expected, there was considerable disagreement between the 2 tests with many of the samples.

Data for a portion of the survey conducted in Biscayne Bay by NOAA and the state of Florida (Long, Sloane et al. 1999) are shown in Figure 8-1. Results of amphipod survival tests of solid-phase sediments with *A. abdita* are compared with those from sea urchin tests of fertilization success and normal embryological development in 100% pore waters done with *A. punctulata*. Based upon results of the 3 tests together, toxicity was less severe and frequent in the portion of the bay between the Port of Miami and Rickenbacker Causeway than in the lower reaches of the Miami River. In the river, 86% of samples (18 of 21) were toxic in at least 1 test, whereas in the bay, 53% (19 of 36) were toxic. Surprisingly, the data did not indicate the formation of a toxic plume near the mouth of the Miami River because no toxicity for any of the 3 tests was observed at the sites in this area.

The data from Biscayne Bay illustrated the differential responses of the 3 tests to the samples. For example, there were only 4 samples (7%, all from the lower Miami River) out of the 57 tested in which toxicity was apparent in all 3 tests and 18 of 57 (32%) in which toxicity was not apparent in all 3 tests. Therefore, perfect agreement among the 3 tests (i.e., all indicated either toxic or nontoxic conditions) occurred in 22 (39%) of the samples. Furthermore, amphipod and urchin development were in agreement on classifying only 13 (23%) of the same samples as toxic, and amphipod survival and urchin fertilization were in agreement in only 4 (7%) of the same samples. As a consequence, the overall incidence of toxicity increased from 20 (35%) in the amphipod tests alone to 38 (67%) when all 3 tests were considered.

Sediment quality was considerably different in the lower Miami River from that of the bay (Figure 8-1). Eighteen of 21 samples tested had highly significant toxicity in the amphipod tests. Only 1 sample collected near the mouth of the river and 2 sandy samples from the Tamiami Canal were nontoxic in the amphipod tests. Concentrations of many chlorinated organic compounds, polycyclic aromatic hydrocarbons (PAHs), and several trace metals were elevated in these samples, and amphipod survival decreased steadily as the concentrations of these substances increased (Long, Sloane et al. 1999). Mean ERM quotients were much higher in the river (average = 0.764, range = 0.172 to 1.980) than in the bay (average = 0.036, range = 0.005 to 0.209). However, only 4 of the 21 samples from the river were toxic in the urchin fertilization tests, whereas 13 were toxic in the urchin embryological develop-

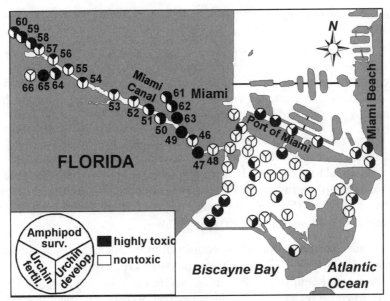

Figure 8-1 Distribution of highly significant responses in 3 toxicity tests of sediments in central Biscayne Bay and the lower Miami River, Florida (from Long, Hameedi, Robertson et al. 1999). Dark shading indicates highly significant responses in each test. Highly significant responses in amphipod survival tests (*A. abdita*) were different from controls ($p \leq 0.05$) and <80% of control response. Highly significant responses in the 2 sea urchin tests (*A. punctulata*) of 100% pore water were different from controls ($p \leq 0.05$) and <80% of control response.

ment tests. The 4 samples that were toxic in the urchin fertilization tests and 12 of the 13 samples that were toxic in the embryological development tests were also toxic in the amphipod test. The 3 samples that were not toxic to amphipods also were not toxic in the urchin fertilization tests (1 of the 3 was toxic in the urchin embryological development test). Therefore, 1 or both of the 2 urchin tests were in agreement in 14 samples and in disagreement in 7 samples from the river, and they were in agreement in 19 samples and in disagreement in 17 samples from the bay. Eighteen of the 19 bay samples in which there was agreement were not toxic in any of the tests. In all 17 of the bay samples in which there was disagreement, the agreement was due to toxicity in one or both of the urchin assays, but not in the amphipod test. There were no correlations between porewater ammonia concentrations and toxicity for the amphipod and fertilization tests, but there was a correlation for the embryo development test when known toxicity thresholds for unionized ammonia were exceeded (7 of the 21 samples).

Percentages of amphipods surviving in the 3 samples from Seybold Canal were among the lowest observed in the entire survey ($n = 226$): 5%, 8%, and 9%. Mean ERM quotients in these samples were 1.5, 1.4, and 1.8, respectively. Percent urchin fertilization was 96%, 14%, and 88%, respectively. Therefore, results of the 2 tests were in agreement in only one of these contaminated samples. However, percent normal embryological development was 0%, 0%, and 3% in tests of 100% pore water in these same samples. The excellent agreement between the amphipod and the embryological development test, but not the fertilization assay, is likely due to differences in mechanisms of toxicity and exposure time. Therefore, they demonstrate the necessity for including a suite of tests with different endpoints, which are differentially sensitive to different types of contaminants.

If the Biscayne Bay survey had been performed with only the amphipod tests, it would have appeared as if there were very little or no toxicity in the bay. Toxic conditions would have appeared to occur only in the river. If, on the other hand, the survey had been run with only the urchin fertilization test, a very different picture would have emerged. Thus, by using the data from all 3 tests, a more comprehensive assessment of conditions in this area was possible.

Results of toxicity tests performed on samples from St. Simons Sound, Georgia, are summarized in Figure 8-2. These data were prepared as part of a survey of 5 estuaries along the South Carolina and Georgia coastline (Long et al. 1998). The results of Microtox tests on organic solvent extracts were in agreement (either toxic or not toxic) with the sea urchin fertilization and embryological development tests on 100% pore waters in 7 and 10 of 20 samples, respectively. The 2 urchin tests of pore waters were in agreement in 10 of the 20 samples. All samples exhibiting toxicity in the fertilization test (4 samples from Brunswick Harbor) were also toxic in the embryological development test. Of the 20 samples collected in the survey, 6 indicated no toxicity in any of the 3 tests, and 1 was toxic in all 3 tests. Two toxicity gradients were apparent in this estuary. The relatively high degree of toxicity in Brunswick Harbor may have been attributable to toxicants entering from storm drains, shipyards, and other maritime facilities. Samples from Terry Creek and adjoining Back River were relatively toxic, as indicated in the Microtox and urchin embryological tests. This area received effluents entering the channel from a nearby industrial mill. One of these samples from Terry Creek was the only sample in this study that was significantly toxic to amphipods. In addition, 5 of the 8 samples collected in the Turtle River were toxic in only the urchin embryological test. Again, as was the case in Biscayne Bay, the spatial distribution or pattern in toxicity would have appeared to be quite different had only 1 test been performed on these samples.

In the third example (Figure 8-3), the results of 3 tests are compared among 66 sediment samples collected from Sabine Lake, an estuarine lagoon on the Texas–Louisiana border, and a portion of the adjoining Intracoastal Waterway (Long et al. 1999). None of the samples from this study had highly significant toxicity in the amphipod test (i.e., survival exceeded 80% of controls). Sixty-two of the 66 samples were significantly toxic in the urchin embryological development tests performed with 100% pore water; therefore, data are shown for tests of 50% porewater concentrations to better illustrate spatial patterns. Results of both urchin tests of pore

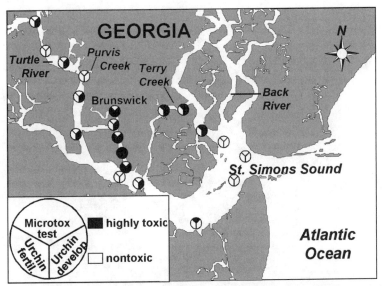

Figure 8-2 Distribution of highly significant responses in 3 toxicity tests of sediments in St. Simons Sound, Georgia (from Long et al. 1998). Dark shading indicates highly significant responses in each test. Highly significant responses in microbial bioluminescence (Microtox) tests were different from combined controls in both Mann-Whitney and Dunnett's t-tests ($p \leq$ 0.05). Highly significant responses in the 2 sea urchin tests (*A. punctulata*) of 100% pore water were different from controls ($p \leq 0.05$) and <80% of control response.

waters are compared with those from cytochrome P-450 reporter gene system (RGS) assays of enzyme induction with organic solvent extracts. Nontoxic conditions were apparent in all 3 tests in 20 samples. Highly significant responses were observed in one or the other urchin tests or both in all 5 samples, with highly significant responses in the P-450 assays. The most severe toxic conditions, as indicated with all 3 tests, were apparent in 1 sample collected in the Neches River near Port Neches and in 2 samples collected in Talor Basin at Port Arthur. Also, 7 of the 9 samples collected in the entrance channel and the Gulf of Mexico had highly significant toxicity in both urchin tests. Highly significant associations were observed between porewater sea urchin embryological development and a wide variety of sediment contaminants and ERM quotients (Long et al. 1999). These associations were not observed with the amphipod data.

The data from these 3 examples and many others illustrate the point made in earlier publications (Ingersoll et al. 1997): that it is important to use multiple tests to identify the presence and absence of toxicity. The use of a single test can easily provide misleading information on the degree, spatial distribution, and extent of toxicity. Because the porewater tests often provide a greater range of response than do the solid-phase survival tests, they often enhance or extend the area identified as toxic.

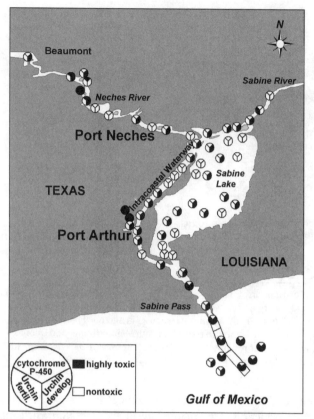

Figure 8-3 Distribution of highly significant responses in 3 toxicity tests of sediments in Sabine Lake, Texas (from Long et al. 2000). Dark shading indicates highly significant responses in each test. Highly significant responses in the cytochrome P-450 enzyme induction assay were >37.1 µg benzo[*a*]pyrene equivalents/g. Highly significant responses in the sea urchin tests (*A. punctulata*) for percent fertilization in 100% pore water were different from controls (*p* ≤ 0.05) and <80% of control response. Highly significant responses in the sea urchin tests (*A. punctulata*) for percent normal embryological in 50% pore water were different from controls (*p* ≤ 0.05) and <80% of control response.

Spatial Extent of Toxicity

The spatial extent of toxicity has been estimated for many U.S. bays and estuaries by weighting results of toxicity tests to the sizes of sampling strata and by expressing the estimates as percentages of the areas surveyed (Long, Robertson et al. 1996; Turgeon et al. 1998). In almost all survey areas, the spatial extent of toxicity was greater for the porewater tests than for the amphipod survival tests of solid-phase sediments. More samples were classified as toxic in the porewater tests, reflecting their greater sensitivity. Also, very little agreement was reported among the different tests, that is, samples that were toxic in both the amphipod and sea urchin fertilization tests represented <4% of the total area surveyed (Long, Robertson et al. 1996). Samples that were toxic in 3 tests (sea urchin fertilization, Microtox, and amphipod) represented only 2.5% of the area surveyed. To further illustrate these points, the percentages of areas sampled that were toxic are contrasted between 2 tests (pore water and solid phase) for 6 regions of the U.S. and for the overall, "national" averages (Table 8-4). Data from several nearby surveys were compiled to provide combined estimates for the bays and estuaries of the northeast, southeast, Southern California, Puget Sound, Texas, Mid-Atlantic, and overall.

Table 8-4 Spatial extent of sediment toxicity in porewater tests and solid-phase tests in 6 estuarine regions of the U.S.[a]

Estuarine regions	Urchin fertilization test of pore waters			Amphipod survival test of sediments		
	Total survey area (km^2)	Toxic area (km^2)	Percent of total area	Total survey area (km^2)	Toxic area (km^2)	Percent of total area
Northeast[b]	56.1	3.8	6.8	491.0	186.1	37.9
Southeast[c]	1964.3	905.2	46.1	1964.3	63.1	3.2
Southern California[d]	99.7	84.9	85.2	105.9	37.2	35.1
Puget Sound[e]	1511.3	44.6	3.0	1511.3	1.0	0.1
Texas bays[f]	1597.0	446.0	27.9	1597.0	0.0	0.0
Mid-Atlantic[g]	4611.8	1024.4	22.2	4611.8	145.4	3.2
National average (1998)[h]	9840.2	2508.9	25.5	10281.2	432.8	4.2

[a] Based on data compiled from NOAA surveys.
[b] Porewater data from Boston Harbor; solid-phase data from Boston Harbor, Hudson-Raritan Estuary, Newark Bay, Long Island Sound.
[c] Porewater data and solid-phase data from Tampa Bay, Biscayne Bay, South Carolina and Georgia bays, Western Florida bays.
[d] Porewater data and solid-phase data from San Pedro Bay, San Diego Bay, and selected coastal lagoons.
[e] Porewater and solid-phase data from northern and central Puget Sound.
[f] Porewater and solid-phase data from Sabine Lake and Galveston Bay.
[g] Porewater and solid-phase data from Delaware and Chesapeake bays.
[h] Porewater data compiled from 1468 samples and solid-phase data compiled from 1696 samples collected nationwide through the 1998 field season.

In the northeast, data from the porewater tests are available only for Boston Harbor, whereas solid-phase tests were performed in Hudson-Raritan estuary, Newark Bay, and Long Island Sound, as well as Boston Harbor (Table 8-4). Otherwise, most of the data were generated from tests of portions of the same samples. There were a few more samples tested for amphipod survival in Southern California than tested for urchin fertilization.

The samples that had highly significant toxicity in the sea urchin fertilization tests of 100% pore water and in amphipod tests of solid-phase sediments represented 7%

and 10%, respectively, of the total survey area in Boston Harbor. However, 100% of the samples had highly significant toxicity in the urchin embryological development test, and 91% and 51% had highly significant toxicity in the tests of 50% and 25% porewater, respectively (Long, Sloane et al. 1996). Additional details on these data are shown in Figures 9-2 and 9-3 of Chapter 9. Samples from Hudson-Raritan estuary and Newark Bay were considerably more toxic, thus raising the combined estimate for the solid-phase tests in the Northeast to 38%. Unfortunately, porewater tests were not conducted in these areas. In nearly all regions, toxicity as determined in the porewater tests was much more widespread than in solid-phase tests. These differences were most apparent in the Southeast, Southern California, Texas bays, and Mid-Atlantic estuaries. Relatively little toxicity was observed in either test in Puget Sound. With all the data combined from NOAA surveys conducted through 1998, the overall combined averages were 26% for the porewater tests (mostly for urchin fertilization) and 4% for the solid-phase tests.

These data confirm the patterns observed in the 3 examples above (Figures 8-1, 8-2, and 8-3): that porewater tests and solid-phase tests infrequently provide duplicative information on the presence of toxicity. Although, they frequently identify the same samples as nontoxic, they only occasionally show toxicity in the same samples and places. Where toxicity is observed with the solid-phase test and not with the urchin fertilization test, significant toxicity is frequently observed with the urchin embryological development assay, thereby demonstrating the contaminant-specific differences for the different tests.

Identification of Reference and Degraded Conditions using the SQT Approach

Data from SQT analyses have been used to identify sites apparently not degraded by pollution, sites with pollution-caused degradation, and sites with mixed results (Chapman 1996). In his sediment classification matrix, Chapman (1996) provided rational explanations for different combinations of results among the 3 elements of the triad. In this section, data from 5 survey areas are shown as examples of the weight of evidence formed to classify the relative quality of sediments where the porewater toxicity tests were included among the toxicity tests. Because the amphipod survival tests rarely indicated toxic conditions in most estuarine samples, the data from the porewater tests often were important in providing toxicological perspective to the chemical and benthic data in these triad studies.

Central Biscayne Bay and Miami River

In the first example, conditions in central Biscayne Bay and the adjoining lower Miami River (stations shown in Figure 8-1) are contrasted using data from chemical analyses, toxicity tests, and benthic analyses reported by NOAA (Long, Sloane et al.

Table 8-5 Mean ERM quotients, results of toxicity tests (adjusted for control results), and measures of benthic abundance in samples from central Biscayne Bay and lower Miami River[a]

Station	Mean ERM quotients	Amphipod survival[b] (%)	Sea urchin fert[c] (%)	Sea urchin devel[c] (% normal)	Benthic infauna				
					Total abundance	Total no. of species	Total no. of arthropods	Total no. of amphipods	Total no. of ampeliscids
Central Biscayne Bay (n = 15)									
Mean	0.04	96	88	58	1569	207	508	183	7
SD	0.03	7	22	47	1220	95	846	317	16
Min	0.01	77	36	0	174	56	17	0	0
Max	0.09	106	102	100	5280	340	3551	1311	62
Lower Miami River									
62	1.36	5[d]	36[e]	0[e]	423	17	22	22	0
51	0.60	9[d]	96	78[e]	3545	23	77	16	0
65	1.98	10[d]	95	74[e]	346	12	0	0	0
58	0.34	32[d]	99	0[e]	758	13	1	0	0
53	0.59	35[d]	102	96	1261	24	101	96	0
46	0.79	41[d]	98	101	2668	26	15	0	0
57	0.30	41[d]	93	97	889	13	0	0	0
Mean	0.85	25	88	64	1413	18	31	19	0
SD	0.57	15	22	41	1132	6	38	32	0

[a] Data from Long, Hameedi, Robertson et al. 1999.
[b] Test species: *A. abdita*.
[c] Test species: *A. punctulata* in 100% pore water; values are control normalized.
[d] Toxicity test results significantly different from controls ($p \leq 0.05$) and sample means < 80% controls.
[e] Toxicity test results significantly different from controls ($p \leq 0.05$) and sample means < detectable significance criteria (Carr and Biedenbach 1999).

1999) (Table 8-5). Based upon the weight of evidence from the triad of measures, the 7 samples from the lower Miami River were considerably degraded relative to the 15 samples from the bay. Mean ERM quotients were <0.1 in all 15 samples from the bay, whereas they ranged from 0.3 to 2.0 in the river (average 0.85). Control-adjusted amphipod survival averaged 96% in the bay. In contrast, amphipod survival ranged from 5% to 41% in the samples from the river. Despite the large differences in chemical concentrations and amphipod survival between the 2 areas, results of the urchin fertilization tests were virtually identical. Results ranged from 36% to 102% fertilization success in both areas and averaged 88% in both the bay and the river. The ranges in response for the urchin embryological development assay also were similar in the 2 regimes: 0% to 100% (average of 58%) in the bay and 0% to 101% (average of 64%) in the river. As illustrated in Figure 8-1, the amphipod test indicated many toxic samples in the river and mostly nontoxic conditions in the bay. In contrast, the urchin fertilization test indicated almost the opposite pattern. The embryo development test was much more sensitive than the fertilization test. It identified 4 of the samples from the river for which there were benthic data as toxic, whereas only one of them was toxic in the fertilization test (Table 8-5). The differences in sensitivity and spatial patterns in toxicity suggested that the 3 tests responded to different contaminants. Correlations between results of the toxicity tests and chemical concentrations confirmed the differential sensitivities among the tests (Long, Sloane et al. 1999).

The composition of the infaunal communities would be expected to change in the river as a result of the discharge of fresh water. Nevertheless, the data acquired for analyses of the benthos are important to help form a weight of evidence regarding relative sediment quality, regardless of the cause of the changes. The primary strength of the SQT approach is to provide data to be used to classify sediments as degraded, uncertain, or undegraded with information from 3 independent lines of analyses, not to assign causality or attribution. The total abundance of all species was similar in the samples from the bay and from the river (Table 8-5). In contrast, the numbers of species and abundance of arthropods, amphipods, and ampeliscid amphipods were severely diminished in the samples from the river. A wide variety of arthropods, echinoderms, polychaetes, mollusks (gastropods and bivalves), and other groups were present in bay samples. A few samples collected in the port channels and/or near the Miami shoreline supported relatively low numbers of species and individuals. Samples with the most diverse and abundant macrofauna generally were collected in seagrass beds, and those with more depauperate assemblages often were collected in navigation channels. The numbers of species in all the Miami River samples were consistently lower than in the samples from the bay. Capitellid worms dominated samples collected near the mouth of the river. Moving farther upstream, the abundance of capitellids diminished, and therefore, total abundance also diminished. The 2 stations sampled in tributaries to the Miami River (Seybold and Tamiami canals) supported the least abundant and least diverse assemblages.

In the lower Miami River, the abundance of all individuals was very high at the 2 downstream stations (46 and 51) because of the abundance of capitellid worms. These samples, however, supported very few species. The polychaetes *Capitella capitata* and *Streblospio benedicti* were dominant in the samples from downstream stations 46, 51, and 53. The cumacean *Cyclaspis* cf *varians* also was abundant at station 46. At station 53, the phyllodocid polychaete *Laeonereis culveri* and the amphipod *Grandidierella bonnieroides* were common along with the capitellids. Farther upstream at stations 57 and 58, the abundance of the capitellids decreased remarkably and the polychaete *Ceratanereis mirabilis* became the dominant species, accompanied by unidentified oligochaetes and the gastropod *Pyrgophorus* sp. The samples from the 2 tributaries to the Miami River supported some of the same species found in the samples from the river. In the samples from station 62 (Seybold Canal), the dominant species was *C. mirabilis*, accompanied by *Pyrogophorus* sp., unidentified oligochaetes, and *G. bonnieroides*. At station 65 (Tamiami Canal), the dominants were the bivalve *Corbicula fluminea* and unidentified oligochaetes.

The triad results, together, classified the sediments at stations 62 and 65 as "highly degraded" as defined in the classification matrix of Chapman (1996). They had the highest mean ERM quotients, they had highly significant toxicity in 2 of the 3 tests, and the benthic assemblages were depauperate (in the case of station 65, supported no amphipods or other arthropods). Sediments collected nearest the mouth of the river (stations 46, 51, and 53) had intermediate ERM quotients (0.6 to 0.8) and had highly significant toxicity to amphipods. However, they were less toxic to sea urchin embryo development than the sediments at stations 62 and 65 and supported intermediate numbers of benthic species and individuals (although dominated by opportunistic capitellids).

Data from some of the riverine stations showed interesting contrasts. Stations 57 and 58 had similar chemical indices (mean ERM quotients of 0.30 and 0.34, respectively), similar results in the amphipod survival tests (41% and 32% survival, respectively), and very similar benthic indices (e.g., both with 13 species, 0 and 1 arthropods, respectively). However, the pore water from station 58 had highly significant toxicity, and that from station 57 was not toxic in the embryo development test. Thus, it appears that the etiological agent or agents, whatever they were, that caused poor embryo development in the sample from station 58 were absent in the sample from station 57. The sample from station 62 was the second most contaminated, indicated the lowest amphipod survival, and had highly significant toxicity in both sea urchin tests. However, the sediments supported relatively high numbers of species, including the relatively sensitive arthropods. Thus, although the station clearly was degraded because of pollution (as per Chapman 1996), it did support some benthic organisms. But because the benthos was dominated primarily by opportunistic worms, all 3 elements of the triad suggested that it was a very degraded location.

Although the porewater tests did not result in classifying more samples from the lower Miami River as toxic than observed in the amphipod tests, they nevertheless added to the weight of evidence used to classify the quality of these sediments, along with information from the chemical and benthic analyses. Several of the samples that were most contaminated and most toxic in the amphipod tests also were toxic in one or both porewater tests and exhibited very depauperate benthic communities.

Delaware Bay

Unpublished data selected from a NOAA survey of Delaware Bay, Delaware provide the second example (Table 8-6). Similar to the situation in Biscayne Bay, samples with mean ERM quotients of <0.01 served as reference samples and, therefore, as a basis for classifying other samples. None of these 23 samples selected to represent reference conditions were toxic in either the amphipod or urchin tests. Amphipod survival averaged 95% and urchin fertilization averaged 101% of controls. There was very little variability in the results of these 2 tests among the 23 samples (standard deviations of 6 and 2, respectively). Enzyme induction in the cytochrome P-450 RGS assays was also very low (average of 2.6 µg B[*a*]p equivalents/g) in these samples. The benthic samples averaged 16 taxa per station and total abundance of 214 animals. All amphipod species and *A. abdita* were present in abundance in most samples.

In contrast to these samples, there were 3 Delaware Bay samples (station numbers 2, 17, and 19) in which mean ERM quotients were relatively high (0.4 to 1.3), amphipod survival was very low (2% to 22%), and the abundance and diversity of the benthos were relatively low (Table 8-6). In these samples, enzyme induction was at least 2 orders of magnitude higher than the average for the 23 reference samples. Also, the numbers of taxa were diminished by about one-half, total abundance was very low in 2 of the 3 samples, and abundance of amphipods (particularly *A. abdita*) was very low. These data provide evidence of pollution-related degradation. However, as in Biscayne Bay, the sea urchin fertilization test did not indicate significant toxicity in these 3 samples.

There were 6 other samples from Delaware Bay in which results of fertilization tests (but not amphipod tests) indicated toxicity when chemical concentrations were only slightly elevated and benthic abundance was reduced in most of them (Table 8-6). In 4 of these samples (21, 23, 25, and 56), urchin fertilization ranged from 6% to 70%, mean ERM quotients were slightly elevated (range: 0.04 to 0.25), and enzyme induction (range: 3 to 20) was slightly above that in the reference samples. There were relatively low numbers of species at these stations, and the abundance of all taxa, all amphipods, and ampeliscid amphipods were depressed. Thus, it appears that, although the sediments were not toxic to amphipods, they were slightly contaminated, had highly significant toxicity to urchin fertilization, and supported depauperate benthic assemblages. Percent urchin fertilization also was depressed in

Table 8-6 Comparisons of mean ERM quotients, toxicity test results, and measures of benthic abundance for 23 reference samples and toxic samples from Delaware Bay[a]

Station	Mean ERM quotients	Mean % amphipod survival[b]	Mean % sea urchin fert[c]	RGS B[a]P eq. (µg/g dry)	No. of taxa	Total abundance	H' diversity	J' evenness	Amphipod abundance	A. abdita abundance
Reference (n = 23)										
Avg.	0.04	95	101	2.6	15.7	214.1	1.7	0.7	97.0	41.3
SD	0.03	6	2	2.4	8.5	314.4	0.5	0.2	100.8	136.5
Min	0.01	82	98	0.2	4	11	1.0	0.4	0	0
Max	0.10	106	103	8.9	42	1265	2.4	1.0	440	650
2	0.37	2[d]	86	298.4	7	72	1.2	0.6	29	0
17	1.32	20[d]	98	344.7	5	75	1.1	0.7	0	0
19	0.73	22[d]	102	183.1	7	203	0.8	0.4	0	0
21	0.25	104	57	7.9	2	16	0.5	0.7	0	0
23	0.25	108	70	18.8	2	116	0.6	0.8	0	0
25	0.17	115	70	20.4	4	5	3			
56	0.04	98	6	2.9	15	324	2.1	0.8	9	9
57	0.11	81	66	3.7	14	1567	1.5	0.6	114	4
60	0.04	104	58	1.8	21	1368	0.7	0.2	240	1161

[a] Unpublished NOAA data.
[b] Test species: A. abdita.
[c] Test species: A. punctulata in 100% pore water; values are control normalized.
[d] Sample means different from control means at $p \leq 0.05$, and differences exceed minimum significant differences.

samples 57 and 60, but chemical concentrations were not particularly high, enzyme induction was relatively low, and the benthos was populated with many animals and species. Overall, the data from the SQT approach suggest that pollution-related degradation occurred at 7 stations among this selected set of samples, 3 with significant toxicity in amphipod tests and 4 with significant toxicity in the urchin tests.

Northern Puget Sound

In the third example, data from a survey of northern Puget Sound, Washington (Long, Hameedi, Robertson et al. 1999) are used to contrast conditions in the Strait of Georgia near the U.S.–Canada border and in nearby Everett Harbor, a highly industrialized port (Table 8-7). There were 10 samples from the Strait of Georgia in which no ERM concentrations were exceeded, mean ERM quotients averaged 0.08, amphipod survival was very high, percent urchin fertilization exceeded that of the controls, enzyme induction was very low, and the benthos was populated with numerous species and individuals. Mollusks in these samples represented about 36% of the total assemblage of biota. In contrast, there were 4 samples from Everett Harbor in which an average of 4 chemicals exceeded an ERM value and mean ERM quotients were roughly an order of magnitude greater than in the samples from the Strait of Georgia. Amphipod survival was high in these samples, but percent urchin fertilization in the porewater tests averaged 21%, and average enzyme induction was elevated by a factor of 31 relative to that for the 10 Strait of Georgia samples. The numbers of species dropped to an average of 7, compared to 52 in the strait, and the abundance of mollusks and all organisms was reduced. There were no echinoderms in the benthic samples from the harbor.

In northern Puget Sound, use of the porewater toxicity tests increased the number of samples classified as toxic relative to the information from the solid-phase amphipod tests. There were 15 samples with highly significant toxicity in fertilization tests of 100% porewater, whereas none had highly significant toxicity in the amphipod tests. In the 14 samples from the Strait of Georgia and Everett Harbor, the results of the urchin fertilization tests agreed quite well with those from the chemical and benthic analyses, in classifying samples as both degraded and undegraded. Statistical correlations between concentrations of classes of PAHs and fertilization success were very high (Spearman-rank, rho > 0.8, $p < 0.001$). Thus, the data from the porewater tests were very helpful in the classification of samples as degraded.

Central Puget Sound

Data from a survey of central Puget Sound (Long et al. 2000) were selected to provide the fourth example (Table 8-8). Chemical, toxicity, and benthic data from 22 stations sampled in relatively undeveloped regions are compared with those from 2 major industrialized bays (Sinclair Inlet near Bremerton and Elliott Bay near

Table 8-7 Averages of ERMs exceeded, mean ERM quotients, toxicity test results, and measures of benthic abundance for samples collected in the Strait of Georgia and Everett Harbor[a]

Average measurements	Strait of Georgia[b] (n = 10)	Everett Harbor[c] (n = 4)
Chemistry		
No. of ERMs exceeded	0	4
Mean ERM quotients	0.08	0.74
Toxicity		
Percent amphipod survival	88	90
Percent urchin fertilization	116	21
Cytochrome P-450 induction (µg/g)	3	94
Benthic effects		
Total abundance	930	69
No. of species	52	7
Arthropod abundance[d]	41	46
Echinoderm abundance[d]	5	0
Mollusk abundance[d]	36	0.7

[a] Data from Long, Hameedi, Robertson et al. (1999).
[b] Stations 10–19.
[c] Stations 86–89.
[d] Expressed as percent of total abundance.

Table 8-8 Averages of mean ERM quotients, control-normalized toxicity test results, and measures of benthic abundance for samples collected in central Puget Sound[a]

Stations	Mean ERM quotients	Percent amphipod[b] survival	Percent urchin[c] fertilization	Benthic abundance	No. of species	Arthropod count
Reference samples (n = 22)	0.05–0.08	80–106	94–118	197–2325	33–176	17–1349
Sinclair Inlet, station 160	0.35	99	2	149	21	3
Elliott Bay, station 115	0.83	97	6	1161	43	9
Elliott Bay, station 182	1.34	98	83	571	88	37
Elliott Bay, station 184	1.18	103	84	731	89	57

[a] Data from Long et al. (2000).
[b] Amphipod test species: *A. abdita.*
[c] Urchin test species: *S. purpuratus.*

Seattle). Among these 22 samples, mean ERM quotients were very low (ranging from 0.05 to 0.08), and both percent amphipod survival and percent urchin fertilization were very high. Measures of benthic community composition were variable among stations, reflecting the natural degree of heterogeneity among the different habitats of Puget Sound. Nevertheless, these benthic metrics exceeded values generally considered as the lower limit for reference conditions (Striplin et al. 1999). The samples from station 160 in Sinclair Inlet and station 115 in Elliott Bay were slightly contaminated, not toxic, in amphipod tests, but had highly significant toxicity in the urchin fertilization tests. Arthropod abundance was extremely low in both samples, whereas total abundance and numbers of species were low in station 160 but not in station 115. It appears, then, that stations 160 and 115 would be classified as degraded, using the classification matrix of Chapman (1996).

The samples from stations 182 and 184 from Elliott Bay, on the other hand, appeared to be the most contaminated (mean ERM quotients of 1.3 and 1.2), but they were not toxic in either test and supported a reasonably abundant and diverse benthos (Table 8-8). Thus, according to Chapman's classification matrix, it would appear that the chemical toxicants at these 2 stations were not highly bioavailable, and therefore, the samples did not exhibit highly significant toxicity in either the laboratory tests or in the field.

Corpus Christi Bay

An SQT study was conducted recently in Corpus Christi Bay, Texas to assess the impacts of stormwater outfalls on this urban estuary (Carr et al. 2000). In addition to stormwater outfall sites, other industrial and domestic outfalls and sites near oil production or dredging activities were also included for comparison. Of the 36 sites sampled, only 1 site exhibited toxicity for the amphipod test, whereas toxicity was observed at 7 and 18 sites for the sea urchin fertilization and embryological development tests, respectively. The 2 sites that were the most toxic in the porewater toxicity tests, but not in the amphipod test, were completely devoid of benthic infaunal organisms. A multivariate analysis was used to create new variables for each component of the triad, which were used in a correlation analysis. Toxicity was significantly correlated with both chemistry and ecological responses. Had only solid-phase tests been conducted in this survey, there would not have been any significant correlations with toxicity because only 1 sample was even marginally toxic in the amphipod test.

In summary, based upon the data from these 5 examples, the use of porewater toxicity tests resulted in classification of more sampling stations as toxic than would have been possible if only the solid-phase test were used. In many instances, the urchin fertilization and/or embryological development tests of pore waters were significantly toxic in samples that were slightly to moderately contaminated. It is desirable to have a test that is able to detect chronic sublethal effects, which are likely to be better predictors of benthic effects than tests with acute toxicity endpoints, as discussed in Chapter 2. Often, they did not indicate toxicity in uncontaminated samples. They sometimes exhibited differential sensitivity (i.e., did not indicate toxicity) in samples that were highly contaminated with complex mixtures of toxicants (e.g., in the lower Miami River). This may be due to the types of contaminants or their binding to dissolved organic carbon (DOC) or other ligands that limit their bioavailability during the short-term exposures. In most cases, one or more measures of benthic abundance and/or diversity were depressed in samples classified as toxic in the porewater tests.

Statistical Analysis of SQT Data

The SQT is an integrated, multivariate indicator of sediment quality (Chapman et al. 1997). The approach integrates results from sediment chemistry, toxicity, and benthic community studies. If results from each component study could be reduced to 1 variable, then at minimum the challenge is to derive concordance among 3 variables. However, the chemistry and benthic studies always generate more than 1 metric, and more than 1 toxicity test may be performed (Carr et al. 2000). Thus, SQT data are a complex of multivariate metrics within a multivariate framework. To ease interpretation, it is necessary to demonstrate concordance among outcomes of the chemistry, toxicity, and benthic data.

The chemical dataset is multivariate because concentrations of many compounds or elements are measured. It is not uncommon to measure concentrations for 100 to 200 chemical species. One simple method to reduce the number of variables is to report sums of classes of contaminants (e.g., total polychlorinated biphenyls [PCBs], DDTs, PAHs, alkanes). There is also evidence that divalent metals have similar toxicity, and they can be summed on a molar basis (Ankley et al. 1994, 1996). Chemical data can be further reduced to 1 indicator variable by ranking or weighting concentrations relative to sediment quality guidelines (SQGs). One method is to use existing criteria, for example, threshold effects level (TEL) or probable effects level (PEL) (MacDonald et al. 1996) or an effects range low (ERL) or ERM (Long et al. 1995), but there are disadvantages to substituting an index for raw data because information is lost.

Benthic community structure data can be difficult for the nonspecialist to interpret. There are 2 basic approaches, depending on whether one wishes to use species-dependent measures or species-independent measures. Species-dependent techniques account for species changes over space or time but are limited to multivariate analyses, which yield results that consist of spatial patterns of similarity, correlation, or concordance of treatments in the study design. Most often, these treatments are stations sampled, so the result is a measure of similarity among stations. Often the results are plotted as tree or cluster diagrams or bivariate plots. The axes of the bivariate plots are new variables derived from the analyses, which can be either qualitative or quantitative and may or may not represent underlying structure in the dataset. Species-independent techniques rely on measures of total community characteristics, for example, abundance, biomass, or species richness or proportional species diversity. Diversity indices can account for proportional representation of species in samples but do not account for individual species that may be pollution sensitive or tolerant, and they discount the importance of rare species. Often, locations have similar diversity indices for totally different reasons.

A more promising approach to synthesizing benthic data is to evolve a benthic index of biotic integrity (BIBI) (Weisberg et al. 1997; Van Dolah et al. 1999; Carr et al. 2000) or the multivariate discriminant-function approach (Engle et al. 1994; Rakocinski 2000). The BIBI approach is powerful because it combines species-dependent and -independent techniques and can incorporate rare or indicator species. Typically, metrics of total community attributes are combined with numbers of indicator species to generate a value that can be compared among samples, stations, or sites. The BIBI approach to reducing benthic community data is promising but is new and still requires research before one approach is adopted universally. Each locale may have different indicator species or indicator scales because of differing background communities, which results from differences in salinity, sediment type, habitat structure, tidal influence, food sources, and keystone predators. A species sensitivity or tolerance must be established. Opportunistic species are often assumed to be disturbance or pollution indicator species (Chapman et al. 1997), and this should be proved with experimentation. It is also not clear which metrics will ultimately prove to be the best indicators of ecosystem health. All studies to date have used different metrics, necessitated by differences in the indigenous fauna in different study locations. Through further research, the goal will be to develop a uniform approach for the development of indicators of ecosystem health.

Although the approach should be uniform, experience has shown that different, natural controlling factors will influence the BIBI variables to varying degrees in different regions. The uniform approach should share common features including 1) the use of multiple attributes combined into a single measure, designed to maximize the ability to distinguish between degraded versus nondegraded benthic conditions, and 2) the biological variability attributable to key, natural controlling factors needs to be accounted for using statistical methods designed for this purpose. An index should be adjusted for the natural controlling factors found to have the strongest influence within a particular region. Similarly, part of the process in developing a BIBI involves performing statistical tests to determine what benthic attributes or combination of attributes do the best job in discriminating between reference and degraded sites (i.e., contaminated and/or toxic sites). Different combinations of attributes will result for different regions. While this approach can be uniform across different regions, the particular attributes and adjustments will vary from region to region.

Given that chemistry and benthic datasets are so richly multivariate, it is a weakness of toxicity studies that often 1 value is generated per location. Porewater toxicity tests have a distinct advantage over solid-phase tests in this regard. Porewater tests can be performed on several endpoints and on a dilution series of samples. The result is a multivariate dataset that allows for analysis to create an index of toxicity that has a rich range of response. This is important because chemistry and benthic data, even when reduced to an index, provide rich ranges of responses. Using several toxicity tests with different species and endpoints is desirable to capture the range of potential toxic responses in an environment. The SQT is strongest when multiple endpoints can be compared (Chapman et al. 1997).

The multivariate nature of the triad data indicates that multivariate techniques are the most reasonable approaches to analyze SQT data (Green et al. 1993). The alternative approach is to use bivariate correlation tests of univariate data. It is tempting to run correlation tests between various metrics of each triad component. However, it is required to run the correlation tests on independent variables. It is often the case that many variables covary in a dependent fashion. It is also necessary to avoid the multiple t-test problem. When using the conventional α rejection rate of 0.05 in a correlation test, the experimenter will make a mistake in rejecting the null hypothesis when it is false 1 in 20 times. It is possible to have a very large matrix of variables from a triad component that would be subjected to multiple t-tests. If using 0.05, the experimenter knows there are errors in the resulting matrix of correlation tests but does not know which tests are correct. The simplest correction is to use the Bonferroni adjustment to the rejection rate, where α is divided by the number of planned comparisons. If a large number of comparisons is planned, this decreases the rejection rate to a level that might not find significant relationships where they actually exist. Correlation tests are best used when each component of the triad is reduced to 1 or 2 variables.

A generic word of caution is necessary about indices. Much has been written about the use and abuse of indices, especially diversity indices. All indices suffer from a loss of information and oversimplification of the data making up the index. Variation in the variables that make up the index is lost, so it is difficult, if not impossible, to know how large a difference among indices is necessary before one is confident that differences truly exist. Indices can have the same values for different reasons and thus not distinguish important trends in the data. In contrast, similar locations can have very different indices if one of the components making up the index has an extreme value, even though that value may be in the natural range of variability. Even though there is a great need for simple, easy to understand indices, it is more likely than not that such indices will be misused.

A variety of multivariate techniques exist, and all are likely to produce the same qualitative results when associations between variables are robust (Chapman et al. 1997). Multivariate analysis can be conducted in a variety of ways to demonstrate concordance between triad components. Variables or indices from each triad component can be compared in the same analysis to classify similar locations.

All triad components can be analyzed separately to create new variables that contain the most pertinent information in the dataset. Multivariate analysis is a transformation of a data matrix to reduce the number of variables where the new variables are in order of decreasing variance and each successive new variable is orthogonal (i.e., not correlated or covarying) with the preceding new variable. Therefore, multivariate analysis of each triad component creates new variables that can be regressed or correlated with each other to demonstrate concordance among triad components.

Principal components analysis has characteristics especially suited for this approach and has been used for this purpose in at least 2 SQT studies that included porewater toxicity tests (Green and Montagna 1996; Carr et al. 2000). The key to interpreting regressions or correlations between new PCA variables created from triad components is to compare PCA variable loadings with PCA observation scores. The data matrix includes columns identifying categorical information of the study design (e.g., sampling date, location, or other treatment effects) and triad responses (i.e., numerical indicators, indices, or measurements). The PCA creates a "load" for each numerical column, indicating its weight in calculating the new PCA variable. Typically, the first new PCA variable (PC1) will indicate overall sample means. The second new variable (PC2) will be orthogonal to the first and therefore will contain no redundant information contained in PC1. The rows of the matrix contain the observations, either as replicates or as means of replicates for each location sampled. The PCA creates a "score" for each observation. The value of the observation score is related to load of triad variables. For example, locations with high concentrations of contaminants will have high chemistry PCA scores, and locations with low survival will have stations with low toxicity PCA scores. Therefore, an inverse correlation would exist between PCA scores for chemistry and PCA scores for toxicity.

The triad components represent multivariate datasets within a multivariate framework. The chemistry and benthic components may contain many variables, and it might be desirable to reduce the number of variables used in the final analysis. One consequence of many redundant or highly correlated variables is that the variables are given a higher weight than they might deserve. For example, it was necessary to reduce more than 100 chemical measurements to 7 to obtain the most accurate description of contaminants near offshore platforms (Kennicutt, Boothe et al. 1996). Because PCA is a variable reduction technique, it can be used iteratively first to generate a list of variables to use in a triad component analysis (i.e., variables with highest loads) and then to create new PCA variables for the concordance analysis.

Confounding factors modify responses measured in triad components. For example, total organic carbon (TOC) and grain size often correlate with concentrations of certain contaminants, salinity and dissolved oxygen (DO) often correlate with benthic abundance and diversity, and ammonia concentrations can correlate with toxicity results. It is essential to measure these and many other ancillary variables to obtain the correct interpretation of an SQT dataset. The ancillary variables are typically handled in 1 of 2 ways. The variables can be normalized, that is, as weights by dividing the triad response variable by the value of the ancillary variable, or the ancillary variable simply can be included in the PCA. Because new variables in a PCA are orthogonal, responses caused by ancillary variables will likely be extracted in PC2 or PC3. In the Gulf of Mexico, PC1 contained high loading for a suite of variables defining the chemical contamination gradient near offshore platforms, and PC2 contained high loading for the sedimentary environment related to the depth at which a platform occurred (Kennicutt, Boothe et al. 1996). In Corpus Christi Bay, Texas, PC1 contained high loadings for organic and metal contaminants, and PC2 contained high loadings for sediment grain size, ammonia, and salinity (Carr et al. 2000). In both cases, toxicity and benthic response to contaminants were separated from response to ancillary confounding factors.

All SQT studies are performed in the context of a field sampling program. It is essential to follow principles of good experimental design to create the sampling plans. The most important guidelines are to avoid pseudoreplication by replicating at the treatment level, to balance the design by having equal numbers of stations in treatment cells, and to include blocks or split plots to control for nuisance variables. Nuisance variables include, but are not limited to, salinity variations or sediment grain size variations. Often, a natural gradient of a nuisance variable runs perpendicular or parallel to a pollution gradient. In these cases, it is important to include the nuisance variable in the design, and not simply resort to a weighting or normalization technique. Including the nuisance variable in the design will allow the experimenter to control for the effects in both univariate and multivariate analyses.

Triad studies should maximize the number of stations, even if this reduces the number of replicate samples (Chapman et al. 1997). The lower number of replicates will lower the power of univariate analyses, but the higher number of stations will increase the power of multivariate analyses. As a general rule, one should try to plan for 10× as many observations (rows) as numeric variables (columns). This guideline is especially difficult to follow if many replicates are taken in each cell and must be reduced to 1 number (or row) per station to perform the multivariate analysis. A large number of reference stations are also needed to identify the natural range of responses so that a degraded response is not incorrectly assigned to values at the low end of the natural distribution of responses.

Synoptic sampling is essential. Ideally, each replicate is tied together as a subsample as well, ensuring that the maximum number of rows will be available for multivariate analyses. The best approach is to take large boxcores and subsample each for chemistry, toxicology, and benthos (Kennicutt, Green et al. 1996). This approach ensures that all measurements are taken simultaneously. It is not always practical to take large boxcores, and occasionally it will be necessary to take multiple small samples to obtain enough sediment for analyses. Different approaches optimize cost versus benefit to obtain a large number of stations. For example, 1 chemical and toxicological measurement can be made of homogenized samples from each station.

A good experimental design is still just a surrogate for the actual conditions at each location. Therefore, a combination of univariate analysis of variance and multivariate analysis will be the best approach to determine concordance of triad components. The station scores resulting for the PCA can be plotted with symbols for the levels of treatments in the experimental design to test hypotheses about toxic and benthic response to actual concentrations of contaminants in the field.

There are several visualization approaches for synthesizing and interpreting SQT data. Indices of triad components can be graphed as 3D response surfaces, triangulation plots, or tertiary diagrams. The PCA scores can be plotted in bivariate plots, or the correlations can be listed in tables. Pie charts, color-coded for response, are effective means to simplify and visualize interpretations of triad data.

Classification of Samples using Multivariate Analysis

When field studies are performed, it is often difficult to interpret toxicity tests and relate the results to the myriad of potential environmental contaminants and natural physicochemical factors. The difficulty in this empirical approach arises primarily from 1) the environmental variability in each measured variable, 2) the tendency of these different variables to covary in relation to similar environmental factors, and 3) the assumption of associating multiple variables with statistical robustness. The most convincing conclusions will be those that are based on demonstrating significant relationships between multiple measures of toxicity correlated with specific contaminants or suites of contaminants. Multivariate statistical analyses can be used to serve this purpose.

Two recent studies used this approach. The approach was first used to assess sediment quality near offshore hydrocarbon production platforms in the Gulf of Mexico (Carr, Chapman et al. 1996; Green and Montagna 1996). Later, the approach was refined in a study to assess sediment quality near stormwater outfalls in Corpus Christi Bay, Texas (Carr et al. 2000). The approach combines measuring toxicity and contaminants and analyzing the data using multivariate analysis. Porewater toxicity employed several species, yielding a multivariate database on toxicity. The physicochemical environment was characterized by measuring key factors (such as sediment grain size, organic carbon content, salinity, DO, etc.) and contaminants present in the environment. The chemical dataset was also multivariate. Multivariate analysis was used to reduce both datasets individually; then correlation analysis was used to compare the reduced datasets.

Multiple indicator variables are often measured at each site and for each sample taken in a study. Thus, datasets produced in all environmental studies have a common multivariate format, that is, a matrix with rows of observations (or samples) and columns of measured responses (of toxicity or chemical factors). A common problem in such a multivariate data matrix is that many variables covary with each other. Therefore, information in each column may not be unique. Another problem is encountered when several univariate tests are employed on such a dataset because this violates assumptions of independence of the variables. A multivariate analysis is required to maintain experiment-wise error rates (i.e., the probability that one or more erroneous statements will be made in an experiment) at the 0.05 level (Kirk 1982).

Principal components analysis is a multivariate method that is also a variable reduction technique. PCA is useful because it transforms the data matrix to create new variables that are 1) mutually orthogonal (i.e., the new variables are not correlated to each other) and 2) extracted in order of decreasing variance (i.e., much of the information, or variance, of the original set of variables is concentrated in the first few principal components [PCs]). The PCs can also be used as predictors in regression analysis because they are orthogonal and co-linearity (i.e., a linear relationship between variables) does not exist.

Results of a PCA are visualized in bivariate plots. Generally, only the first 2 PC factors (PC1 and PC2) are used in the plots. The results are visualized in 2 ways: as factor patterns and as loading scores. Each new dataset is simply a matrix, that is, rows of observations versus columns of variables. The factor patterns are the PC coefficients for each variable or column. These vector patterns are used to interpret what PC1 and PC2 represent by plotting the column heading as the symbol for each point. Next, the loading scores for each observation are plotted using the site name as the symbol for each point.

The plot of loading scores allows the analyst to visualize the relationships or correlation among the sampling units, that is, the stations in the present study. The PCA is performed once for each dataset (i.e., the toxicity and chemical datasets) to create the new datasets of PC scores. The most interesting new variable in the new dataset is invariably PC1 because it will represent the overall average response and retain most of the variance in the entire original dataset. Because the PCs from the toxicity and chemistry datasets are collinear, they can be subjected to further regression analysis. A simple convention for naming PCs derived from different datasets is to append a prefix representing the name of the originating dataset. For example, PC1 from the chemical dataset is ChemPC1. Regression of the new PC variables reveals the power of this approach (Figure 8-4). Because toxicity should respond as a function of the chemical background, it is best to plot ToxPC on the ordinate and ChemPC on the abscissa. The resulting plot has 4 quadrants. If contaminants cause toxicity at specific locations, then an inverse relationship should exist between ToxPC and ChemPC.

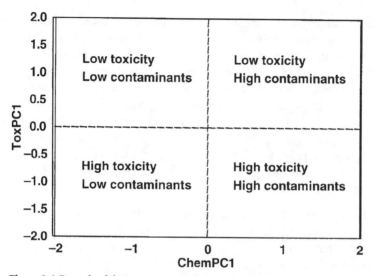

Figure 8-4 Example of the interpretation of regression of PC variables derived from separate toxicity (ToxPC1) and chemical (ChemPC1) analyses

Stations (or sites) with low levels of contaminants and no toxicity should cluster in the upper left-hand quadrant. Sites with high levels of contaminants and low survival should cluster in the lower right-hand quadrant. Scatter would be expected, especially near the origin. Stations located in the upper right or lower left quadrant do not exhibit a consistent response. This can occur for at least 3 reasons:

1) There is simply a great deal of variance in the response or concentrations of chemicals present.

2) The toxic response could be caused by a chemical or environmental factor not measured (for stations in the lower left quadrant).

3) The chemicals are not bioavailable or not toxic (for stations in the upper right quadrant).

The following discussion represents a reanalysis of data from 2 recent reports (Carr, Chapman et al. 1996; Carr et al. 2000) to demonstrate the approach outlined above. All multivariate analyses were performed using the FACTOR procedure contained in SAS software (SAS 1991). Multivariate analysis is complex, employing many different options. The following programming statements are provided as an example of how options were used to analyze the chemical dataset.

TITLE1 'Principal Components Analysis on the chemical dataset (CHEM.SD2)';

PROC FACTOR DATA=CHEM METHOD=PRINCIPAL ROTATE=VARIMAX COV SCREE NFACTORS=5 NPLOT=3 OUT=SCORES OUTSTAT=PCSTATS;

```
VAR TTOTPAH TUCM TTOTALK TSAND TSILT TCLAY TTOC TTIC REDOX TFE
TCD TAL TBA TNO2 TSIO3 TSAL TTEMP;

RUN;

TITLE2 'SCORES.SD2: Principal Component Scores for stations';

PROC PRINT DATA=SCORES NOOBS;

VAR STA FACTOR1 FACTOR2 FACTOR3;

RUN;

PROC PLOT DATA=SCORES;

PLOT FACTOR1*FACTOR2 $ STA='*' / PLACE=(S=RIGHT LEFT: H=2 -2);

RUN;

TITLE2 'PCSTATS.SD2: Principal Component Statistics';

PROC PRINT DATA=PCSSTAT;

RUN;

TITLE2 'LOADS.SD2: Principal Component Loadings form Chemical Variables';

DATA STATS;

SET PCSTATS;

WHERE _TYPE_='PATTERN';

RUN;

PROC TRANSPOSE DATA=STATS OUT=LOADS;

ID _NAME_;

RUN;

PROC PRINT DATA=LOADS;

RUN;

PROC PLOT DATA=LOADS;

PLOT FACTOR1*FACTOR2 $ _NAME_='*' / PLACE=(S=RIGHT);

RUN;
```

The above programming demonstrates 8 manipulations of the primary dataset (CHEM). Each manipulation is represented by a PROC or DATA step. First, the PCA is run to create 2 new datasets (SCORES and PCSTATS). Then, PC scores for stations are printed and plotted. Next, variable loadings are extracted from the statistics dataset, transposed, printed, and plotted. The options that affect the final analysis most are the method, rotation, and choice of matrix. Here, the PC method is used with VARIMAX rotation. Rotation is very useful because it aligns PC1 parallel to the axis. There are many rotation options available. By definition, PC2 is perpendicular to PC1, so it is also parallel to the second axis. The PCA can be run on the correlation matrix (which is the default) or the covariance matrix (COV option). When performing PCA on the covariance matrix, the analysis does not treat all the variables as if they have the same variance. This is useful because nondetect or zero values often are encountered at many stations. Co-occurring zeroes have a perfect correlation but zero covariance. Prior to multivariate analysis, measurement data were log transformed, that is, TUCM=LOG(UCM+1), and percentage data were arcsin transformed, that is, TTOC = ARSIN(SQRT(TOC/100)).

Results of the PCA are visualized in bivariate plots. Generally, only the first 2 PC factors (PC1 and PC2) are used in the plots. The results are visualized in 2 ways: as variable (response) loading patterns and as observation (sample) scores. Each dataset is simply a matrix, that is, rows of observations versus columns of variables. The loading patterns are the PC coefficients for each new variable or column. These vector patterns are used to interpret the meaning of PC1 and PC2 by plotting the column heading as the symbol for each point. Next, the scores for each observation are plotted using the site, station, or sample name as the symbol for each point. The plot of scores allows visualization of relationships or correlation among the observations. To assess porewater quality among observations (i.e., sites), the chemical and toxicity PC scores are plotted (as in Figure 8-4) using the observation name as the plot symbol.

In the study of Corpus Christi Bay outfalls (Carr et al. 2000), bulk sediment contaminant concentrations and toxicity were measured. The chemistry dataset contained 11 trace metals, 44 PAHs, 29 pesticides, and 61 PCBs. A PCA can be influenced by the number of variables in a dataset; therefore, it was necessary to reduce the number of variables that went into the chemistry PCA (Kennicutt, Boothe et al. 1996). These variables were first reduced by summing the constituents of families of compounds into 5 categories: total NS&T Program PAHs (NSTPAHs), total chlordanes, total DDTs, total hexachlorocyclohexanes (HCHs), and total PCBs. Then, PCA was performed to choose representative metals. The final chemistry PCA was performed on selected chemical and other abiotic variables that described the contaminant and hydrographic background at each site. In addition, sediment grain size, TOC, salinity, DO, and unionized ammonia (UAN) in pore water were included in the chemistry dataset. There were 5 separate toxicity tests for 3 species (amphipod and mysid solid-phase, sea urchin porewater fertilization, and embryological development tests). The PCA was performed on the chemistry (Figure 8-5) and toxicity datasets (Figure 8-6).

The PCA for bay outfall chemistry is easy to interpret (Figure 8-5). The first axis, ChemPC1, represents metals associated with fine sediments and high ammonia

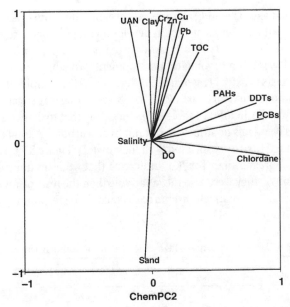

Figure 8-5 Principal component loading pattern for chemical variables (ChemPC) in the Corpus Christi Bay outfall study. Data extracted from Carr et al. 2000.

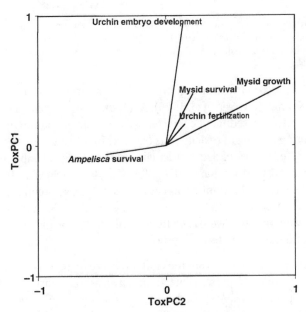

Figure 8-6 Principal component loading pattern for toxicity variables (ToxPC) in the Corpus Christi Bay outfall study. Data extracted from Carr et al. 2000.

levels. The second axis, ChemPC2, represents organic contaminants. High values on both axes represent high levels of urchin development, polychaete reproduction, or copepod survival. The toxicity PCA is less easy to interpret (Figure 8-6). Both axes represent high levels of embryological development, growth, or survival. Apparently urchin embryo tests (ToxPC1) responded differently to the samples than mysid growth or amphipod survival (ToxPC2). There were strong statistical associations only for toxicity with ChemPC1 (Table 8-9); indicating that toxicity was associated with high levels of metals or ammonia but not hydrocarbons. A total of 12 stations out of 36 exhibited toxicity associated with high metals (Figure 8-7). The correlation with ToxPC2 was greater than ToxPC1, indicating that while urchin embryo tests were more sensitive, they were also more variable than the tests of mysid growth or amphipod survival. Alternatively, urchin toxicity could be responding to something not included in the chemical PCA.

Table 8-9 Correlations between PC axes for the chemical and toxicity datasets[a,b]

Station		Chem PC1	Chem PC2
Corpus Christi Bay, TX (36 stations)	Tox PC1	-0.44 ($p = 0.0076$)	-0.11 ($p = 0.5070$)
	Tox PC2	-0.77 ($p = 0.0001$)	$+0.14$ ($p = 0.4202$)
Gulf of Mexico (12 stations)	Tox PC1	-0.58 ($p = 0.0491$)	-0.16 ($p = 0.6179$)
	Tox PC2	-0.68 ($p = 0.0145$)	-0.26 ($p = 0.4148$)

[a] Expressed as Pearson product correlation coefficients and (in parentheses) the probability that the correlations equal 0.
[b] Data from Gulf of Mexico (Carr, Chapman et al. 1996) and Corpus Christi Bay (Carr et al. 2000).

In the Gulf of Mexico, a hydrocarbon production platform on the continental shelf was studied (Kennicutt, Green et al. 1996). A full suite of sediment chemical contaminants, ancillary sediment characteristics, and water column nutrients were measured. During 1 cruise, 3 toxicity tests (sea urchin development to pluteus larvae, polychaete number of eggs per female, and harpacticoid nauplii survival) were performed at 1 platform named HI-389A (Carr, Chapman et al. 1996). The total number of chemical variables was enormous (>200), so preliminary PCA was performed to choose representative subsets of the variables (Kennicutt, Boothe et al. 1996). Those same variables were used for a chemical PCA on just the data collected synoptically with the toxicity tests (Figure 8-8).

The PCA for the chemical data from the Gulf study is easy to interpret (Figure 8-8). The first axis, ChemPC1, represents metals and hydrocarbons associated with coarse sediments near platforms, as opposed to background metals and fine sediments found farther away from the platforms. The second axis, ChemPC2, represents hydrographic differences in the water column near and away from platforms. High values for only PC1 represent high contaminant loads. Contaminated sediments

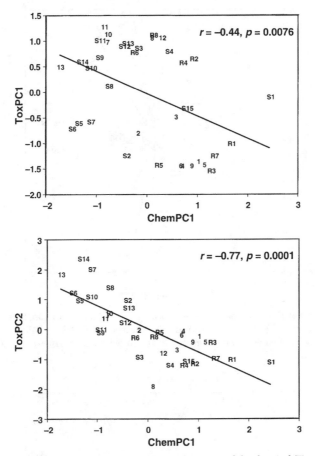

Figure 8-7 Regression of station scores from PCA of the chemical (Figure 8-5, ChemPC1) and toxicity (Figure 8-6, ToxPC1) datasets from the Corpus Christi Bay outfall study. Stations with prefix "S" are stormwater outfall stations; "R" are reference stations; all others were other stations suspected of having elevated levels of contaminants. See Carr et al. (2000) for details of station locations.

near offshore platforms are different from contaminated sediments in bays. In bays, contaminants are deposited with fine sediments, but near these offshore platforms, contaminants are deposited with coarse drill cuttings. The toxicity PCA is also easy to interpret (Figure 8-9). High values on both axes represent high levels of urchin development, polychaete reproduction, or copepod survival. All 3 measures are highly correlated, so they did not separate cleanly into 2 distinct components. The general vector pattern is not parallel with the axes, but is on a 45° angle, despite use of the rotation option of the PCA. There are strong statistical associations only for toxicity with ChemPC1 (Table 8-9), indicating that toxicity was associated with high levels of metals and hydrocarbon contaminants. A total of 5 stations out of 12 exhibited toxicity associated with high concentrations of contaminants (Figure

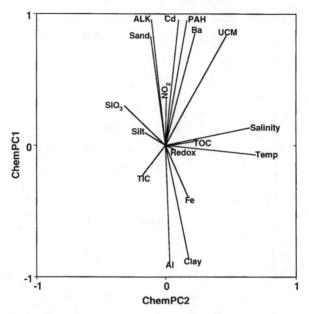

Figure 8-8 Principal component loading pattern for chemical variables (ChemPC) in the Gulf of Mexico hydrocarbon production platform study (Kennicutt, Boothe et al. 1996; reprinted with permission)

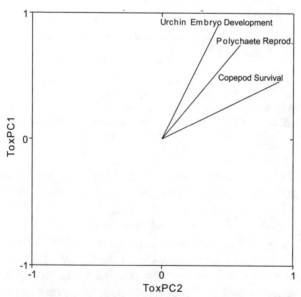

Figure 8-9 Principal component loading pattern for toxicity variables (ToxPC) in the Gulf of Mexico hydrocarbon production platform study (Carr, Chapman et al. 1996; reprinted with permission)

8-10). Four of the 5 stations (with a prefix 2) lie within 50 m of the platform. The strength of the correlation with ToxPC2 is similar to that of ToxPC1 because the loadings lie on an angle, rather than parallel to the PC1 axis.

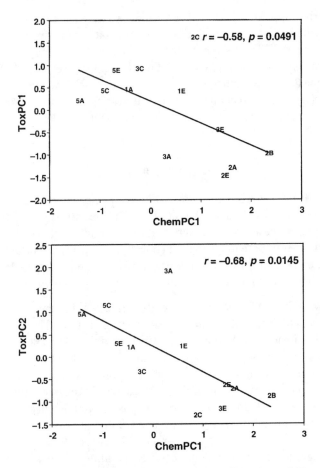

Figure 8-10 Regression of station scores from PCA of the chemical (Figure 8-8, ChemPC1) and toxicity (Figure 8-9, ToxPC1) datasets from the Gulf of Mexico hydrocarbon production platform study. Stations prefixes designate distance from the platform: 5 = 3000 m from platforms, 4 = 500 m, 1 = 200 m, 3 =100 m, and 2 = 50 m. For station locations see Kennicutt, Green et al. (1996).

Comparison of the results of the 2 studies (1 in a bay and the other on the shelf) indicates that the PCA technique is very powerful at detecting toxicity related to suites of contaminants in 2 very different environments (Table 8-9). The bay study indicated that different suites of contaminants can be identified as the primary sources of concern. The shelf study indicated that a large number of stations (likely 30 or more) are necessary to detect differences among stations with statistical certainty. The correlation was high in the offshore study, but the significance was low. The approach also identified stations where uncertainty exists (Figure 8-4).

The multivariate approach described here can be extended. For example, both studies described above also contained data on benthic community structure. When benthic data are available, the approach can be used to perform an SQT analysis. The statistical approach for SQT should follow that of Green et al. (1993) and Green and Montagna (1996). A further extension of the approach is to regress new variables derived from the chemical PCA with univariate measures of toxicity, such as would be required when only 1 toxicity measurement is made.

Summary

Data compiled and reviewed from many sediment quality surveys have demonstrated the importance of including multiple tests of toxicity as part of the analytical plan. Similar to the approach used to evaluate chemical contamination with analyses of numerous potentially toxic chemicals, it is equally important to measure toxicity with more than 1 test. Furthermore, it is equally important to develop a weight of evidence with information from analyses of the resident infauna.

Experience with the tests of porewater toxicity has shown that they often identify more samples as toxic than do the solid-phase tests of survival. The incidence of toxicity and frequency distributions of results showed the relatively high sensitivity of sea urchin tests of porewater samples. Often, when results of amphipod survival tests indicated none of the samples was toxic, the porewater tests were useful in identifying both the spatial extent and spatial patterns in toxicity. Often, results of these tests also were instrumental in classification of both reference and degraded conditions, based upon the triad of data, when amphipod tests showed nontoxic conditions. Classifications of samples in multivariate analyses demonstrated the statistically significant correspondence between a number of chemical variables and results of the toxicity tests. Thus, it appears that the porewater toxicity tests can provide valuable information on the relative quality of sediments in a comprehensive assessment that follows a triad approach.

References

Ankley GT, Di Toro DM, Hansen DJ, Berry WJ. 1996. Technical basis and proposal for deriving sediment quality criteria for metals. *Environ Toxicol Chem* 15:2056–2066.

Ankley GT, Di Toro DM, Hansen DJ, Mahoney JD, Berry WJ, Swartz RC, Hoke RA, Thomas NA, Garrison AW, Allen HE, Zarba CS. 1994. Assessing potential bioavailability of metals in sediments: A proposed approach. *Environ Manag* 18:331–337.

Carr RS, Biedenbach JM. 1999. Use of power analysis to develop detectable significance criteria for sea urchin toxicity tests. *Aquat Ecosyst Health Manag* 2:413–18.

Carr RS, Chapman DC, Presley BJ, Biedenbach JM, Robertston L, Boothe P, Kilada R, Wade T, Montagna P. 1996. Sediment porewater toxicity assessment studies in the vicinity of offshore oil and gas production platforms in the Gulf of Mexico. *Can J Fish Aquat Sci* 53:2618–2628.

Carr RS, Long ER, Chapman DD, Thursby G, Biedenbach JM, Windom H, Sloane G, Wolfe DA. 1996. Sediment quality assessment studies of Tampa Bay, Florida. *Environ Toxicol Chem* 15:1218–1231.

Carr RS, Montagna PA, Biedenbach JM, Kalke R, Kennicutt MC, Hooten R, Cripe G. 2000. Impact of storm-water outfalls on sediment quality in Corpus Christi Bay, Texas, USA. *Environ Toxicol Chem* 19:561–574.

Chapman PM. 1996. Presentation and interpretation of sediment quality triad data. *Ecotoxicology* 5:327–339.

Chapman PM, Anderson B, Carr S, Engle V, Green R, Hameedi J, Harmon M, Haverland P, Hyland J, Ingersoll C, Long E, Rodgers Jr J, Salazar M, Sibley PK, Smith PJ, Swartz RC, Thompson B, Windom H. 1997. General guidelines for using the sediment quality triad. *Mar Pollut Bull* 34:368–372.

Engle VD, Summers JK, Gaston GR. 1994. A benthic index of environmental condition of Gulf of Mexico estuaries. *Estuaries* 17:373–384.

Green RH, Boyd JM, MacDonald JS. 1993. Relating sets of variables in environmental studies: The sediment quality triad as a paradigm. *Environmetrics* 4:439–457.

Green RH, Montagna P. 1996. Implications for monitoring: Study designs and interpretation of results. *Can J Fish Aquat Sci* 53:2629–2636.

Kennicutt II MC, Boothe PN, Wade TL, Sweet ST, Rezak R, Kelly FJ, Brooks JM, Presley BJ, Wiesenburg DA. 1996. Geochemical patterns in sediments near offshore production platforms. *Can J Fish Aquat Sci* 53:2554–2566.

Kennicutt II MC, Green RH, Montagna P, Roscigno PF. 1996. Gulf of Mexico Offshore Operations Experiment (GOOMEX) Phase I: Sublethal responses to contaminant exposure –introduction and overview. *Can J Fish Aquat Sci* 53:2540–2553.

Kirk RE. 1982. Experimental design. 2nd ed. Monterey CA, USA: Brooks/Cole. 911 p.

Ingersoll CG, Ankley GT, Baudo R, Burton GA, Lick W, Luoma S, MacDonald DD, Reynoldson TB, Solomon KR, Swartz RC, Warren-Hicks WJ. 1997. Work group summary report on uncertainty evaluation of measurement endpoints used in sediment ecological risk assessment. In: Ingersoll CG, Dillon T, Biddinger GR, editors. Ecological risk assessment of contaminated sediments. Pensacola FL, USA: Society of Environmental Toxicology and Chemistry (SETAC). p 297–352.

Long ER. 2000. Degraded sediment quality in U.S. estuaries: A review of magnitude and ecological implications. *Ecol Appl* 10:338–349.

Long ER, Hameedi MJ, Harmon M, Sloane GM, Carr RS, Biedenbach J, Johnson T, Scott KJ, Mueller C, Anderson JW, Wade TL, Presley BJ. 1999. Survey of sediment quality in Sabine Lake, Texas and vicinity. Silver Spring MD, USA: National Oceanic and Atmospheric Administration. NOAA Technical Memorandum NOS NCCOS CCMA 137. 27 p.

Long ER, Hameedi J, Robertson A, Dutch M, Aasen S, Ricci C, Welch K, Kammin W, Carr RS, Johnson T, Biedenbach J, Scott KJ, Mueller C, Anderson J. 1999. Sediment quality in Puget Sound. Year 1 Report - Northern Puget Sound. Olympia WA, USA: National Oceanic and Atmospheric Administration (NOAA) Technical Memorandum NOS NCCOS CCMA Nr 139 and Washington State Department of Ecology (WDOE) Publication No. 00-03-055. 347 p.

Long ER, Hameedi J, Robertson A, Dutch M, Aasen S, Ricci C, Welch K, Kammin W, Carr RS, Johnson T, Biedenbach J, Scott KJ, Mueller C, Anderson J. 2000. Sediment quality in Puget Sound. Year 2 Report - Northern Puget Sound. Olympia WA, USA: NOAA NOS NCCOS CCMA Technical Memo Nr 147 and WDOE Publication No. 99-347.

Long ER, MacDonald DD, Smith SL, Calder FD. 1995. Incidence of adverse biological effects within ranges of chemical concentrations in marine and estuarine sediments. *Environ Manage* 19:81–97.

Long ER, Robertson A, Wolfe DA, Hameedi J, Sloane GM. 1996. Estimates of the spatial extent of sediment toxicity in major U.S. estuaries. *Environ Sci Technol* 30:3585–3592.

Long ER, Scott GI, Kucklick J, Fulton M, Thompson B, Carr RS, Biedenbach J, Scott KJ, Thursby G, Chandler GT, Anderson JW, Sloane GM. 1998. Magnitude and extent of sediment toxicity in selected estuaries of South Carolina and Georgia. Silver Spring MD, USA: National Oceanic and Atmospheric Administration. NOAA Technical Memorandum NOS ORCA 128.

Long ER, Sloane GM, Carr RS, Scott KJ, Thursby GB, Wade T. 1996. Sediment toxicity in Boston Harbor: Magnitude, extent, and relationships with chemical toxicants. Silver Spring MD, USA: National Oceanic and Atmospheric Administration. NOAA Tech Memo NOS ORCA 96, 133 p.

Long ER, Sloane GM, Scott GI, Thompson B, Carr RS, Biedenbach J, Wade TL, Presley RJ, Scott KJ, Mueller C, Brecken-Fols G, Albrecht B, Anderson JW, Chandler GT. 1999. Magnitude and extent of chemical contamination and toxicity in sediments of Biscayne Bay and vicinity. Silver Spring MD, USA: National Oceanic and Atmospheric Administration. NOAA Technical Memorandum NOS ORCA 141.

MacDonald DD, Carr RS, Calder FD, Long ER, Ingersoll CG. 1996. Development and evaluation of sediment quality guidelines for Florida coastal waters. *Ecotoxicology* 5:253–278.

O'Connor TP, Daskalakis KD, Hyland JL, Paul JF, Summers JK. 1998. Comparisons of sediment toxicity with predictions based on chemical guidelines. *Environ Toxicol Chem* 17:468–471.

Paul JF, Scott KJ, Holland AF, Weisberg SB, Summers JK, Robertson A. 1992. The estuarine component of the US EPA's environmental monitoring and assessment program. *Chem Ecol* 7:93–116.

Rakocinski D, Brown SS, Gaston GR, Heard RW, Walker WW, Summers JK. 2000. Species-abundance-biomass responses by estuarine macrobenthos to sediment chemical contamination. *J Aquat Ecosyst Stress Recovery* 7:210–214.

SAS Institute Inc. 1991. SAS/STAT® User's guide, Version 6. 4th ed. Volume 2. Cary NC, USA: SAS Institute Inc. 846 p.

Striplin Environmental Associates, Roy F. Weston Inc. 1999. Puget Sound reference value project. Task 3: Development of benthic effects sediment quality standards. Olympia WA, USA: Striplin Environmental Associates, Inc.

Turgeon DD, Hameedi J, Harmon MR, Long ER, McMahon KD, White HH. 1998. Sediment toxicity in U. S. coastal waters. Silver Spring MD, USA: National Oceanic and Atmospheric Administration. NOAA Special Report.

Van Dolah RF, Hyland JL, Holland AF, Rosen JS, Snoots TR. 1999. A benthic index for assessing sediment quality in estuaries of the southeastern United States. *Mar Environ Res* 48:269–283.

Weisberg SB, Ranasinghe JA, Dauer DM, Schaffner LC, Diaz RJ, Frithsen JB. 1997. An estuarine benthic index of biotic integrity (B-IBI) for Chesapeake Bay. *Estuaries* 20:149–158.

CHAPTER 9

Uses of Porewater Toxicity Tests in Sediment Quality Triad Studies

R Scott Carr (Workgroup Leader), Edward R Long, Julie A Mondon, Paul A Montagna, Pasquale F Roscigno

The Sediment Quality Triad (SQT) approach to sediment quality assessments was born out of the realization that integrated information was necessary to accurately portray the relative quality of sediments (Long and Chapman 1985). The SQT approach relies upon the integration of data from chemical analyses, toxicity tests, and infaunal benthic structure analyses. Data from the 3 components of the triad can be used by sediment assessors to form a weight of evidence to compare and rank the relative quality of sediment samples and regions of a study area (Chapman et al. 1987; Long 1989).

Tabular decision matrices and graphical methods commonly are used to convey information from the triad components (Chapman 1996). Multiple toxicity tests commonly are applied as a part of SQT studies, and guidance is available with which to use these data in judging the relative toxicity of sediments (Chapman et al. 1997). Use of a single, unit-less index to portray the toxicity of sediments is not encouraged; rather, the use of the data from all individual test results has been recommended (Chapman et al. 1997). Thus, the toxicity data from tests of pore waters should be used along with the parallel data from tests of other sediment phases to form a weight of evidence and to determine concordance among the triad components.

Although toxicity tests of pore waters have been used primarily in large-scale (e.g., estuary-wide) triad surveys (Carr, Chapman, Howard, Biedenbach 1996; Fairey et al. 1998; Carr et al. 2000), they can be used for related purposes. Most applications of these tests in SQT studies involve various kinds of determinations of spatial status and temporal trends. The types of hypotheses tested, the kinds of tests selected, and the methods used during data interpretation are dependent upon the objectives of the study.

Porewater Toxicity Testing: Biological, Chemical, and Ecological Considerations. R. Scott Carr and Marion Nipper, editors.
© 2003 Society of Environmental Toxicology and Chemistry (SETAC). ISBN 1-880611-65-1

Applications and Objectives of Sediment Quality Triad Studies

Seven possible applications of porewater toxicity tests in triad studies include but are not restricted to the following:

1) Identifying spatial trends in sediment quality within a specified study area (i.e., an assessment of status and trends)

2) Determining the magnitude (e.g., degree and severity) of degraded sediment quality

3) Identifying and justifying designation of hotspots with unacceptable sediment quality

4) Determining sediment quality adjacent to a designated upland waste site (e.g., off-site migration of contaminants)

5) Determining the biological significance of known contamination

6) Determining temporal trends in sediment quality within a specified area

7) Development of sediment quality guidelines (SQGs) or validation of their predictive ability in the field.

Identifying spatial trends in sediment quality within a specified study area: An assessment of status and trends

The SQT approach is often used to identify spatial trends, gradients, or other patterns in sediment quality. A weight of evidence can be used to determine the relative quality of sediments within a bay, lake, estuary, or harbor. In this application, the data from porewater toxicity tests are used, along with those from other toxicity tests, chemical analyses, and benthic structure analyses, to identify patterns in sediment quality within all reaches of the study area. Sediment quality would be expected to deteriorate closest to toxicant sources, especially in depositional zones. In past surveys done by National Oceanic and Atmospheric Administration (NOAA), the tests conducted on pore waters have identified areas with toxicity that were nontoxic in the solid-phase amphipod tests, often when chemical concentrations were elevated only slightly above background or detectable levels. This situation was reported in Tampa Bay, Florida, (Long, Carr et al. 1995; Carr, Long et al. 1996), northern Puget Sound, Washington (Long, Hameedi et al. 1999), and southern Biscayne Bay, Florida, USA (Long, Sloane et al. 1999). In these 3 areas, the incidence of significant results in the porewater tests was much higher than in the solid-phase amphipod survival tests.

Examples of applications of the triad approach in which pore waters were tested for toxicity were the sediment quality assessments performed as components of the National Status & Trends (NS&T) program (Long et al. 1996). In these NOAA studies, the triad of measurements was conducted on samples collected throughout major marine bays and estuaries in the U.S. Other examples include the Gulf of Mexico Offshore Operations Monitoring Experiment (GOOMEX) studies done in the Gulf of Mexico (Carr, Chapman, Presley et al. 1996); the assessment of stormwater effects in Corpus Christi Bay, Texas, (Carr et al. 2000); a sediment quality survey of San Diego Bay, California and vicinity (Fairey et al. 1998); and a survey of sediment quality in Boston Harbor, Massachusetts, USA (Hyland and Costa 1995).

Determining the magnitude (degree and severity) of degraded sediment quality

Using results from status and trends surveys, the magnitude of toxicity can be expressed for each survey area. Magnitude is a function of the severity of the toxicological response (e.g., percent survival or fertilization success) and the incidence of significant toxicity. Magnitude also can be expressed as the percentage of the survey area that is significantly toxic. By developing data from stratified-random sampling designs, NOAA has been able to compare the percentages of the surface areas that were toxic in each of several tests, including urchin fertilization tests, among many different bays (Long et al. 1996). In most cases, the spatial extent of toxicity observed in the urchin fertilization tests of pore waters greatly exceeded the spatial extent of toxicity in solid-phase tests of amphipod survival, indicative of the higher sensitivity of the porewater tests.

Identifying and justifying designation of hotspots with unacceptable sediment quality

Another outcome of status and trends surveys can be the discovery or verification of "hotspots" (i.e., small areas with unacceptable sediment quality). If surveys are conducted to embrace areas either known or suspected to be contaminated, the data can be used to justify a determination that they are unacceptable, for example, contaminated with toxic chemicals, toxic in one or more assays, and containing altered benthic populations. The strongest evidence for pollution-caused degradation of sediments is based upon significant results in all 3 components of the triad (Chapman et al. 1997; Long and Wilson 1997). Several areas in San Diego Bay had chemical concentrations well above screening guidelines, were toxic in either amphipod survival or urchin embryo development tests, or both, and had adversely altered benthic infauna (Fairey et al. 1998). Data from San Diego Bay stations that were least contaminated, nontoxic, and populated with an abundant infauna were used to determine a reference envelope with which to compare conditions at putative hotspots (Long and Wilson 1997). Agreement on the presence of toxicity in both the amphipod and urchin development tests occurred in 22 of 164 samples, and lack of agreement in toxicity occurred in 27 samples, demonstrating the importance of performing more than 1 test in assessment surveys (Fairey et al.

1998). Because of the frequent lack of agreement in results among toxicity tests, it is important to determine a priori the weight or differential significance of each test.

Determining sediment quality adjacent to a designated upland waste site: Off-site migration of contaminants

Field surveys of sediment quality often involve collection of samples near upland waste disposal sites to determine the degree to which resources in a bay or lake have been adversely affected. For example, Swartz et al. (1994) measured chemical concentrations of DDT and other substances, toxicity, and amphipod abundance in benthic samples near a waste site in San Francisco Bay, California. In a large survey of the New York and New Jersey Harbor, USA, sediments were tested for toxicity near an upland waste site adjacent to the Passaic River, New Jersey (Long, Wolfe et al. 1995). Such assessments, therefore, may encounter potentially toxic conditions that have resulted from leaching or runoff of toxicants from adjacent upland sites. Porewater toxicity tests can be used in such areas to help form a weight of evidence regarding the relative quality of sediments that have been affected by the upland site.

Determining the biological significance of known contamination

In many cases, SQT studies are conducted in areas for which some sediment quality data exist from previous studies. Because many studies develop primarily chemical data, the biological significance of chemical contamination is unknown. Although SQGs can be used to predict the probabilities of adverse effects such as toxicity, empirical evidence of effects provides verification of the predictions based upon use of the guidelines. Studies based upon the SQT approach, and in which porewater toxicity tests were performed, have been conducted in many areas such as San Diego Bay (Fairey et al. 1998); San Pedro Bay, California (Anderson et al. 2001); and Tampa Bay, Florida (Carr, Long et al. 1996).

Determining temporal trends in sediment quality within a specified area

The most comprehensive evidence of biologically significant temporal changes in sediment quality is developed with SQT studies conducted in the same area with the same methods. Although porewater toxicity tests were not used, the U.S. Environmental Protection Agency (USEPA) conducted SQT studies in an area offshore of California over the course of several years to determine temporal trends in sediment quality (Swartz et al. 1986). In these surveys, they were able to quantify the degree to which chemical contamination, toxicity, and benthic abundance and diversity changed during the time period. Comprehensive chemical analyses, 4 toxicity tests, and species-level benthic analyses were conducted in Puget Sound, and the spatial extent of degraded and undegraded conditions was determined as a basis for estimating changes in quality in future years (Long et al. 2002).

Developing SQGs or validating their predictive ability in the field

Using one of several empirical approaches to the development of SQGs, many different agencies and programs have developed numerical values associated with observations of toxicity or other adverse effects. Often, these guidelines have been based upon statistical analyses of matching chemical and toxicological data from portions of the same samples. Such analyses could be based upon toxicological data from any large database developed with consistent protocols. For example, a large amount of data from amphipod survival and urchin fertilization tests exists along with the necessary chemical information (Field et al. 1999). Field validation of the predictive ability of numerical guidelines was conducted with data ($n = 1086$) compiled from marine and estuarine studies, many of which included information from porewater toxicity tests and benthic analyses (Long et al. 1998). The ability of these guidelines to accurately predict adverse benthic effects was determined with data from a portion of these same samples (Hyland et al. 1999). In a study initiated in Sydney Harbour, Australia, the predictive ability of North American SQGs has been determined using endpoints from benthic community structure analyses and tests of amphipod survival, microbial bioluminescence, and urchin fertilization (Dr. Gavin Birch, Department of Geology, University of Sydney, personal communication).

In the following discussion, the potential uses of the porewater toxicity tests in SQT studies will be described. In all cases, we assumed that the results from these tests would be used to form a weight of evidence regarding relative sediment quality and to determine the statistical relationships among the triad components. However, in all cases, it is important to remember that best professional judgment is necessary in the interpretation of these data, and short, quick, easy answers to questions regarding sediment quality are not likely (Chapman et al. 1997).

Desirable Attributes of Porewater Toxicity Tests Used in SQT Studies

Ingersoll et al. (1997) have discussed the uncertainty associated with various types of sediment toxicity tests and test endpoints (see Tables 18-1 and 18-2 in Ingersoll et al. 1997). The following section provides an update of that discussion with a specific focus on porewater toxicity tests.

Insensitivity to confounding factors

When a sediment porewater toxicity test is to be used, several confounding factors that are common to all sediment toxicity testing should be considered. Because of anoxic conditions that exist a few millimeters below the surface layer of aquatic sediments, most sediments are naturally toxic to many forms of aquatic organisms.

Bioturbation can oxygenate sediments, but free molecular oxygen decreases rapidly with depth. Ferrous iron and various sulfide species, in particular, immediately react with oxygen to produce a redoxycline of increasing reduced-chemical species with depth. Levels of ammonia, a normal constituent of nutrient fluxes found in sediment, also are influenced by anthropogenic inputs. Consequently, low dissolved oxygen (DO), high sulfide, and elevated ammonia can cause mortality in sediment, which masks the toxicity that could have been attributed to a particular anthropogenic contaminant of concern (Jones-Lee and Lee 1993). Ammonia concentrations in sediments can be considered confounding when they result from normal sediment processes. Further, depending on sample collection, storage, and processing, increased ammonia concentrations in pore water may produce artifacts. It is important, therefore, to know the tolerance ranges for parameters such as pH, ammonia, and sulfide for a particular test species in order to interpret the test results (see Chapter 7).

In sea urchin fertilization tests of pore water and amphipod survival tests of solid-phase sediments in marine bays, toxicologically significant concentrations of ammonia occur very rarely in the test chambers. Long et al. (1998) reported that concentrations of unionized ammonia exceeded the levels of effective or lethal concentration to 50% of the test population (EC50 or LC50) in only 3% samples for both tests.

Ecological relevance

Sediments are the ultimate repository and integrator of chemical contamination in aquatic environments. Chemicals entering the environment have an affinity for particles and will bind to sediment and its constituents to levels above background. Therefore, the benthic community is exposed to contaminant concentrations that produce a series of responses, from acute toxicity to more subtle chronic impacts. The organisms in sediment play key roles in biogeochemical transformations that are important in the coupling of benthic and pelagic environments. The SQT approach provides for a pragmatic assessment of the benthic environment's interactions with the in-place contaminants.

Using a sediment porewater toxicity test further provides information about these interactions because it uses the interstitial water that is available to benthic organisms at concentrations that represent the organism's exposure to the bioavailable contaminants. In areas where environmental variability is great or the level of contamination is moderate, or where sediment mineralogy is complex, use of sediment porewater toxicity testing provides insight into an environment where contaminant correlations to changes in the environment are difficult to observe and interpret.

The SQT approach provides a reliable assessment for evaluating the effect of contamination on the sediment environment. The SQT provides a mechanism to test whether the benthic environment has been impaired by the introduction of contaminants and provides a measure of whether this effect is within the variability experienced within that benthic environment. Many water bodies have naturally toxic sediments that support abundant biological resources. For example, hypoxic Gulf of Mexico sediments support a robust shrimp fishery. The coupling of sediment toxicity with benthic community responses can be more easily achieved with sediment porewater toxicity tests that link them within the framework of the SQT.

In many ways, the SQT approach, using porewater toxicity, challenges the trend in the regulatory community to use numeric approaches in which a statistic is produced, simplifying the decision-making process. The dynamic benthic environment does not lend itself to simple numerics of significant adverse impacts. The SQT, as it is being applied to more and more benthic environments, may eventually find patterns, trends, or levels of classifications that are applicable to many more environments. Rather than focusing on differences in benthic environments, some underlying commonality may emerge to provide the specificity and simplicity that decision-makers seek.

The SQT framework allows an investigator to use best professional judgment to determine the weight of evidence needed for an integrative assessment. The appropriate chemical characteristics of the sediment are judged against the assemblages found in the benthos and compared with habitats and communities that have not experienced impacts. The SQT assumes that biological responses found in benthic studies and toxicity testing are a function of the concentration of chemicals sorbed to the sediment (Chapman 1986).

Information obtained from the SQT is best used in a descriptive mode, in which the relative quality of sediments among sampling sites is evaluated and classified. Sites can be ranked independently for each of the SQT components to determine both geographic and temporal trends in sediment quality. Further, predictive models can be developed in which the relationships between collected biological and chemical data can be used to estimate the relative degree of contamination that may be associated with biological effects. Finally, the SQT data can be used to rank sites for remedial action (Long 1989a).

Sensitivity to target contaminants of concern

In order to predict the community responses in the field, which is the ultimate objective of sediment toxicity tests, the test organism or endpoint must be similar in sensitivity to the most sensitive life stages of the most sensitive organisms that would naturally occur in a particular marine ecosystem. Ideally, the test organism will be sensitive to a wide variety of contaminants or respond only to a particular class of chemicals. One of the major problems with the commonly employed solid-phase sediment toxicity tests is that they measure acute lethality in an adult animal. Such tests may not be predictive of population and community responses in the field. Benthic organisms are subject to chronic exposures that affect growth and reproductive processes that are undetected by these acute assays.

Range in response

In addition to sensitivity, a desirable attribute of any sediment toxicity test is the ability to differentiate among samples, all of which may be highly contaminated or moderately contaminated. Unlike amphipod solid-phase tests, porewater samples can be tested in a dilution series test design that allows highly toxic samples to be differentiated at the lower dilutions. The commonly employed amphipod solid-phase test has been observed to have a very narrow range in response to moderately contaminated areas, whereas the porewater tests with sea urchins often exhibited a wider range in response among samples (Carr, Long et al. 1996; Long et al. 1997). This range in response provides data that are more useful in SQT studies, particularly when multivariate statistical analyses are performed on the SQT dataset. When all the samples are either nontoxic or so toxic that the response is the same for all samples (as has sometimes been observed with the undiluted porewater samples but not at lower dilutions), it is very difficult to show concordance or associations between toxicity and the other triad parameters. Test data that show a wide range in response are, therefore, highly desirable.

Most Appropriate Porewater Toxicity Tests for Use in SQT Studies

Sediment porewater toxicity testing is used to evaluate potential in situ effects of contaminated sediments on aquatic organisms (Ankley et al. 1991). Early life history testing of sediment pore water can result in greater discrimination between sites. Embryos and gametes exhibit greater sensitivity to contaminants than adult life stages, providing information on contaminant bioavailability and toxicity during critical developmental stages. Many test species can be potential candidates for porewater toxicity tests: embryo or larval stages of mollusks, polychaetes, crustaceans, echinoderms, and fish. Even though porewater toxicity testing may be an order of magnitude more sensitive than whole-sediment toxicity testing, such sensitivity allows for further investigations of those sediments that may be producing more complex changes in the benthic community (Carr and Chapman 1992; Carr, Chapman, Howard, Biedenbach 1996; Carr, Long et al. 1996). However, caution should be exercised when interpreting these results because of potential alterations in chemical pathways.

Within the SQT, several different toxicity tests should be used to reduce the uncertainty and limit the occurrence of false negative results. This is especially important in moderately contaminated areas where traditional sediment toxicity tests usually fail to exhibit a response. Multiple endpoints should be measured to ensure an integrated assessment of sediment contamination (Carr 1998). Candidate species and endpoints recommended for use in porewater toxicity tests are shown in Table 12-3 (Chapter 12).

Analysis for Weight-of-Evidence Interpretation

The SQT was first proposed in the 1980s (Long and Chapman 1985) as an approach to integrate data from synoptic samples analyzed for sediment chemistry, benthic community structure, and toxicity tests to assess relative sediment quality or determine pollution-induced degradation. The main challenge facing investigators who have applied this approach to aquatic impact assessments over the past 2 decades is how to interpret this array of data from these different measurement endpoints. There have been numerous suggestions put forth in the literature (Chapman 1996; Chapman et al. 1997; Long and Wilson 1997; Long 1998), including summary indices (Chapman 1990, 1992a; Alden 1992), tabular decision matrices (Chapman 1992b, 1996; Carr 1993; Carr, Long et al. 1996; Chapman et al. 1996), scaled ranking factors (Canfield et al. 1996; Carr, Long et al. 1996; Carr et al. 2000), or multivariate analyses (Green 1993a, 1993b; Green et al. 1993; Chapman et al. 1996; Green and Montagna 1996; Carr et al. 2000). All of these approaches require appropriate reference stations. In addition, they all involve a weight-of-evidence interpretation or a way to draw conclusions based on congruent or conflicting lines of evidence. As with multivariate techniques for data analysis, where there is no one "right" way to relate sets of variables (Green 1993a), a recurring theme in these papers (Chapman 1996; Chapman et al. 1997) is that there is no one "best" way to depict or use the SQT.

Looking at the SQT from an integrated systems approach, the ecologist is given a particularly difficult task. Faced with a potential wealth of environmental information, the ecologist must decide what information is the most relevant, how to obtain it, how to integrate what is obtained, and how to relate it to what are often nebulous or ill-defined categories (Faust 1986a). While the SQT provides a basic framework for the classes of data to be collected, there still is an enormous range of choice in each of the 3 legs of the triad for the sampling design (e.g., number of stations, number of replicates, type of sampler, time of year), the types of analyses to be performed (contaminants of concern for chemical analyses, sieve size or level of identification for benthic community analyses, type of tests and target organisms for toxicity testing), and the method of interpretation (for each of the data groups as well as the integration of the entire dataset). This wide range of choices and results presents its own set of challenges when it comes to interpretation (Germano 1999).

While some investigators may experience greater intellectual freedom employing the SQT because "it is neither constrained nor defined by any particular method of data analysis or presentation" (Chapman et al. 1997), investigators still need to exercise caution in both their sampling design and data interpretation methods. Because the universal answer ("It depends....") to all questions is equally applicable to investigators wondering how to both design and interpret SQT investigations, it is imperative that investigators be clear about what questions are being asked before any samples are collected. Because the SQT by definition will include competing lines of evidence, both the sampling design and analysis framework should be established a priori (Chapman 1996). For example, depending upon the objective of a study, there may or may not be a need for porewater toxicity testing. If the assessment requires no information on the toxicity of chemicals bioavailable in the pore water, then there would be no need to include such tests. Any one of a number of excellent references will help with overall sampling design (e.g., Green 1979, 1984; Eberhardt and Thomas 1991; Rose and Smith 1992). The following discussion will concentrate on the appropriateness of including porewater toxicity tests as one of the lines of evidence in SQT studies as well as potential methods of interpretation.

The basic conundrum facing investigators who deal with SQT studies is that no one line of evidence by itself is sufficient to determine environmental degradation, and the data are not evaluated at equivalent levels of information hierarchy. For example, the sediment chemistry analyses give bulk sediment chemical concentrations of various contaminants of concern. In an isolated system, these chemical concentration numbers are meaningless by themselves; if there were no biological system or organism available with which they could interact, it would not matter what concentration was measured or how different any particular chemical concentration was at a reference station. What matters is their effect or impact on biological systems, which can be predicted with effects-based numerical guidelines or measured empirically with toxicity tests.

Unfortunately, chemical concentration alone is not a sufficient indicator of bioavailability (if it were, there would be no need to collect the lines of evidence from the other legs of the SQT). The currently accepted method of determining whether a sediment contaminant is bioavailable is through acute or chronic bioeffects testing. Because it is impossible to analyze for all possible chemicals of concern that may be present, bioeffects tests are performed with the intent that a) they will indicate whether any "elevated" concentrations of chemicals measured are potentially bioavailable, and b) they will indicate whether any unknown or unmeasured contaminants are present in sediments that may have an adverse environmental effect. While there are inherent interpretive difficulties with this logical construct (e.g., Spies 1989), the array of bioeffects tests that has been developed is accepted by the scientific and regulatory community as the best method available for indicating whether sediment chemical concentrations are bioavailable or have the potential to cause what would be considered an adverse environmental effect.

If one had to construct an information hierarchy from the matrix of SQT data available for interpretation, the toxicity test results appear as the "middle rung," indicating whether the sediment sampled at a particular location has the potential to adversely affect the particular taxon selected as the test organism in the bioeffects test. The implication is that, if the sediment is shown to be harmful to this particular test species by any one of a suite of measurement endpoints, there is a potential that it would also be harmful to other aquatic species. The next jump in interpretive integration in the SQT is at the population level with the results of the benthic community analyses. While sediment chemical concentrations may be elevated and toxicity test results may indicate the potential bioavailability of these chemicals, the evidence of degradation in the field at a higher systems level is assessed in an SQT investigation by determining whether or not the benthic community has been negatively altered. This is another line of evidence that can be assessed by a wide range of metrics, depending on the particular objective of the study.

Given the array and systems-level hierarchy of data available from an SQT investigation, if an investigator has chosen to use a porewater toxicity test either by itself or along with other toxicity tests, such as a solid-phase amphipod or mysid test, how the porewater test data are evaluated will depend in large part on the evaluation method chosen. The 2 most common approaches are either a tabular decision matrix (e.g., Chapman 1996) or a multivariate approach (e.g., Green and Montagna 1996). The typical interpretive framework for the tabular decision matrix is shown in Table 9-1.

Table 9-1 Tabular decision matrix for SQT data[a]

Chemistry	Toxicity	Benthos	Possible conclusion
+	+	+	Strong evidence for pollution-induced degradation
−	−	−	Strong evidence against any pollution-induced degradation
+	−	−	Contaminants not bioavailable
−	+	−	Unmeasured contaminants or confounding factors have the potential to cause degradation
−	−	+	Alteration is not due to bioavailable contaminants
+	+	−	Contaminants are bioavailable, but in situ effects are not demonstrated
−	+	+	Unmeasured bioavailable contaminants or confounding factors are causing degradation
+	−	+	Contaminants are not bioavailable, alteration is not due to bioavailable contaminants

[a] Reprinted from Chapman et al. 1996, copyright 1996, with permission from Elsevier Science.

Several cautionary statements should be made at this point about the above interpretations, before we even consider how to incorporate porewater toxicity tests into the overall framework. Chapman et al. (1991) noted that one of the weaknesses of the SQT was the "lack of statistical criteria for the combined Triad components." The tabular decision matrix applications of the SQT have descriptive statistical evaluations that are individually applied to each component (toxicity and benthic community results compared to reference results to designate a "+" or "−") rather than as criteria for the combined components. Because of the potential for a range of background conditions for both the toxicity and benthic community data, the use of a "reference envelope" approach rather than an absolute value for reference conditions has been advocated as a basis for comparisons (Long and Wilson 1997).

The evaluation of the chemistry data as to whether or not they are designated with a "+" or "−" (which has substantial implications for the "Possible conclusions" column) has been done in a number of different ways, using either effects range median (ERM) values of Long, MacDonald et al. (1995), probable effects level (PEL) values of MacDonald et al. (1996), ERM quotients (Long and Wilson 1997), PEL quotients (Carr et al. 2000), or even effects range low (ERL) values (Carr, Chapman, Howard, Biedenbach 1996). Because all of these sediment chemical guidelines were determined by co-occurrence comparisons with toxicity tests, aside from the problem of circular reasoning, the possibility of a false positive or negative in that particular column in the decision matrix would necessarily require the investigator to also consider alternative conclusions as potential explanations for the observed patterns. Another problem noted by Green and Montagna (1996) with a tabular decision matrix for evaluating SQT results is that the experiment-wide error rate is not controlled in this type of approach. However, with the recent evaluation of the predictive abilities of these guidelines (Long et al. 1998), the analyst can now estimate the probabilities that samples would be toxic in acute tests by comparing their chemical data for the site with those evaluated in an independent database (Long and MacDonald 1998).

Given all of the above qualifications, the investigators have the choice of incorporating porewater toxicity tests as the sole toxicity measure in an SQT, if they feel that is appropriate (e.g., Green and Montagna 1996), or in combination with other toxicity tests (e.g., Long et al. 1990; Carr, Chapman, Howard, Biedenbach 1996; Carr, Long et al. 1996; Carr et al. 2000). When used in combination with other toxicity tests, investigators will naturally feel more confident in their conclusion (as is the case with the overall SQT interpretation) if all the various tests show the same result (either all "hits" or all "no hits"). The quandary comes when the porewater tests show a response different from that of the other tests performed. A possible solution when one is faced with this situation is to continue the weight-of-evidence tabular decision matrix approach within the suite of toxicity tests applied. Determination as to whether or not one assigns an overall "+" or "−" to the "Toxicity" column in the SQT decision table could become increasingly difficult and more subjective because of the various possible permutations as more tests are added. Clearly, if more than one type of porewater test is used in combination with one or more solid-phase tests, a tabular decision matrix would not be recommended as a means of evaluation. With such a large data array, one could employ an unvalidated ranking procedure for the data or the more preferable approach of using multivariate analyses (principal component analysis [PCA], multiple dimensional scaling

[MDS], canonical correlation analysis [CCA], etc.). Such an approach was recently reported where PCA was used to categorize toxicity, chemistry, and benthic indices of biotic integrity (BIBIs) as a "+" or "−" for the decision matrix (Carr et al. 2000).

Multivariate approaches have their own advantages and disadvantages; while these are covered in more detail in the following section, one would need to evaluate whether the additional data arrays generated by porewater tests because of the use of dilution series, as compared with solid-phase tests, give equal weight to the overall "toxicity" determinations of the various stations sampled. Some investigators have chosen to use a cutoff of 50% (i.e., judging the sample as being toxic if 50% or more of the assays performed tested positive) when combining the results of multiple toxicity tests (Hyland et al. 2000). This approach has the advantage of 1) providing a mechanism to make spatial and temporal toxicity comparisons in cases where the number of assay types has varied across sampling locations and years and 2) helping compensate for the under- or oversensitivity of any particular assay.

Psychologists have shown that having more data does not necessarily allow an investigator to make a better decision (e.g., Goldberg 1968; Dawes 1979; Faust 1986b). Improvements in judgmental accuracy are usually more an exercise in exclusion of variables than inclusion of more measurements (Faust 1989). A limited set of validated predictors that are simply added together is as predictive as optimally weighted variables (Goldberg 1968; Dawes 1979; Faust 1986a, 1986b). Including additional variables that have not been validated just increases the likelihood of decreasing the overall accuracy of any overall assessment or conclusion. Before any toxicity tests are routinely included as part of SQT investigations, we would be better off spending the resources to field validate these tests as a predictive tool by accurately assessing covariation (Arkes 1981; Germano 1999), determining whether their inclusion increases diagnostic accuracy (Faust 1986b), and finally, determining whether the information they provide is redundant with other variables or instead produces incremental validity for the final assessment (Faust 1986a).

Use of Porewater Toxicity Tests to Differentiate between Degraded and Nondegraded Conditions

As suggested by Chapman (1996), preparing strong evidence either for or against pollution-induced degradation is relatively straightforward once the thresholds for designating a significant effect for each component of the triad have been established (the difficulty is in determining these threshold values). Samples in which there is strong evidence of degraded conditions would be those in which effects-based, numerical guidelines are exceeded for many substances and/or by large degrees, in which toxicity is demonstrated in all or most of the tests, and in which the benthic infauna are severely depressed in abundance and diversity relative to reference areas (Chapman et al. 1997). Sites that might qualify as hotspots could be those with similar characteristics (Long and Wilson 1997). In contrast, sites that might be classified as reference locations or could be used in developing a reference envelope would be those in which chemical concentrations are less than all numerical guidelines, results of all toxicity tests are negative (nontoxic), and a rich and diverse infauna is supported (Chapman et al. 1997; Long and Wilson 1997).

In an SQT survey of Puget Sound, undegraded samples had very low chemical concentrations (mean ERM quotients < 0.1), very high percent amphipod survival and urchin fertilization (>80%), and very abundant and diverse benthos (Table 8-8). The large range in the benthic data is reflected in these data because of the large range in depths, textures, and other natural environmental factors among sampling locations. In contrast, chemical concentrations were much higher in 4 other samples (mean ERM quotients 0.35 to 1.34). In 2 of these samples, urchin fertilization was very low, indicating significant toxicity, and total abundance, species richness, and arthropod abundance were very low. In the remaining 2 samples, however, urchin fertilization was relatively high, indicating nontoxicity, and metrics of infaunal abundance and diversity were relatively high. The sample from station 160 was classified as a "triad hit" and represented less than 0.1% of the total surface area of the survey (Long et al. 2002). The percentages of the 2 survey areas in Puget Sound with unanimous degraded conditions, unanimous undegraded conditions, and mixed results are contrasted by Long et al. (2003) (Table 9-2).

Table 9-2 Estimated spatial extent of degraded sediments in Puget Sound, based on results of triad studies, expressed as percentages of survey areas

Indices of sediment quality	Percent of Northern Puget Sound (Total area: 774 km^2)	Percent of Central Puget Sound (Total area: 731 km^2)
Contaminated, toxic, altered benthos	1.5	<0.1
Contaminated, toxic, abundant benthos	4.5	13.6
Mixed chemical and toxicity results, abundant benthos	69.3	33.2
Uncontaminated, not toxic, abundant benthos	25.7	53.1

In a portion of Biscayne Bay near the Port of Miami ($n = 15$), mean ERM quotients were less than 0.1, amphipod survival exceeded 76%, urchin fertilization exceeded 75% in most samples, and total number of species ranged from 56 to 340. In contrast, samples ($n = 7$) from the lower Miami River often were very degraded; mean ERM quotients ranged from 0.3 to 2.0, amphipod survival was 5% to 41%, and total number of species was 12 to 26. Urchin fertilization in the lower Miami River samples remained relatively high (93% or higher) in 6 of the 7 samples (Long, Sloane et al. 1999).

The task of classifying samples as degraded or undegraded is much more difficult when results of the different tests and analyses are not unanimous. As shown in the data from Puget Sound, the majority of samples provided mixed results among the different SQT tests and assays. Often, the large middle class is difficult to characterize and justify, frequently requiring professional experience in the interpretation of the data.

In the many SQT studies in which porewater toxicity tests were performed, every possible combination and permutation of results probably have been observed. Generally, results of different toxicity tests performed with the same samples show overlapping but different spatial patterns in the survey areas, indicative of relatively low agreement among tests (Long, Sloane et al. 1999).

Samples with high contamination and high total organic carbon (TOC) content might be expected to be nontoxic in solid-phase and porewater tests but toxic in Microtox tests because, in this case, the toxicants that were otherwise not bioavailable were eluted with solvent extractions. In some cases, we have observed toxicity in amphipod survival tests in samples with high levels of organic compounds, but not in urchin fertilization tests. The causes of these differences in responses are unknown. Disagreement in results may be a factor of patchiness within the sample.

Sperm cells may not be highly responsive to the presence of organics, especially over the short time of the exposures. More frequently, urchin fertilization tests are significantly toxic in moderately contaminated samples, especially sandy samples with low TOC content, in the absence of toxicity in the Microtox and amphipod tests. Echinoderm tests performed with embryological development as the endpoint are highly sensitive to the presence of ammonia in the pore waters, much more so than the fertilization test. In any case, professional experience is necessary during interpretation of results. Simple and easy classifications of sediment quality should not be expected.

Spatial Patterns in Sediment Quality Observed in SQT Studies

In studies of marine bays and estuaries of the U.S., a variety of spatial patterns in sediment quality have been observed (Turgeon et al. 1998). These patterns probably were a function of proximity to sources and a combination of physical processes that resulted in dispersal and sedimentation of toxicants. In future SQT studies, sediment assessors might expect to observe one of the following spatial patterns in sediment quality:

1) General lack of degraded conditions. These situations generally occur in systems that are adjacent to relatively rural areas and/or where physical processes rapidly disperse toxicants and therefore act to dilute their concentrations.

2) Highly localized areas of degraded conditions. Degraded conditions most often occur in relatively small peripheral bayous, harbors, boat basins, coves, or inlets adjoining the main basin of a large system. In these situations, toxicants probably were released into the peripheral body of water, deposition of the chemicals occurred near the sources because of poor flushing of the peripheral area, and chemicals that were transported to the main basin of the system were rapidly dispersed by physical processes.

3) Spatial gradient of increasing sediment quality with increasing distance from pollutant sources. It is not unusual to encounter systems in which degraded conditions are clearly worst nearest the pollutant sources and gradually and incrementally improve with distance from the sources as a result of dispersal and dilution of the toxicants.

4) Patchy patterns in degraded conditions. Very heterogeneous conditions may be encountered in areas where dispersal of toxicants and their deposition are interrupted or complicated by the presence of islands, peninsulas, jetties, or other physical obstructions, or where there are multiple sources, or where erosion and scouring occur in some regions of the system. Patchy results may occur in only one or a minority of the triad analyses and not the others.

5) Widespread distribution of degraded conditions. Degraded conditions may be relatively ubiquitous in areas with significant toxicant inputs, multiple widely distributed sources, and poor physical dispersal and transport. In some surveys, degraded conditions were found from the head of the system to its entrance channel.

In some survey areas, it is possible to foretell what the results of an SQT survey will show, and in other areas, it is impossible to correctly guess the outcome of such a survey. In addition, conditions may change with time, so the pattern observed in 1 year or season may not be observed in another.

Concordance between Porewater Toxicity Tests and Sediment Contamination

During the 1990s, numerous sediment quality assessment surveys using porewater toxicity tests were conducted in bays and estuaries of the east, west, and Gulf coasts of the U.S. as well as in some areas of the continental shelf. In some of these studies, there appears to be excellent concordance between the concentration of bulk sediment contaminants and the incidence and degree of toxicity observed with porewater toxicity tests. A good example of a high degree of concordance between sediment chemistry and porewater toxicity was a comprehensive SQT survey performed in Tampa Bay (Carr, Long et al. 1996). This study, which was conducted over a 2-year period and included 55 sites with 3 stations per site (165 stations total), spread over the entire system from the highly industrial port area of northern Hillsboro Bay to the barrier islands near the mouth of the estuary. The SQG values of Long, MacDonald et al. (1995) were used to calculate a cumulative ERM that incorporated a measure of the amount and degree of the measured contaminants for which an ERM has been calculated. Comparing these cumulative ERM values with the porewater toxicity data demonstrated a high degree of concordance between the contaminant levels and toxicity (Figure 9-1).

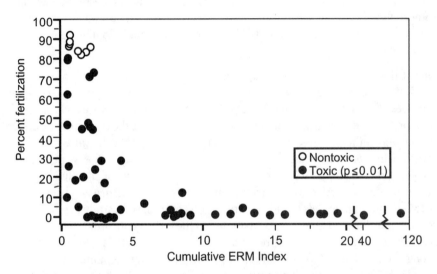

Figure 9-1 Relationship between sea urchin fertilization in 100% pore water and the cumulative ERM index for Tampa Bay (Carr, Long et al. 1996; reprinted with permission from the Society of Environmental Toxicology and Chemistry [SETAC])

The degree of concordance between contaminant concentrations and porewater toxicity was sufficiently strong in this dataset that all samples higher than a cumulative ERM index of 2.5 were significantly toxic for 100% of the samples. One of the major reasons for the excellent concordance between porewater toxicity data and sediment chemistry in this study was the wide range in contaminant concentrations in this study area and the wide range of responses in porewater toxicity observed. This concordance was not observed between the ERM cumulative index and the amphipod data because only 2% of the samples exhibited significant toxicity with that test.

Many of the sediment quality assessment studies that have been conducted in recent years have used a probabilistic, randomized sampling design to select the specific sites to be included in a regional survey. Using this type of approach allows spatial estimates of toxicity to be done for a particular region, but the range in contaminant levels for a particular study may be considerably less than in a design that focuses on hotspots and contaminant gradients, as was often done in the past. A good example of this randomized sampling design is the survey conducted in 4 bays along the northwestern coast of Florida (Long et al. 1997). Although porewater toxicity was observed at a number of sites throughout these bays, the range in contaminant concentrations was small enough that little concordance was observed between sediment contaminant concentrations and porewater toxicity for any one bay system (see Tables 14, 20, and 26 in Long et al. 1997). When all the bays were compared, however, there was enough of a range and a sufficient sample size that some highly significant associations were observed between porewater toxicity and polycyclic aromatic hydrocarbons (PAHs), particularly (Table 28 in Long et al. 1997).

One of the realities of sediment quality assessment studies is that only a small fraction of the chemicals and their degradation products that occur in the complex mixtures of contaminants typically found in urbanized estuaries are included in the "comprehensive" chemical analyses. Another reality is that the response of different species and endpoints used in toxicity tests is chemical specific, because they measure different mechanisms of toxicity. Yet another reality is that we do not yet fully understand all the factors that control the bioavailability of sediment-associated contaminants. All of these factors can affect the concordance we observe between contaminant levels and toxicity. A good example of a study in which these realities may have substantially contributed to the lack of concordance between sediment contamination and porewater toxicity was conducted in Boston Harbor (Long, Sloane et al. 1995). Boston Harbor is one of the most contaminated harbors in the U.S. Using a probabilistic random sampling design, 26 of 30 sites sampled exceeded at least 1 ERM, and 13 of those 26 exceeded 10 or more ERMs (Long, Sloane et al. 1995). For reasons likely related to the realities and limitations of sediment quality assessment studies mentioned previously, the sea urchin (*Arbacia punctulata*) fertilization assay did not detect the presence of the measured contaminants (Figure 9-2). The sea urchin embryo development assay, however, was extremely sensitive to the pore water from these sites, with 89% of the samples significantly toxic at a 50% dilution (Figure 9-3). Even with these high levels of contaminants and the high proportion of toxic samples, there was a low degree of concordance between sediment contaminant levels and porewater toxicity in the sea urchin embryological development assay (Long, Sloane et al. 1995). In this instance, the high levels of contaminants at the majority of sites (resulting in a small range although high concentrations), the high degree of covariance among the contaminants of concern, and the high proportion of toxic sites (also resulting in a small range of response) contributed to the lack of statistical concordance in this study.

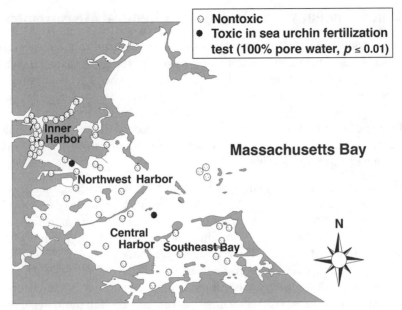

Figure 9-2 Results of sea urchin fertilization test with undiluted pore water from Boston Harbor

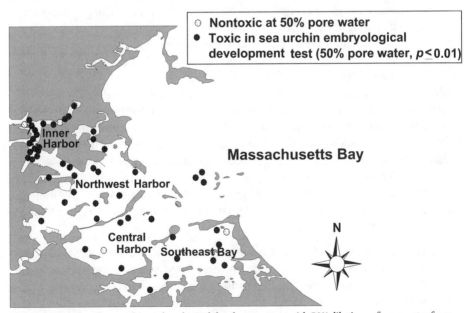

Figure 9-3 Results of sea urchin embryological development test with 50% dilutions of pore water from Boston Harbor

Bulk Sediment versus Porewater Chemical Measurements

The vast majority of data from comprehensive sediment quality assessment surveys that could be used to examine concordance between porewater toxicity test data and sediment-associated contaminants are for bulk sediment chemistry concentrations only. Because of the additional effort and cost required to collect enough pore water to perform chemical analyses and the uncertainty surrounding the potential artifacts and losses occurring during the extraction procedure, porewater chemical measurements have been performed in conjunction with porewater toxicity tests on a very limited number of samples.

One study in which porewater chemical measurements were made at some select sites was part of the Minerals Management Service (MMS)-sponsored GOOMEX program, designed to evaluate the impacts of offshore petroleum production platforms in the Gulf of Mexico (Carr, Chapman, Presley et al. 1996; Kennicutt, Green et al. 1996). At one platform near the Flower Gardens Marine Sanctuary, located approximately 120 miles off the coast of Texas, all discharges were required to be shunted to within 10 m of the bottom to minimize possible contamination of the Flower Gardens reefs. This requirement, in conjunction with the low energy environment near the bottom of this deep-water platform (~140 m), resulted in very high levels of metal contamination (primarily zinc, cadmium, lead, and barium) in the immediate vicinity of the platform (Kennicutt, Booth et al. 1996). The sea urchin (*A. punctulata*) embryological development test, the polychaete (*Dinophilus gyrociliatus*) reproduction test, and the harpacticoid nauplii (*Longipedia americana*) survival test using pore water all detected toxicity at the stations in the immediate vicinity of the platform (Figure 9-4), a finding that corresponded with the presence of high concentrations of metals at these stations (Carr, Chapman, Presley et al. 1996). Chemical measurements of the pore water from these same sites were also performed for both metals and organics. The organic contaminants were not detectable in the pore water, but the metals concentrations were high enough to be responsible for the observed toxicity, and there was good concordance between the concentrations of these porewater metals and the observed toxicity for all 3 species (Carr, Chapman, Presley et al. 1996). At least part of the reason that good concordance was observed between sediment porewater chemistry and toxicity with this limited dataset is that there was a strong contaminant gradient that tracked well with the observed toxicity.

Another example, in which a select group of porewater samples was analyzed chemically, was in a study conducted in northern Puget Sound (Long, Hameedi et al. 1999). A subset of 10 porewater samples along a gradient from contaminated Everett Harbor out into the Sound, which had previously been shown to exhibit a toxicity gradient with the sea urchin (*Strongylocentrotus purpuratus*) fertilization test, was examined chemically. A number of PAHs and organochlorines were detected in these porewater samples, which did not correspond particularly well with the

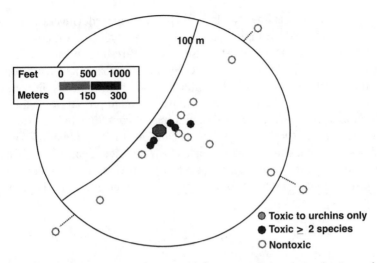

Figure 9-4 Results of sea urchin embryo development, polychaete reproduction, and copepod survival toxicity tests in the vicinity of a petroleum production platform (Carr, Chapman, Presley et al. 1996; reprinted with permission)

porewater toxicity. The reason for the lack of correlation between specific contaminants and the observed toxicity is likely related to the small sample size and the small range in contaminant concentrations. This study did show that measurable quantities of organic contaminants can be present in pore water extracted by oxic techniques (Carr and Chapman 1995) that have been routinely employed in sediment quality assessment surveys using porewater toxicity tests.

Desirable Attributes of Future Porewater Toxicity Tests

The absence of knowledge relating to the mechanisms of expression of test endpoints inhibits the ability of the researcher to accurately interpret in situ environmental conditions and to predict potential toxicity response. Considerable interest in the cytochrome P450 mixed function oxygenases (MFO) system in fish and other organisms has resulted in detailed investigations identifying the mechanisms and the chemical classes that elicit induction of the CYP1A detoxification response (Stegeman et al. 1992; Livingstone 1993; Di Giulio et al. 1995). Attention has also been focused on investigation of heavy metal-induced biosynthesis of metallothionein in aquatic organisms and its potential use in environmental monitoring (Di Giulio et al. 1995). Conversely, no equivalent "mechanism of effects" data exist for the commonly known porewater toxicity tests. Understanding the mechanisms responsible for the expression of porewater toxicity endpoints would be advantageous in identifying the specific toxicants that may contribute to the measured response.

A suite of porewater assessment tools would be useful for reducing the uncertainty factor associated with determining causality. The toxicity identification evaluation (TIE) procedure, which involves a stepwise process of testing and analysis of chemical characteristics to identify potentially causative toxicants, is currently the best method for determining causality using porewater toxicity tests (Burgess et al. 1996; Chapters 6 and 7). Ideally, the optimum assessment procedure would involve identification of the specific chemical significance of contaminated sites after an initial screening. Nonspecific assays that are currently in use could be used as exploratory tools to identify toxicity. Single-chemical test endpoints could then be applied to identify the chemicals responsible for the toxicity and to infer whether the community degradation is likely to be a result of a particular chemical or source.

Conclusions

1) Porewater toxicity tests may be able to satisfy a number of different objectives in SQT studies, mostly involving estimations of status and trends.

2) Toxicity in porewater tests, together with toxicity in other tests, elevated chemical contamination, and altered infaunal populations, can be used to classify sites as degraded.

3) Observations of toxicity in porewater toxicity tests alone may not constitute a sufficient weight of evidence to classify sediments as degraded.

4) Often, toxicity in porewater tests occurs in samples with levels of chemical contamination insufficient to cause toxic responses in other tests.

5) The response of different species and endpoints used in toxicity tests is chemical specific because they measure different mechanisms of toxicity.

6) Only a small fraction of the potentially toxic chemicals and their degradation products that occur in the complex mixtures of contaminants typically found in urbanized estuaries is included in the "comprehensive" chemical analyses.

7) The factors that control the bioavailability of sediment-associated contaminants are not completely understood. Concordance between porewater toxicity data and sediment chemistry is most likely to be observed in studies where there is a wide range in contaminant concentrations and a wide range of responses in porewater toxicity.

Recommendations

1) Covariation should be better assessed and base rates determined for the toxicity tests used in SQT studies.

2) Samples determined to be toxic in less than 100% pore water should be classified as more toxic than those determined to be toxic only in 100% pore water.

3) A multivariate analysis is the best statistical approach to demonstrate concordance among SQT components.

4) SQT studies should be designed to maximize the number of stations, even if some reduction in the number of replicate samples is necessary.

5) Confounding factors are best accounted for by normalization in univariate tests and by simply including the confounding variables in a multivariate analysis.

6) Field sampling programs should follow good design principles and include synoptic sampling for triad components.

7) A suite of tests with different species and endpoints should be used to reduce uncertainty and help identify causality.

8) A uniform approach should be used in developing BIBIs that incorporate adjustments for the natural controlling factors and an optimized combination of benthic metrics based on region-specific data for differentiating between degraded and nondegraded sites.

Research Needs

1) More synoptic porewater chemistry and toxicity data are needed.

2) Dose–response relationships need to be known for confounding factors in existing porewater toxicity tests.

3) A better understanding of the relationships between concentrations of chemical mixtures in sediments and pore water and porewater toxicity is needed.

4) Region-specific BIBIs and indices of ecological health need to be developed.

5) Generalized multiparameter patterns in ecological structure and function, as opposed to species differences in different locations, relative to porewater toxicity need to be determined.

6) Development of chemical-specific porewater toxicity tests would be desirable.

7) Research on the toxicological mechanism at the cellular level for porewater test organisms is needed.

References

Alden II RW. 1992. Uncertainty and sediment quality assessments: I. Confidence limits for the Triad. *Environ Toxicol Chem* 11:637–644.

Anderson BS, Hunt JW, Phillips BM, Fairey RJ, Oakden JM, Puckett HM, Stephenson M, Tjeerdema RS, Long ER, Wilson CJ, Lyons M. 2001. Sediment quality in Los Angeles Harbor: A triad assessment. *Environ Toxicol Chem* 20:359–370.

Ankley GT, Schubauer-Berigan MK, Dierkes JR. 1991. Predicting the toxicity of bulk sediments to aquatic organisms with aqueous test fractions: Pore water vs. elutriate. *Environ Toxicol Chem* 10:1359–1366.

Arkes HE. 1981. Impediments to accurate clinical judgment and possible ways to minimize their impact. *J Consulting Clinic Psychol* 49:323–330.

Burgess RM, Ho KT, Morrison GE, Chapman G, Denton DL. 1996. Marine toxicity identification evaluation (TIE). Phase I guidance document. Washington DC, USA: U.S. Environmental Protection Agency. EPA/600/R-96/054. 70 p.

Canfield TJ, Dwyer FJ, Fairchild JF, Haverland PS, Ingersoll CG, Kemble NE, Mount DR, LaPoint TW, Burton GA, Swift MC. 1996. Assessing contamination in Great Lakes sediments using benthic invertebrate communities and the Sediment Quality Triad approach. *J Great Lakes Res* 22:565–583.

Carr RS. 1993. Sediment quality assessment survey of the Galveston Bay system. Galveston TX, USA: Galveston Bay National Estuary Program. GBNEP-30. 101 p.

Carr RS. 1998. Marine and estuarine porewater toxicity testing. In: Wells PG, Lee K, Blaise C, editors. Microscale testing in aquatic toxicology: Advances, techniques, and practice. Boca Raton FL, USA: CRC. p 523–538.

Carr RS, Chapman DC. 1992. Comparison of whole sediment and pore-water toxicity tests for assessing the quality of estuarine sediments. *Chem Ecol* 7:19–30.

Carr RS, Chapman DC. 1995. Comparison of methods for conducting marine and estuarine sediment porewater toxicity tests: Extraction, storage, and handling techniques. *Arch Environ Contam Toxicol* 28:69–77.

Carr RS, Chapman DC, Howard CL, Biedenbach JM. 1996. Sediment quality triad assessment survey of the Galveston Bay, Texas system. *Ecotoxicology* 5:341–364.

Carr RS, Chapman DC, Presley BJ, Biedenbach JM, Robertson L, Boothe P, Kilada R, Wade T, Montagna P. 1996. Sediment porewater toxicity assessment studies in the vicinity of offshore oil and gas production platforms in the Gulf of Mexico. *Can J Fish Aquat Sci* 53:2618–2628.

Carr RS, Long ER, Chapman DC, Thursby G, Biedenbach JM, Windom H, Sloane G, Wolfe DA. 1996. Toxicity assessment studies of contaminated sediments in Tampa Bay, Florida. *Environ Toxicol Chem* 15:1218–1231.

Carr RS, Montagna PA, Biedenbach JM, Kalke R, Kennicutt MC, Hooten R, Cripe G. 2000. Impact of storm-water outfalls on sediment quality in Corpus Christi Bay, Texas, USA. *Environ Toxicol Chem* 19:561–574.

Chapman PM. 1986. Sediment quality criteria from the Sediment Quality Triad: An example. *Environ Toxicol Chem* 5:957–964.

Chapman PM. 1990. The Sediment Quality Triad approach to determining pollution-induced degradation. *Sci Total Environ* 97/98:815–825.

Chapman PM. 1992a. Sediment quality triad approach. In: Sediment classification methods compendium. Washington DC, USA: U.S. Environmental Protection Agency. EPA-823-R-92-0006. p 10-1–10-18.

Chapman PM. 1992b. Pollution status of North Sea sediments: An international integrative study. *Mar Ecol Prog Ser* 91:253–264.

Chapman PM. 1996. Presentation and interpretation of Sediment Quality Triad data. *Ecotoxicology* 5:327–339.

Chapman PM, Anderson B, Carr S, Engle V, Green R, Hameedi J, Harmon M, Haverland P, Hyland J, Ingersoll C, Long E, Rodgers J Jr, Salazar M, Sibley PK, Smith PJ, Swartz RC, Thompson B, Windom H. 1997. General guidelines for using the Sediment Quality Triad. *Mar Pollut Bull* 34:368–372.

Chapman PM, Dexter RN, Long ER. 1987. Synoptic measures of sediment contamination, toxicity and infaunal community composition (the Sediment Quality Triad) in San Francisco Bay. *Mar Ecol Prog Ser* 37:75–96.

Chapman PM, Paine MD, Arthur AD, Taylor LA. 1996. A triad study of sediment quality associated with a major, relatively untreated marine sewage discharge. *Mar Pollut Bull* 32:47–64.

Chapman PM, Power EA, Dexter RN, Andersen HB. 1991. Evaluation of effects associated with an oil platform, using the Sediment Quality Triad. *Environ Toxicol Chem* 10:407–424.

Dawes, RM. 1979. The robust beauty of improper linear models in decision making. *Am Psychol* 34:571–582.

Di Giulio RT, Benson WH, Sanders BM, Van Veld PA. 1995. Biochemical mechanisms: Metabolism, adaptation, and toxicology. In: Rand GM, editor. Fundamentals of aquatic toxicology: Effects, environmental fate and risk assessment. Washington DC, USA: Taylor & Francis. p 523–561.

Eberhardt LL, Thomas JM. 1991. Designing environmental field studies. *Ecol Monogr* 61:53–73.

Fairey R, Bretz C, Lamerdin S, Hunt J, Anderson B, Tudor S, Wilson CJ, LeCaro F, Stephenson M, Puckett M, Long ER. 1998. Assessment of sediment toxicity and chemical concentrations in the San Diego Bay region, California, USA. *Environ Toxicol Chem* 17:1570–1581.

Faust D. 1986a. Research on human judgment and its application to clinical practice. *Prof Psychol Res Pract* 17:420–430.

Faust D. 1986b. Learning and maintaining rules for decreasing judgment accuracy. *J Personality Assess* 50:585–600.

Faust D. 1989. Data integration in legal evaluations: Can clinicians deliver on their premises? *Behavior Sci Law* 7:469–483.

Field LJ, MacDonald DD, Norton SB, Severn CG, Ingersoll CG. 1999. Evaluating sediment chemistry and toxicity data using logistic regression modeling. *Environ Toxicol Chem* 18:1311–1322.

Germano JD. 1999. Ecology, statistics, and the art of misdiagnosis: The need for a paradigm shift. *Environ Rev* 7:167–190.

Goldberg LR. 1968. Simple models or simple processes? Some research on clinical judgments. *Am Psychol* 23:483–496.

Green RH. 1979. Sampling design and statistical methods for environmental biologists. New York NY, USA: J Wiley.

Green RH. 1984. Some guidelines for the design of biological monitoring programs in the marine environment. In: White HH, editor. Concepts in marine pollution measurements. College Park MD, USA: Maryland Sea Grant College. p 647–655.

Green RH. 1993a. Relating two sets of variables in environmental studies. In: Rao CR, editor. Multivariate analysis: Future directions. Amsterdam, The Netherlands: Elsevier. p 151–165.

Green RH. 1993b. Relating two sets of variables in environmental studies. In: Patil GP, Rao CR, editors. Multivariate environmental statistics. New York NY, USA: Elsevier. p 140–163.

Green RH, Boyd JM, MacDonald JS. 1993. Relating sets of variables in environmental studies: The Sediment Quality Triad as a paradigm. *Envirometrics* 4:439–457.

Green RH, Montagna P. 1996. Implications for monitoring: Study designs and interpretation of results. *Can J Fish Aquat Sci* 53:2629–2636.

Hyland JL, Balthis WL, Hackney CT, Posey M. 2000. Sediment quality of North Carolina estuaries: An integrative assessment of sediment contamination, toxicity, and condition of benthic fauna. *J Aquat Ecosyst Stress Recovery* 8: 107–124.

Hyland JL, Costa H. 1995. Examining linkages between contaminant inputs and their impacts on living marine resources of the Massachusetts Bay ecosystem through application of the Sediment Quality Triad method. Cambridge MA, USA: A. D. Little, Inc. Report MBP-95-03 to the Massachusetts Bay Program.

Hyland JL, Van Dolah RF, Snoots TR. 1999. Predicting stress in benthic communities of Southeastern U.S. estuaries in relation to chemical contamination of sediments. *Environ Toxicol Chem* 18:2557–2564.

Ingersoll CG, Dillon T, Biddinger GR. 1997. Ecological risk assessment of contaminated sediments. Pensacola FL, USA: Society of Environmental Toxicology and Chemistry (SETAC). 390 p.

Jones-Lee A, Lee GF. 1993. Potential significance of ammonia as a toxicant in aquatic sediments. First International Specialized Conference on Contaminated Sediments: Historical Records, Environmental Impact, and Remediation, Proc. London, UK: International Assoc Water Quality. p 223–232.

Kennicutt II MC, Booth PN, Wade TL, Sweet ST, Rezak R, Kelly FJ, Brooks JM, Presley BJ, Wiesenberg DA. 1996. Geochemical patterns in sediments near offshore production platforms. *Can J Fish Aquat Sci* 53:2554-2566.

Kennicutt II MC, Green R, Montagna P, Roscigno P. 1996. Gulf of Mexico Offshore Operations Monitoring Experiment (GOOMEX), Phase I: Sublethal responses to contaminant exposure: Introduction and overview. *Can J Fish Aquat Sci* 53:2540–2553.

Livingstone DR. 1993. Biotechnology and pollution monitoring: Use of molecular biomarkers in the aquatic environment. *J Chem Technol Biotech* 57:195–211.

Long ER. 1989. Use of the Sediment Quality Triad in classification of sediment contamination. In: Contaminated marine sediments: Assessment and remediation. Washington DC, USA: National Research Council, Marine Board. p 78–99.

Long ER. 1998. Sediment quality assessments: Selected issues and results from the National Oceanic and Atmospheric Administration's National Status and Trends Program. Chapter 6, Ecological risk assessment: A meeting of policy and science. Pensacola FL, USA: Society of Environmental Toxicology and Chemistry (SETAC). p 111–132.

Long ER, Buchman MF, Bay SM, Breteler RJ, Carr RS, Chapman PM, Hose JE, Lissner AL, Scott J, Wolfe DA. 1990. Comparative evaluation of five toxicity tests with sediments from San Francisco Bay and Tomales Bay, California. *Environ Toxicol Chem* 9:1193–1214.

Long ER, Carr RS, Thursby GA, Wolfe DA. 1995. Sediment toxicity in Tampa Bay: Incidence, severity, and spatial extent. *Fla Sci* 58:163–178.

Long ER, Chapman PM. 1985. A Sediment Quality Triad: Measures of sediment contamination, toxicity and infaunal community composition in Puget Sound. *Mar Pollut Bull* 16:405–415.

Long ER, Dutch M, Aasen S, Welch K, Ricci C, Hameedi MJ. 2003. Chemical contamination, acute toxicity in laboratory test, and benthic impacts in sediments of Puget Sound: A summary of results of the joint, 1997–99 Ecology/NOAA survey. Olympia WA, USA: Washington State Dept of Ecology Technical Report. (in preparation)

Long ER, Field LJ, MacDonald DD. 1998. Predicting toxicity in marine sediments with numerical sediment quality guidelines. *Environ Toxicol Chem* 17:714–727.

Long ER, Hameedi J, Robertson A, Dutch M, Aasen S, Ricci C, Welch K, Kammin W, Carr RS, Johnson T, Biedenbach J, Scott KJ, Mueller C, Anderson JW. 1999. Sediment quality in Puget Sound: Year 1 - Northern Puget Sound. Olympia WA, USA: Washington State Dept. of Ecology. NOAA NOS NCCOS CCMA, Technical Memo Nr 139, Washington State Dept. of Ecology Pub. No. 99-347. 221 p, 8 appendices.

Long ER, Hameedi J, Robertson A, Dutch M, Aasen S, Ricci C, Welch K, Kammin W, Carr RS, Johnson T, Biedenbach J, Scott KJ, Mueller C, Anderson JW. 2000. Sediment quality in Puget Sound: Year 2 - Central Puget Sound. Olympia WA, USA: Washington State Dept. of Ecology. NOAA NOS

NCCOS CCMA Technical Memo Nr 147. Washington State Dept of Ecology (WDOE) Pub No. 00-03-055.

Long ER, MacDonald DD. 1998. Recommended uses of empirically-derived sediment quality guidelines for marine and estuarine ecosystems. *Human Ecol Risk Assess* 4:1019–1039.

Long ER, MacDonald DD, Smith SL, Calder FD. 1995. Incidence of adverse biological effects within ranges of chemical concentrations in marine and estuarine sediments. *Environ Manage* 19:81–97.

Long ER, Robertson A, Wolfe DA, Hameedi J, Sloane GM. 1996. Estimates of the spatial extent of sediment toxicity in major U.S. estuaries. *Environ Sci Technol* 30:3585–3592.

Long ER, Sloane GM, Carr RS, Johnson T, Biedenbach J, Scott KJ, Thursby GB, Crecelius E, Peven C, Windom HL, Smith RD, Loganathon R. 1997. Magnitude and extent of sediment toxicity in four bays of the Florida panhandle: Pensacola, Choctawhatchee, St. Andrew and Apalachicola. Silver Spring MD, USA: National Oceanic and Atmospheric Administration, Coastal Monitoring Bioeffects Assessment Division. NOAA Technical Memorandum NOS ORCA 117.

Long ER, Sloane GM, Carr RS, Scott KJ, Thursby GB, Wade T. 1995. Sediment toxicity in Boston Harbor: Magnitude, extent, and relationships with chemical toxicants. Silver Spring MD, USA: National Oceanic and Atmospheric Administration, Coastal Monitoring and Bioeffects Assessment Division. NOAA Technical Memorandum NOS ORCA. 85 p, 31 figures, 4 appendices.

Long ER, Sloane GM, Scott GI, Thompson B, Carr RS, Biedenbach J, Wade TL, Presley RJ, Scott KJ, Mueller C, Brecken-Fols G, Albrecht B, Anderson JW, Chandler GT. 1999. Magnitude and extent of chemical contamination and toxicity in sediments of Biscayne Bay and vicinity. Silver Spring MD, USA: National Oceanic and Atmospheric Administration. NOAA Technical Memorandum NOS NCCOS CCMA 141.

Long ER, Wilson CJ. 1997. On the identification of toxic hot spots using measures of the Sediment Quality Triad. *Mar Pollut Bull* 34:373–374.

Long ER, Wolfe DA, Scott KJ, Thursby GB, Stern EA, Peven C, Schwartz T. 1995. Magnitude and extent of sediment toxicity in the Hudson-Raritan estuary. Silver Spring MD, USA: National Oceanic and Atmospheric Administration. NOAA Tech Memo NOS ORCA 88. 230 p.

MacDonald DD, Carr RS, Calder FD, Long ER, Ingersoll CG. 1996. Development and evaluation of sediment quality guidelines for Florida coastal waters. *Ecotoxicology* 5:253–278.

Rose KA, Smith EP. 1992. Experimental design: The neglected aspect of environmental monitoring. *Environ Manag* 16:691–700.

Spies RB. 1989. Sediment bioassays, chemical contaminants and benthic ecology: New insights or just muddy water? *Mar Environ Res* 27:73–75.

Stegeman JJ, Brower M, Di Giulio RT, Forlin L, Fowler B, Sanders B, Van Veld PA 1992. Molecular responses to environmental contamination: Proteins and enzymes as indicators of contaminant exposure and effects. In: Huggett RJ, Kimerle RA, Merle PM, Bergman HL, editors. Biomarkers: Biochemical, physiological and histological markers of anthropogenic stress. Boca Raton FL, USA: Lewis. p 235–335.

Swartz RC, Cole FA, Lamberson JO, Ferraro SP, Schults DW, DeBen WA, Lee II H, Ozretich RJ. 1994. Sediment toxicity, contamination and amphipod abundance at a DDT- and dieldrin-contaminated site in San Francisco Bay. *Environ Toxicol Chem* 13:949–962.

Swartz RC, Cole FA, Schults DW, DeBen WA. 1986. Ecological changes on the Palos Verdes Shelf near a large sewage outfall: 1980–1983. *Mar Ecol Prog Ser* 31:1–13.

Turgeon DD, Hameedi J, Harmon MR, Long ER, McMahon KD, White HH. 1998. Sediment toxicity in U.S. coastal waters. Silver Spring MD, USA: National Oceanic and Atmospheric Administration. NOAA Special Report.

CHAPTER 10

Experiences with Porewater Toxicity Tests from a Regulatory Perspective

Linda Porebski

T he use of porewater toxicity tests in a regulatory context is not common. This chapter will look at the current status of porewater toxicity testing for regulatory use and share the experience of Environment Canada's Disposal at Sea Program as it moves towards regulatory use of toxicity tests in 2001. The program has developed a tiered testing approach to evaluating marine sediments for ocean disposal permits. Chemical-specific screening limits are evaluated and, if triggered, will necessitate additional evaluation using a battery of marine toxicity tests. The proposed toxicity test battery includes an amphipod survival and a bivalve bioaccumulation test in whole sediment, a Microtox solid phase, and an echinoid fertilization test with sediment pore water (Porebski and Osborne 1998).

Background

About 15 years ago, the use of sediment toxicity tests as indicators or predictors of potential effects began to find favor. It was theorized that whole organisms could be used to actually test bioavailability of contaminants, both measured and unmeasured, and produce an integrated and reproducible response, representative of what similar organisms might be experiencing in the field. The regulatory focus was on whole sediment tests (Sergy 1987; Puget Sound Dredged Disposal Analysis [PSDDA] 1989; U.S. Environmental Protection Agency [USEPA] 1991). There was, however, an interest in adapting for sediment use some of the existing aqueous-phase tests, many of which had the advantage of being smaller, faster, more sensitive, and less expensive than their whole sediment counterparts. The aqueous-phase tests also tended to cover a broader range of endpoints, including reproduction, fertilization, growth and development, and genotoxicity. In many dredging arenas, the choice was to do these aqueous-phase tests using elutriates in order to mimic what the pelagic organisms would potentially be exposed to during disposal (USEPA 1991). Sediment extracts were another phase for testing and were seen as a way to mimic organism digestion, or concentrate low levels of contaminants to a worst case scenario (Centre for Environment, Fisheries and Aquaculture Science [CEFAS] 1998). Interstitial or pore water was also considered for use in sediment toxicity assessments because this matrix was believed to control the bioavailability of sediment-associated contaminants to interstitial-dwelling organisms (Adams et al. 1985; Di Toro 1991).

Status of Regulatory Uses of Porewater Toxicity Tests

Evaluations of regulatory potential for porewater testing have been made by various regulatory agencies. In Canada, an Aquatic Effects Technology Evaluation (AETE) program was established to facilitate cooperative industry and government review of appropriate technologies for assessing the impacts of mine effluents on the aquatic environment (ESG International 1999). A technical evaluation of porewater toxicity and chemistry concluded that porewater toxicity testing in the laboratory was reasonable if samples were collected and processed properly and the test exposures were realistic. A field demonstration project was recommended (Burton 1998). The AETE synthesis report, however, did not recommend the use of porewater tests for routine monitoring of mining but suggested porewater chemistry and toxicity may be useful for more detailed investigation on a site-specific basis (ESG International 1999).

The Netherlands National Institute for Coastal and Marine Management was also looking for a battery of regulatory tests to evaluate sediments for disposal at sea. One of the test methods evaluated was a rotifer (*Brachionus plicatilis*) toxicity test done in pore water. A standard method was available, and the test was "repeatable" (meaning the method could be executed successfully) but showed high variance in interlaboratory tests and had a low score in meeting a set of acceptability criteria. Solid-phase amphipod and Microtox tests were selected over the porewater rotifer test for regulatory use (Shipper and Stronkhorst 1999).

South Africa is also in the process of developing a procedure for regulating the disposal of dredged sediments. The first step in this procedure is a chemical analysis, which at present is limited to a range of heavy metals. The second step, which is necessary only for those sediments that "fail" the chemical screening, involves the use of bioassays. Following a technology transfer session with Canada, a battery of 3 tests is currently being considered. These include a Microtox test, an amphipod survival test, and a sea urchin fertilization test using pore water extracted from the sediments of concern.

The sea urchin fertilization test with the South African urchin *Tripneustes gratilla* has been routinely used for effluent testing over many years, and the procedure is well documented. However, it has only recently been used on a trial basis for sediment assessments based on pore water (extracted using the Canadian protocol). In these cases, the tests have generally shown only minimal inhibition of fertilization, even on sediments from areas that appear to be quite severely contaminated on the basis of chemical analyses. On the other hand, in some cases, toxicity (>25% inhibition of fertilization) was registered in sediments that did not fail the metal concentration criteria, indicating that substances other than heavy metals were influencing the result (personal communication, May 2001, Colin Archibald, CSIR, South Africa). As a result, the South African authorities, in a submission to the 24th Meeting of the Scientific Group of Contracting Parties to the London Convention (South Africa 2001), expressed their concern over the reliability of using porewater extractions for sediment assessments. They also indicated that it was their intention to expand the list of contaminants for which chemical analyses would be required.

The USEPA, the U.S. Army Corps of Engineers (USACE), and the American Society for Testing and Materials (ASTM) do not appear to have produced standard methods for toxicity testing with pore water, although elutriate or extract methods exist with many of the same species and endpoints that are used for porewater testing (USEPA 1991; ASTM 1999). There is a porewater testing method described in *Standard Methods* (Carr 1998b). The U.S. Geological Survey (USGS) has conducted a great deal of porewater testing in a monitoring or assessment mode (Carr 1998b). No programs currently use, or plan to use, porewater toxicity tests in a formal regulatory context, although interpretation options have been developed (Carr and Biedenbach 1999) and methods standardized for echinoid fertilization and development tests. Studies have also been conducted with porewater-dwelling species (Carr et al. 1989; Carr, Long et al. 1996).

In the United Kingdom, the Joint Assessment and Monitoring Program (JAMP) under the Oslo and Paris Commission (OSPAR) included porewater toxicity testing in their monitoring guidelines. This program advocates the use of the oyster embryo development test but also allows for the use of other species such as small polychaetes and copepods that inhabit pore water. This program assesses the extent of spatial effects and the organizational levels of these effects on a case-by-case basis. There is no formal regulatory use of these tests linked to permitting or cleanup action (OSPAR 1998). New Zealand evaluated porewater toxicity tests with an indigenous sand dollar species, *Fellaster zelandiae* (Nipper et al. 1997, 1998).

Other tests and endpoints in pore water have also been developed and have potential for regulatory use. Carr's (1998a) list includes bacteria, algae, polychaetes, mollusks, crustaceans, echinoderms, and teleosts. Most require better standardization and field validation. In Chapter 12, the workgroup assembled a list of potential species, with their known state of development. Future regulatory assessments may focus on filling the data gaps for their use.

Disposal at Sea Program in Canada

In Canada, disposal at sea is regulated under the Canadian Environmental Protection Act of 1999 (CEPA 1999), which passed 31 March 2000, replacing the previous

CEPA. One of its new requirements is that before any open water disposal permit is granted, the material is evaluated according to a waste assessment framework (CEPA 1999). One of the steps within this assessment process is the characterization of the waste's physical, chemical, and biological properties. Environment Canada is now finalizing a tiered approach for this waste characterization process and must have new regulations in place by 2001. Among the requirements of the act is an obliga- tion, following new international principles, to establish an upper action level above which disposal at sea cannot be permitted (Porebski and Osborne 1998).

The proposed assessment process has Tier 1 screening levels or lower action levels, which will be chemical numbers at or below which no adverse biological effects are expected. Levels above these criteria would trigger Tier 2 investigations of sediment quality, in the form of toxicity and possibly bioaccumulation testing. The toxicity test battery selected includes whole sediment tests (amphipod survival, Microtox, polychaete growth, bivalve bioaccumulation) and a porewater test (echinoid fertilization) (Environment Canada 1992a, 1992b, 1998; USEPA 1993). If the sediments or waste materials pass the biological tests, open water disposal can be considered. Failure in more than one of the biological tests (or of the acute test alone) disqualifies the material for open water disposal. Interpretive criteria for the toxicity tests have been proposed for use but require field validation (Porebski and Osborne 1998).

Canada's Decision to Use Porewater Tests

A battery of tests was chosen for regulatory development about 10 years ago. To be suitable for regulatory use, each test in the battery had to demonstrate these characteristics:

- Its species and endpoint are relevant.
- The species is easily accessed and maintained in the laboratory.
- The expected routes of exposure are known.
- It has consistent responses to contaminants of concern.
- Confounding factors are understood and accounted for.
- It can be easily handled in the laboratory.
- Its variability is within reasonable bounds (inter- and intralaboratory).
- The method available is standardized, clear, and repeatable.
- Interpretation is available.
- Cost and time to conduct the test are not prohibitive for commercial use.

For assessing sediments, whole sediment tests were favored and make up most of the test battery. However, there was already a set of inexpensive water-based tests that could be adapted and would provide additional assessment of another component of the environment (i.e., pore water). Adaptation of water-based tests was also considered if these tests addressed sublethal endpoints (e.g., growth, reproduction, development).

Once Environment Canada had accepted the use of water-based tests in the waste assessment framework, a decision was necessary on the appropriate aqueous phase for sediment testing. The general biological test protocol offered the choice of testing liquids from extracts (using solvents), elutriates (sediment slurries), or pore water (interstitial water).

In the U.S. and the UK, elutriates are used for toxicity testing of dredged material because they mimic the potential exposure during open water disposal of dredged sediment. As noted in Environment Canada (1994), elutriates do not measure the toxicity of pore waters or bedded sediments. There may also be a problem with particulates in the elutriate affecting the test results and confounding their interpretation (CEFAS 1998).

Extracts could be used to enable a concentration of contaminants in an area of low contamination to simulate a worst-case scenario (CEFAS 1998). Potential for solvent-related effects and the unclear relationship between the toxicity of the extract and the actual toxicity of the original sample made this approach less desirable for disposal at sea assessments.

Pore water appeared to be the most suitable aqueous phase for sediment assessment. Water, especially pore water, is thought to be the primary route of uptake for most benthic invertebrates and most lipophilic chemicals (Adams 1987). The equilibrium partitioning theory suggests that concentrations of most substances in the pore water approach equilibrium with the solid phase (USEPA 1999); consequently, pore water has been collected for toxicity testing to approximate the relative toxicity of contaminated sediments and/or to assess contaminant levels (Environment Canada 1994).

In general, sublethal porewater tests are reported as being more sensitive, by an order of magnitude or more, than whole sediment tests (lethal and sublethal) (Carr 1998b). Many studies show a good relationship between porewater toxicity test responses and contaminant levels (Carr, Chapman, Howard 1996; Carr, Chapman, Presley et al. 1996; Carr, Long et al. 1996; Carr et al. 2000). The next sections deal with the limited experience that Canada has had with the tests it has selected and is considering for regulatory use in assessing sediments for disposal at sea.

Environment Canada's echinoid fertilization general test method document (Environment Canada 1992a) includes the following species:

- *Strongylocentrotus droebachiensis*, green sea urchin
- *S. purpuratus*, purple sea urchin*

- *Dendraster excentricus*, eccentric sand dollar*
- *Arbacia punctulata*, Arbacia
- *Lytechinus pictus*, white sea urchin.*

Those with an asterisk are commonly used for porewater testing, largely due to availability of these species.

The bulk of Canadian experience to date has been in monitoring and research programs, with 1 or 2 studies done each year in disposal at sea site monitoring and several more done in a risk assessment context. On a policy basis, those wishing to dispose of sediments at sea may choose to do the echinoid fertilization test in pore water (as part of a larger toxicity test battery) and can be granted an ocean disposal permit if they are judged to have passed all tests. To date, however, most applicants, once they exceed the chemical-specific lower action levels, choose to forego disposal at sea and seek land-based disposal, rather than do the toxicity tests. Thus, data from these permit applications has been limited.

The Big Issues

Variability

For regulatory purposes, particularly if a test is to be conducted by different laboratories, including contract laboratories, repeatability and inter- and intralaboratory variability must be known and within acceptable levels (high repeatability, low variability). A literature review done in 1992 for the general method document suggested echinoid interlaboratory variability averaged about 63% for EC50 (effective concentration to 50% of the test population) values based on round-robin testing, which was similar to the precision found in an interlaboratory comparison of chemical analyses (i.e., 60%) (Environment Canada 1992a). Within-laboratory precision also appeared to be acceptable. USEPA (1988) listed coefficients of variation for *A. punctulata* from 30% to 48%. Environment Canada's within-laboratory variability (control chart results for reference toxicants) are presented in Table 10-1 for the commonly used test species.

Variability in some studies appeared to be greater in the assessment of more contaminated sediments (Zajdlik et al. 2000). For example, Figure 10-1(A) shows plotted raw data from a study conducted in an organic pollution gradient using sediments from Sydney Harbour, Nova Scotia, Canada. The variability (distance between maximum and minimum responses) between stations for amphipod (*Amhiporeia virginiana*) survival remains fairly even as sediment contamination decreases from station 1 to 9 and in the control and 2 reference stations (12 and St Ann's). Table 10-2 shows mean contaminant levels. Figure 10-1(B) shows echinoid fertilization where there is higher variability of response in the more contaminated

Table 10-1 Within-laboratory variability of bacteria and echinoid
species tested with a reference toxicant[a]

Species	Reference toxicant	Coefficient of variation (%)
Vibrio fisheri	Zn	27
L. pictus	Cu	45
D. excentricus	Cu (at 10 °C)	28
D. excentricus	Cu (at 15 °C)	10 and 41
S. purpuratus	Cu	16

[a] Sources: Environment Canada control charts; Pacific Environmental Science
Centre, Vancouver BC, Canada; Environmental Toxicology Lab, Moncton NB,
Canada.

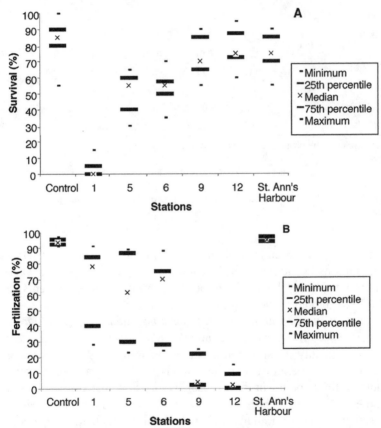

Figure 10-1 A) Study of Sydney Harbour sediments. Amphipod (*Amphiporeia virginiana*)
survival results when all replicate data (field and laboratory replicates at a station) are plotted. Polycyclic aromatic hydrocarbon (PAH) and polychlorinated biphenyl (PCB) contamination decreases from stations 1 to 9. Stations 12 and St Ann's are reference stations. Here,
survival increases as contamination decreases, and variability is fairly even between stations.
B) Echinoid (*D. excentricus*) fertilization success results when all replicate data are plotted.
Stations are as in 10-1A. Fertilization decreases as contamination decreases, and variability is
markedly higher in the more contaminated stations (1, 5, and 6).

Table 10-2 Summary of selected mean chemical levels at stations in Sydney Harbour [a]

Variables	Stations					
	1	5	6	9	12	St Ann
Main sediment variables[b,c,d]						
PAH	**212**	**86**	**36**	3.1	*0.5*	*0.4*
PCB	**2.1**	**1.2**	**0.7**	*>0.07*	*>0.06*	*>0.03*
Hg	**0.7**	0.5	0.3	*0.04*	*0.02*	*0.06*
Pb	**286**	**214**	**133**	32	*21*	37
Zn	**516**	**865**	**281**	*91*	*56*	*84*
NH$_3$–N	43.1	36.8	38.1	32.6	13.6	45.6
Main porewater variables						
PAH[e]	21.1	11.1	6.1	1.5	<0.3	<0.7
PCB[e]	0.76	0.97	0.46	<0.48	<0.51	<0.44
Hg[f]	0.12	0.60	na	<0.05	0.06	<0.05
Zn[f]	<0.2	<0.2	na	<0.2	<0.2	2.1
Cr[f]	0.7	<0.6	na	0.6	<0.6	1.3
NH$_4$ (total)[f]	11.2	10.2	11.3	35.6	33.0	9.5

[a] Source: Reprinted with permission from Zajdlik et al. 2000.
[b] In mg/g.
[c] Bold numbers exceed Canadian Council of Ministers of the Environment (CCME) Probable Effects Levels (PELs).
[d] Italicized numbers do not exceed CCME Threshold Effects Levels (TELs) for marine sediments.
[e] In µg/L.
[f] In mg/L.

sediments (stations 1, 5, and 6) and lower variability in the control or reference samples (stations 12 and St. Ann's). To a lesser extent, the same pattern was seen in other tests in this study that were run concurrently on the same sediments (data not shown). Two echinoid species (*D. excentricus* and *L. pictus*), 3 amphipod species (*Rhepoxinius abronius, Eohaustorius washingtonianus,* and *Eohaustorius estuarius*), and 2 polychaete species (*Polydora cornuta* and *Bocardia proboscidea)* all showed some increase in variability in response at the more contaminated stations (stations 1, 5, and 6). This may be because contaminant levels in this study are approaching a threshold for these species, or that exposure is somehow "patchy," so that some are being stressed to the point of response, while others are not. From this perspective, an increase in response variability could be seen (and potentially used) as another indicator of increased stress on the population. There is concern, however, when the variability is so large as to obscure differences. For example, in this study, within-sample variability (replicates at 1 station) for *D. excentricus* was greater than the variability among samples (replicates compared from different stations) at the 3 more contaminated stations (Zajdlik et al. 2000). Given the heterogeneity of sediments, this may be a result of physical or chemical differences in field replicates rather than a lack of consistency within the test, but alternately, it implies that in using this test, the differences between stations or between exposure and control sites may be obscured by the variability within a site.

Tolerance levels to noncontaminants

An area where there is an obvious need for further study, for both porewater and whole sediment tests, is in the understanding of how various factors in the environment, which may not be regarded as contaminants, affect the test organisms. A commonly recommended practice in toxicity testing is to compare results from test samples to those from a reference sample. The reference sediment is defined in the Canadian test method documents as being "a field collected sample of presumably clean (uncontaminated) sediment, selected for properties (e.g. particle size, compactness, total organic content) representing sediment conditions that closely match those of the sample(s) of test sediment except for the degree of chemical contamination" (Environment Canada 1998). Finding these sites, however, and matching all the noncontaminant factors such as ammonia, sulfide, pH, and Eh, has proven to be very difficult, in our experience. A regulator, who needs to make timely and consistent decisions, cannot perform an endless search for a perfect reference site. Standard reference sites could be chosen and extensively characterized as is done in some jurisdictions (PSDDA 2000). The other route is to gain an understanding of what factors, in addition to the contaminants of concern, are likely to affect the toxicity test results. One way to address this issue, at least partially, is to know species-specific tolerance levels to the major confounding factors (noncontaminants) (Environment Canada 1998). This facilitates selection of an appropriate test species and improves the interpretation.

As seen in Chapters 5 and 7, ammonia is a very common confounding factor that may correlate to responses in porewater toxicity tests. Information on the species-specific tolerance to ammonia of the echinoid species used for disposal at sea is therefore invaluable for test interpretation (Table 10-3). These figures, however, are based on a small dataset from a single set of experiments. Clearly, a more comprehensive dataset is needed to increase the confidence in these numbers. Ideally, this dataset should be generated using porewater exposures, as has been done for *A. punctulata* (Figure 10-2) (Carr and Biedenbach 1996). There have been some examples of *A. punctulata* tolerances to ammonia in pore water being higher than the seawater-only exposures (Carr and Biedenbach 1996). It should be determined whether this finding is also valid for the species currently used in Canada.

Sulfide is another common confounding factor. Environment Canada studies did not find any strong correlations between sulfide and fertilization in *L. pictus* in 2 sets of toxicity tests done to assess ammonia and sulfide effects. This likely speaks to the volatile nature of sulfides, rather than a lack of toxicity (Tay et al. 1998). The workshop participants compiled a list of other potential confounding factors (Eh, pH, etc., see Chapter 12). Given the obvious implications of interpreting the tests

Table 10-3 Ammonia tolerance levels in water for 3 echinoid fertilization tests

Parameter	Species					
	L. pictus[a]		*S. purpuratus*[b]		*D. excentricus*[b]	
	NOEC (pH 7.7)	IC25[c] (pH 7.6)	NOEC (pH 8.0)	IC25 (pH 8.0)	NOEC (pH 8.0)	IC25 (pH 8.0)
Total NH_4 (mg/L)	33.6	98.4	10.0	>100	10.0	16.5
NH_3 (mg/L) (undissociated/ unionized)	0.66	1.53	0.17	>1.69	0.21	0.34

[a] Tay et al. 1998, based on 20-min fertilization test.
[b] Bailey et al. 1995, based on 40-min fertilization test.
[c] IC = inhibition concentration.

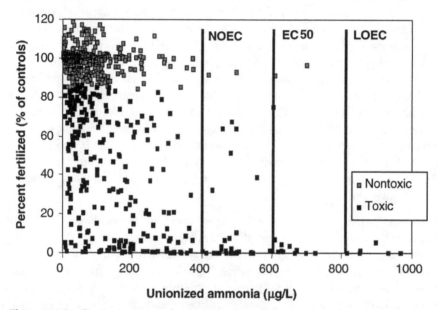

Figure 10-2 Fertilization in *A. punctulata* (100% pore water) versus unionized ammonia concentration (*n* = 865) (from Carr and Biedenbach 1996; reprinted with permission).

without understanding species tolerances and sensitivities to these factors, the systematic determination of tolerance levels for major noncontaminants, or confounding factors, is necessary in order to develop formal regulatory reference methods.

It should be noted that factors such as ammonia are not always regarded as confounding factors or noncontaminants. In fact, one instance in which porewater tests have provided clear results on toxicity is in the monitoring of salmon net pens on the Canadian east coast. In a study to determine whether results of chemical analyses and toxicity tests of sediments reflected the New Brunswick Department of Fisheries and Aquaculture's ratings of its aquaculture sites for low to high impact, *L. pictus* fertilization was judged to be (along with Microtox solid phase) the most sensitive toxicity indicator. In this case, the fertilization levels were inversely related to the ammonia levels, which decreased with increased distance from the pens (Burridge et al. 1999).

Relationship to contaminants

There is a body of literature that demonstrates a relationship between porewater toxicity and contaminant concentrations (some examples include Long, Carr et al. 1995; Carr, Long et al. 1996; Carr, Chapman, Howard 1996; Carr, Chapman, Presley et al. 1996; Long et al. 1999; Carr et al. 2000). This research focuses mainly on *A. punctulata* tests, with either a fertilization or embryological development endpoint. Environment Canada has produced a smaller body of fertilization data on *L. pictus*, *D. excentricus*, and *S. purpuratus* (Lee et al. 1995; Tay et al. 1997, 1998; Burridge et al. 1999; Hamel et al. 1999; Sullivan 1999). In much of this work, the relationship between echinoid response and contaminants has been less clear. Many of these studies, however, deal with sediments that have relatively low levels of contamination, so that any observed toxicity may stem from noncontaminants or unmeasured factors.

In addition to protocol development work, which looked at such things as general sensitivity to contaminants, optimum exposure time, and sperm/egg ratios (Beak Consultants Ltd. 1992; Environment Canada 1992a), the program has conducted 2 pollution gradient studies with the complete battery of biological tests plus sediment and porewater chemistry and benthic community analysis. The object was to look at the performance of tests and the proposed interpretation.

Gradient studies

The first study was done over a metals gradient in Belledune Harbour, New Brunswick, with smelter outfall as the main source of cadmium (0.4 to 10 mg/kg), lead (26 to 749 mg/kg), zinc (132 to 1240 mg/kg) and arsenic (11 to 35 mg/kg). The bulk sediment chemical levels for these metals ranged from greater than probable effects levels (PELs) to less than threshold effects levels (TELs) (Canadian Council of Ministers of the Environment [CCME] 1999). Porewater metals were generally below method detection levels, but because of small sample volumes, these levels may not have been low enough to be within the normal range of sea water for more than a few elements (Porebski et al. 1999). The ongoing difficulty of getting enough sample to perform adequate chemical analysis will need to be addressed. Calculation of acid volatile sulfide/simultaneously extractable metals (AVS/SEM) ratios (Allen et al. 1993) suggested that only the most contaminated station should show responses based on divalent metals. Amphipod and Microtox tests showed no responses, but lead bioaccumulated in *Macoma balthica* tissue, increasing as sediment levels increased (r^2 = 0.86), so that the tissue levels exceeded Health Canada tolerance levels for lead (of 0.5 μg/g wet weight), where the sediment lead levels exceeded the PEL for bulk sediment. Benthic community structure suggested that the more contaminated stations were more impacted than the chosen reference station and that the correlation was likely not due to differences in physical factors at the site, including ammonia. Several of the porewater toxicity tests suggested toxicity, but with all 3 species (*L. pictus, D. excentricus,* and *S. purpuratus*), the fertilization decreased at the midpoints of the gradient but was greater at the reference site and at the most contaminated sites. Only 1 of the 3 species (*L. pictus*), however, would have "passed" at the more contaminated sites using the current interpretation criteria, which specify that a sediment fails where there is at least a 25% decrease in fertilization between control and test pore waters and that difference is statistically significant (Porebski et al. 1998). A better understanding is needed of what factors or additional contaminants are depressing fertilization at certain stations.

The second study was done over a gradient of primarily polycyclic aromatic hydrocarbons (PAHs) and polychlorinated biphenyls (PCBs) in Sydney Harbour, Nova Scotia. Mercury, lead, and zinc were also present along the gradient. The porewater chemistry indicated a gradient for PAHs and PCBs, while bulk metals were generally below detection limits (Table 10-2). SEM showed a gradient for zinc and cadmium concentrations. Here, amphipod mortality, Microtox light reduction, and PAH bioaccumulation in clams all increased with increasing chemical levels in the sediment. The benthic community structure also showed lower abundance and richness at the more contaminated stations. The porewater fertilization tests with *D. excentricus* produced less clear results. The tests failed according to interim interpretation criteria at all stations, except for 1 reference and the negative control stations. This might have been expected, given that even the reference stations had some contaminants over TELs. Statistical analysis, however, found no correlation between toxicity and any measured contaminants but rather found a relationship with ammonia, sulfide, pH, and moisture (Zajdlik et al. 2000). Plotting the data (Figure 10-1B) also showed an almost inverse relationship with the contaminant gradient. The *D. excentricus* assay came very close to passing in the most contaminated station. Indeed, 2 of the 3 field replicates would have passed. The *L. pictus* data showed a similar shape and similar lack of correlation to contaminants. If permit applicants had chosen this species, they could have technically "passed" this assay at the most contaminated station. The correlations to the noncontaminants are of less concern here than is the apparent lack of sensitivity to contaminants. PAHs, PCBs, and metals that were measured both in the sediment and pore water did correlate to responses in solid-phase assays and bioaccumulation tests. Benthic community analysis also suggested the more contaminated stations were impacted. High levels of dissolved organic carbon (DOC) or other ligands may be binding contaminants in pore water, thereby reducing their bioavailability (Wiener and Giesy 1979; Caron 1989; Playle et al. 1993; Ingersoll 1995; Chapter 7). Long, Sloane et al. (1995) noticed a similar response in contaminated Boston Harbor sediments using the fertilization test with *A. punctulata*. In this case, however, a second porewater test with *A. punctulata*, using embryological development as the endpoint, was extremely sensitive to these samples, with 89% exhibiting significant toxicity at the 50% porewater dilution. It remains unclear, therefore, what factors or mechanisms are responsible for the different responses. Possibilities include methodological factors, route or extent of exposure changing depending on endpoint, or actual differences in sensitivity of that stage of development to particular contaminants.

It is recognized that the dataset here is small, and the porewater tests are generally found to be more, rather than less, sensitive to contaminants (Carr, Long et al. 1996; Carr, Chapman, Howard 1996; Carr et al. 2000), but given the implications, further research on this point will be essential to better define the performance capabilities of the test and the conditions under which its regulatory use would be suitable.

Reference sites

Another issue that continues to be of importance in the regulatory use of toxicity tests, in general, is the selection of a negative control or natural reference site against which to compare the toxicity response. In the Sydney Harbour study, for instance, 2 sites that were free (below screening levels or TELS) of PAHs and PCBs were chosen as "clean reference sites" (N.B.: full chemistry later revealed one or more other contaminants over TEL). The first was in Sydney Harbour but appeared to be more coarse grained and less organically enriched. The second, taken from nearby St. Ann's Harbour, was geographically removed from the test sites but more closely matched grain size, total organic carbon (TOC), ammonia, etc. Neither site was a perfect geochemical match for the test sites (the same but "substantially free of contaminants"), but both could be considered suitable reference sites. Filtered seawater was also used as a negative control and served as a measure of organism health. In the past, Environment Canada had great difficulty in locating suitable reference sites when using porewater tests for monitoring studies, or when proponents chose to use them for permit applications. It was for this practical reason that Canada's interim interpretation criteria for ocean disposal require a comparison of the test results to the seawater control rather than to a reference pore water. Proponents and peer reviewers have argued that a reference site would be a more meaningful or relevant comparison.

The Sydney Harbour study gives a graphic example of the consequences of reference site selection. In the porewater tests, in particular, there was a very marked difference in the way each species responded to the different reference sites, so that *L. pictus* showed high fertilization in the control and Sydney Harbour references while *D. excentricus* showed high fertilization in the control and the St. Ann's Harbour references. Each species did poorly in the other reference, although the reasons for this are unclear. Thus, when compared to control or to one or the other reference, as shown in Table 10-4, the results of the test can be changed quite dramatically, even though the actual fertilization in the test pore waters is the same in each case (Zajdlik 2000). In reality, expert judgment would have been used to disqualify the reference sediment with poor performance, but in cases where the response is more marginal, better guidance on what constitutes an acceptable reference sediment pore water is needed. Additional research is needed to provide guidance for selecting a reference site for porewater testing or for developing alternate means of interpreting the test.

Table 10-4 Effect of changing reference sites on porewater test results

Species	Station[a]					
	1	5	6	9	Sydney	St. Ann's
Test pore water versus control water[b]						
D. excentricus	F	F	F	F	F	P
L. pictus	P	F	P	F	P	F
Test pore water versus Sydney reference pore water[b]						
D. excentricus	P	P	P	P	na[c]	P
L. pictus	P	F	P	F	na	F
Test pore water versus St Ann's reference pore water[b]						
D. excentricus	F	F	F	F	F	na
L. pictus	P	P	P	P	P	na

[a] Stations numbers range from 1, which is most contaminated, to the Sydney and St. Ann's reference stations, which are least contaminated.
[b] F = fail, P = pass, according to interim disposal at sea criteria.
[c] na = not available.

Responses at low contaminant levels

After permits are issued, toxicity testing may also be used at the disposal sites to monitor the long-term effects of disposal. It is not uncommon to find that the porewater toxicity tests are more sensitive than the other solid-phase tests in the battery. One example of this occurred at a disposal site in Cap-aux Meules, Québec that routinely receives dredged material. The site was being assessed for possible PAH contamination. Assessment followed Canadian disposal-site monitoring guidelines and was assessed much in the same way as was done in the gradient studies, with chemistry, toxicity tests, and benthic assessments (Hamel et al. 1999). No contamination above detection levels was found. Solid-phase toxicity tests and benthic assessment suggested no impact, but porewater tests failed when compared to a seawater control. Ammonia and TOC were correlated to the response, but ammonia levels did not exceed tolerance levels for *L. pictus* fertilization and were not significantly different across stations, and therefore, they did not satisfactorily explain the results. Fertilization was greatest at the disposal site and decreased toward the reference site. Anoxia was suggested as a possible cause because the sand content was higher on the disposal mounds, possibly increasing the oxygen levels. Further investigation of how to deal with sediments being anoxic and what the laboratory oxygenation process does to porewater toxicity is also needed (see Chapter 7). The report recommended that the results be compared to the reference pore water rather than the seawater control, which would have brought the results in line with the other toxicity results. Whereas this appears to be the right approach for these data, the practical difficulties remain: locating suitable reference sites and ensuring that they are both well matched and clean (see Chapter 12).

Relevance and Appropriateness of Sampling Methods

Another area where our general protocol documents are not precise enough for regulatory use is in the collection and extraction methods for pore water. Because these methods are not fixed, each laboratory makes choices, including whether to filter, what speed to centrifuge and for how long, or whether to exclude air (see Chapters 5 and 7). A look at past monitoring and research reports, just within Environment Canada, revealed several inconsistencies, such as single versus double centrifuging, different exposure times, or different test temperatures across laboratories and projects, which could potentially affect the results (Lee et al. 1995; Tay et al. 1997; Zajdlik et al. 2000). Clear guidance needs to be incorporated into a standard reference method for each test. Recommendations on collection and handling of pore water for toxicity testing coming from this workshop summarize areas of agreement on best practices and identify where further research is needed.

The draft interpretation criteria that the disposal at sea program uses to pass or fail the test in Canada should also be looked at in light of the analysis done on *A. punctulata*, whereby interpretation could be based on detectable significance criteria (Carr and Biedenbach 1999). If the Canadian program species respond similarly to *A. punctulata*, it is possible that the current interpretation criteria, which allow a 25% difference between control and test pore waters (and allow the difference to be significant), are too lenient. Minimum significant difference is a function of test method design and must be determined by a large amount of data collected in exactly the same manner. The minimum significant difference for *A. punctulata* fertilization averaged about 17% (Carr and Biedenbach 1999). The inclusion of *A. punctulata* in the list of species recommended for routine use in Canada may also be advisable and should be further investigated.

Conclusions

Whereas porewater testing has not been adopted globally as a regulatory tool, it remains a reasonable option for regulatory use in Canada, provided additional research and standardization are able to address the concerns identified. As a complement to other solid-phase tests, porewater tests provide an additional means of integrating what typical organisms might be "seeing" in a sediment, both measured and unmeasured. They allow the implementation of a precautionary approach, whereby appropriate action can be taken to limit harm when there is reason to believe that effects are likely, based on a weight of evidence, even when the specific links between the cause and the effects have not been proven. Assessments in the U.S. (Florida, Texas, Washington), Canada (New Brunswick), Mexico (Nipper and Carr 2001), and other locations have shown that the tests are sensitive and practical. Work now should focus on identifying and dealing with noncontaminant factors and factors that explain lack of responses in contaminated sediments. Standardizing test methods and their interpretation and reaching decisions on the best use of reference pore waters are also important. Environment Canada will be working over the next years to close data gaps and generate an echinoid porewater standard reference method suitable for regulatory use.

References

Adams WJ. 1987. Bioavailability of neutral lipophilic organic chemicals contained on sediments: A review. In: Dickson KL, Maki AW, Brungs WA, editors. Fate and effects of sediment-bound chemicals. New York NY, USA: Pergamon. p 219–244.

Adams WJ, Kimerle RA, Mosher RG. 1985. Aquatic safety assessment of chemicals sorbed to sediments. In: Cardwell RD, Purdy R, Bahner RC, editors. Aquatic toxicology and hazard assessment: Seventh Symposium. Philadelphia PA, USA: American Society for Testing and Materials. ASTM STP 854. p 429–453.

Allen HE, Fu G, Deng B. 1993. Analysis of acid-volatile sulfide (AVS) and simultaneously extracted metals (SEM) for the estimation of potential toxicity in aquatic sediments. *Environ Toxicol Chem* 12:1441–1453.

[ASTM] American Society for Testing and Materials. 1999. Standard guide for conducting static acute toxicity tests with echinoid embryos: Designation E1536. Volume 11.04, ASTM annual book of standards. Philadelphia PA, USA: ASTM.

Beak Consultants Limited. 1992. Fertilization assay with echinoids: Interlaboratory evaluation of test options. Ottawa ON, Canada: Prepared for Environment Canada, Test Method Development and Application Division. Unpublished report TS-23.

Burridge LE, Doe K, Haya K, Jackman PM, Lindsay G, Zitko V. 1999. Chemical analyses and toxicity tests on sediments under salmon net pens in the Bay of Fundy. Canadian Technical Report of Fisheries and Aquatic Sciences No. 2291. 33 p.

Burton Jr GA. 1998. Assessing aquatic ecosystems using pore waters and sediment chemistry. Ottawa ON, Canada: Prepared for Canada Centre for Mining and Energy Technology, Natural Resources Canada, Aquatic Effects Technology Evaluation (AETE) Program. Project 3.2.2.a. 107 p.

Caron G. 1989. Modeling the environmental distribution of nonpolar organic compounds: The influence of dissolved organic carbon in overlying and interstitial water. *Chemosphere* 19:1473–1482.

Carr RS. 1998a. Marine and estuarine porewater toxicity testing. In: Wells PG, Lee K, Blaise C, editors. Microscale aquatic toxicology: Advances, techniques and practice. Boca Raton FL, USA: CRC Lewis. p 523–538.

Carr RS. 1998b. Sediment porewater testing. In: Clesceri LS, Greenberg AE, Eaton AD, editors. Standard methods for the examination of water and wastewater, Section 8080. 20th ed. Washington DC, USA: American Public Health Association. p 8-37–8-41.

Carr RS, Biedenbach JM. 1996. Influence of ammonia on sea urchin porewater toxicity tests. 23rd Annual Aquatic Toxicity Workshop, Proceedings; 1996 Oct 6–9; Calgary AB, Canada.

Carr RS, Biedenbach JM. 1999. Use of power analysis to develop detectable significance criteria for sea urchin porewater toxicity tests. *Aquat Ecosyst Health Manag* 2:413–418.

Carr RS, Chapman DC, Howard CL. 1996. Sediment quality triad assessment survey in the Galveston Bay complex. *Ecotoxicology* 5:1–25.

Carr RS, Chapman DC, Presley BJ, Biedenbach JM, Robertson L, Boothe P, Kilada R, Wade T, Montagna P. 1996. Sediment porewater toxicity assessment studies in the vicinity of offshore oil and gas production platforms in the Gulf of Mexico. *Can J Fish Aquat Sci* 53:2618–2628.

Carr RS, Long ER, Chapman DC, Thursby G, Beidenbach JM, Windom HL, Sloane GM, Wolfe DA. 1996. Toxicity assessment studies of contaminated sediments in Tampa Bay, Florida. *Environ Toxicol Chem* 15:1218.

Carr RS, Montagna PA, Biedenbach JM, Kalke R, Kennicutt MC, Hooten R, Cripe G. 2000. Impact of storm-water outfalls on sediment quality in Corpus Christi Bay, Texas. *Environ Toxicol Chem* 19:561–574.

Carr RS, Williams JW, Fragata CTB. 1989. Development and evaluation of a novel marine sediment pore water toxicity test with the polychaete *Dinophilus gyrociliatus*. *Environ Toxicol Chem* 8:533–543.

[CCME] Canadian Council of Ministers of the Environment. 1999. Canadian environmental quality guidelines. Winnipeg MB, Canada: CCME.

[CEFAS] Centre for Environment, Fisheries and Aquaculture Science. 1998. Monitoring and surveillance of non-radioactive contaminants in the aquatic environment and activities regulating the disposal of wastes at sea. Burnham-on-Crouch, Essex, UK: CEFAS. 1995 and 1996 Science Series, Aquatic Environment Monitoring Report No. 51.

[CEPA] Canadian Environmental Protection Act. 1999. Statutes of Canada, Chapter 33. Ottawa, ON, Canada: Federal Government of Canada.

Di Toro DM, Zarba CS, Hansen DJ, Berry WJ, Swartz RC, Cowan CE, Pavlou SP, Allen HE, Thomas NA, Paquin PR. 1991. Pre-draft technical basis for establishing sediment quality criteria for non-ionic organic chemicals using equilibrium partitioning. Washington DC, USA: U.S. Environmental Protection Agency, Office of Water.

Environment Canada. 1992a. Biological test method: Fertilization assay using echinoids (sea urchins and sand dollars). Ottawa ON, Canada: Environment Canada. EPS 1/RM/27.

Environment Canada. 1992b. Biological test method: Toxicity test using luminescent bacteria (*Photobacterium phosphoreum*). Ottawa ON, Canada: Environment Canada. EPS 1/RM/24.

Environment Canada. 1994. Guidance document on the collection and preparation of sediments for physicochemical characterization and biological testing. Ottawa ON, Canada: Environment Canada. EPS 1/RM/29.

Environment Canada. 1998. Biological test method: Reference method for determining acute lethality of sediment to marine or estuarine amphipods. Ottawa ON, Canada: Environment Canada. EPS 1/RM/35.

ESG International. 1999. The aquatic effects technology evaluation (AETE) program synthesis report of selected technologies for cost-effective environmental monitoring of mine effluent impacts in Canada. Ottawa ON, Canada: Canada Centre for Mineral and Energy Technology (CANMET), Natural Resources Canada. 116 p.

Hamel P, Chabot R, Provencher M, St-Laurent D. 1999. Suivi environnemental au site d'immersion CM-7 des déblais de dragage du havre de Cap-aux-Meules, Îles-de-la-Madeleine, Québec (1996). Montreal QB, Canada: Environment Canada. Quebec regional report. 65 p.

Ingersoll CG. 1995. Sediment tests. In: Rand GM, editor. Fundamentals of aquatic toxicology: Effects, environmental fate, and risk assessment. Washington DC, USA: Taylor & Francis. p 231–255.

Lee DL, Yee SG, Fennel M, Sullivan DL. 1995. Biological assessment of three ocean disposal sites in Southern British Columbia. North Vancouver BC, Canada: Environment Canada. Pacific and Yukon Regional Report 95-07. 39 p, appendices.

Long ER, Carr RS, Thursby GA, Wolfe DA. 1995. Sediment toxicity in Tampa Bay: Incidence, severity, and spatial extent. _Fla Sci_ 58:163–178.

Long ER, Hameedi J, Robertson A, Aasen S, Dutch M, Ricci C, Welch K, Kammin W, Carr RS, Johnson T, Biedenbach J, Scott KJ, Mueller C, Anderson JW. 1999. Survey of sediment quality in Puget Sound, Year 1 - Northern Puget Sound. Seattle WA, USA: National Oceanic and Atmospheric Administration. NOAA Technical Memorandum. 249 p, 11 appendices.

Long ER, Sloane GM, Carr RS, Scott KJ, Thursby GB, Wade T. 1995. Sediment toxicity in Boston Harbor: Magnitude, extent, and relationships with chemical toxicants. Silver Spring MD, USA: National Oceanic and Atmospheric Administration, Coastal Monitoring and Bioeffects Assessment Division. NOAA Technical Memorandum NOS ORCA. 85 p, 31 figures, 4 appendices.

Nipper MG, Carr RS. 2001. Porewater toxicity testing: A novel approach for assessing contaminant impacts in the vicinity of coral reefs. _Bull Mar Sci_ 69:407–420.

Nipper MG, Martin ML, Williams EK. 1997. The optimization and validation of a marine toxicity test using the New Zealand echinoid, _Fellaster zelandiae_. _Australas J Ecotoxicol_ 3:109–115.

Nipper MG, Roper DS, Williams EK, Martin ML, VanDam LF, Mills GN. 1998. Sediment toxicity and benthic communities in mildly contaminated mudflats. _Environ Toxicol Chem_ 17:502–510.

[OSPAR] Oslo and Paris Commission. 1998. Joint Assessment and Monitoring Programme (JANMP) guidelines for general biological effects monitoring. Oslo, Norway and Paris, France: OSPAR. 15p.

Playle RC, Dixon DG, Burnison K. 1993. Copper and cadmium binding to fish gills: Modification by dissolved organic carbon and synthetic ligands. _Can J Fish Aquat Sci_ 50:2667–2677.

Porebski LM, Doe KG, Lee D, Pocklington P, Zajdlik B. 1999. Evaluating the techniques for a tiered testing approach to dredged sediment assessment: A study over a metal concentration gradient. _Environ Toxicol Chem_ 18:2600–2610.

Porebski LM, Doe KG, Lee D, Pocklington P, Zajdlik B, Atkinson G. 1998. Interpretative guidance for bioassays using pollution gradient studies: First year report. Ottawa ON, Canada: Prepared for Environment Canada, Marine Environment Division. Unpublished report WM-20.

Porebski LM, Osborne JM. 1998. The application of a tiered testing approach to the management of dredged sediments for disposal at sea in Canada. _Chem Ecol_ 14:197–214.

[PSDDA] Puget Sound Dredged Disposal Analysis. 1989. Management plan report (MPR) - Phase II. Seattle WA, USA: U.S. Army Corps of Engineers, Seattle District; U.S. Environmental Protection Agency, Region X; Washington State Dept of Natural Resources; Washington State Dept of Ecology.

[PSDDA] Puget Sound Dredged Disposal Analysis. 2000. User's manual. Seattle WA, USA: U.S. Army Corps of Engineers, Seattle District; U.S. Environmental Protection Agency, Region X; Washington State Dept of Natural Resources; Washington State Dept of Ecology.

Sergy GA. 1987. Recommendations on aquatic biological tests and procedures for environmental protection, C&P, DOE. Edmonton AB, Canada: Prepared for Environment Canada, Conservation and Protection, Technology Development and Technical Services Branch. Internal Technical Report TS-20. 102 p.

Shipper C, Stronkhorst J. 1999. Quality assurance of five sediment toxicity tests. Society of Environmental Toxicology and Chemistry (SETAC) 20th Annual Meeting, Proceedings; 1999 Nov 14–18; Philadelphia PA, USA.

South Africa. 2001. Waste assessment guidance: Application of biological assessment techniques. Progress in the assessment of dredged material from South African Ports. Submitted to the 24th meeting of the Scientific Group of the London Convention. London, UK: International Maritime Organization. Submission: LC/SG 24/2/3.

Sullivan DL. 1999. Ucluelet Inlet survey 1996. Vancouver BC, Canada: Environment Canada, Pacific and Yukon Region Report (unpublished). 23 p.

Tay KL, Doe KG, MacDonald AJ, Lee K. 1997. Monitoring of the Black Point ocean disposal site, Saint John Harbour, New Brunswick. Saint John NB, Canada: Environment Canada, Atlantic Region. Ocean Disposal Report #9. ISBN: 0-662-25655-7. 133 p.

Tay KL, Doe KG, MacDonald AJ, Lee K. 1998. The influence of particle size, ammonia, and sulfide on toxicity of dredged materials for ocean disposal. In: Wells PG, Lee K, Blaise C, editors. Microscale aquatic toxicology: Advances, techniques and practice. Boca Raton FL, USA: CRC Lewis. p 559–574.

[USEPA] U.S. Environmental Protection Agency. 1988. Sea urchin (*Arbacia punctulata*) fertilization test method 1008. In: Weber CI, Horning WB II, Klemm DJ, Neiheisel TW, Lewis PA, Robinson EL, Menkendick JR, Kessler FA, editors. Short-term methods for estimating the chronic toxicity of effluents and receiving waters to marine and estuarine organisms. Cincinnati OH, USA: USEPA. EPA/600/4-87/028.

[USEPA] U.S. Environmental Protection Agency. 1991. Evaluation of dredged material proposed for ocean disposal: Testing manual. Washington DC, USA: USEPA. EPA-503/8-91/001.

[USEPA] U.S. Environmental Protection Agency. 1993. Guidance manual: Bedded sediment bioaccumulation tests. Washington DC, USA: USEPA. EPA/600/R-93/183.

[USEPA] U.S. Environmental Protection Agency. 1999. Methods for the derivation of site specific equilibrium partitioning sediment guidelines (ESGs) for the protection of benthic organisms. Washington DC, USA: USEPA.

Wiener JG, Giesy JP. 1979. Concentrations of Cd, Cu, Mn, Pb, and Zn in fishes in a highly organic softwater pond. *J Fish Res Board Can* 36:270–279.

Zajdlik BA, Doe KG, Porebski LM. 2000. Interpretive guidance for biological toxicity tests using pollution gradient studies: Sydney Harbour. Ottawa ON, Canada: Environment Canada. EPS 3/AT/2, 150 p.

Chapter 11

Comparison of Sediment Quality Guideline Values Derived Using Sea Urchin Porewater Toxicity Test Data with Existing Guidelines

R Scott Carr, James M Biedenbach, Don MacDonald

The most commonly employed marine sediment quality guidelines (SQGs) (Long et al. 1995; MacDonald et al. 1996) are based on data predominantly from laboratory studies that use the 10-day solid-phase toxicity test with amphipods. The amphipod test measures survival of adult organisms and has been shown repeatedly to be considerably less sensitive to contaminant effects than other tests that have been used routinely in monitoring and assessment programs along the Atlantic and Gulf coasts of the U.S. (Carr, Chapman, Howard, Biedenbach 1996; Carr, Long et al. 1996; Carr et al. 1997, 1998, 2000; Long et al. 1997; Long, Scott et al. 1998; Long, Sloane et al. 1998). Numerous sediment quality assessment surveys have recently been conducted along the Atlantic and Gulf coasts of the U.S., using the sea urchin (*Arbacia punctulata*) fertilization and embryological development tests with pore water. The areas that have been sampled include locations in the states of Massachusetts, South Carolina, Georgia, Florida, and Texas and 200 stations in the vicinity of offshore oil and gas production platforms in the Gulf of Mexico.

All of these sea urchin porewater toxicity tests were conducted at the U.S. Geological Survey (USGS) (formerly the U.S. Fish and Wildlife Service [USFWS] and the National Biological Service [NBS]) Marine Ecotoxicology Research Station, using identical procedures. None of these porewater toxicity data have been included in the database used by Long et al. (1995) and MacDonald et al. (1996) to generate their guideline values. From the studies cited above, there are sufficient chemistry and toxicity data to calculate sediment effects concentrations (SECs), as was done by Ingersoll et al. (1996) for the freshwater amphipod *Hyalella azteca* and the midge *Chironomus riparius*. Ingersoll et al. (1996) calculated SECs using co-occurrence data from individual stations rather than mean values derived from specific studies, as was done by Long et al. (1995) and MacDonald et al. (1996). The purpose of this chapter is to compare SECs derived from porewater toxicity test data with existing SQGs.

Porewater Toxicity Testing: Biological, Chemical, and Ecological Considerations. R. Scott Carr and Marion Nipper, editors.
© 2003 Society of Environmental Toxicology and Chemistry (SETAC). ISBN 1-880611-65-1

Methods

Sample collections and chemical analyses

Sediment samples from surveys conducted in the following areas were included in the database for the development of the SECs: Boston Harbor, Massachusetts (NBS 1994a; Long et al. 1995); Charleston Harbor, Winyah Bay, and Savannah River, South Carolina and St. Simon Sound, Georgia (NBS 1993, 1995; Long et al. 1998); Biscayne Bay, Tampa Bay, Choctawhatchee Bay, Apalachicola Bay, St. Andrew Bay, and Pensacola Bay, Florida (USFWS 1992; NBS 1994b, 1995a, 1995b; Long et al. 1995, 1997; USGS 1997a; Long, Scott et al. 1998; Long, Sloane et al. 1998); Galveston Bay, Lavaca Bay, and Sabine Lake, Texas (Carr 1993, 1999; NBS 1996; USGS 1997b; Carr et al. 2001); and 200 stations in the vicinity of offshore oil and gas production platforms in the Gulf of Mexico (Kennicutt 1995; Carr, Chapman, Presley et al. 1996). Although different sampling methods were used in the different studies (i.e., modified Van-Veen grab, box cores, and hand cores), precautions were always taken to minimize any sample contamination. The depth of the samples varied among the studies from 2 to 10 cm. Overlying water was always decanted before the sediment sample was obtained. Sediment samples were placed in precleaned polyethylene containers, and the samples were stored on ice or held refrigerated (4 °C) until the pore water was extracted.

Porewater toxicity testing

Sediment porewater extraction procedure

A pressurized pneumatic extraction device was used to extract pore water from the sediments. This extractor is made of polyvinyl chloride (PVC) and uses a 5 μm polyester filter. It is the same device used in previous sediment quality assessment surveys (USFWS 1992; Carr 1993; NBS 1993, 1994, 1995a, 1995b; USGS 1997a, 1997b, 1998; Carr et al. 1998, 2000, 2001; Carr and Nipper 1998). The apparatus and extraction procedures have been described previously (Carr and Chapman 1995).

Sediment samples were held refrigerated (4 °C) until the pore water was extracted. Samples were always processed within 5 days of the time of collection, except for the study in the Gulf of Mexico when samples were held refrigerated for up to 2 weeks before they were processed. Pore water was extracted using a pneumatic extraction method (Carr and Chapman 1995). After extraction, the porewater samples were centrifuged in polycarbonate bottles at 1200 ×g for 20 minutes to remove any suspended particulate material and were then frozen in precleaned amber glass bottles.

Porewater toxicity testing with sea urchins

Two days before the start of a toxicity test, the samples were moved from the freezer to a refrigerator at 4 °C. One day prior to testing, samples were completely thawed in a tepid water bath. Temperature of the samples was maintained at 20 ± 1 °C. Sample salinity was measured and adjusted to 30 ± 1‰, if necessary, using reagent-grade purified water or concentrated brine. Other water quality measurements (dissolved oxygen [DO], pH, sulfide and ammonia concentrations) were made. Temperature and DO were measured with YSI meters; salinity was measured with a Reichert or American Optical refractometer; and pH, sulfide (as S^{-2}), and total ammonia expressed as nitrogen (TAN) were measured with Orion meters and their respective probes. Unionized ammonia, expressed as nitrogen concentrations (UAN), was calculated for each sample using the respective salinity, temperature, pH, and TAN values. Any samples containing less than 80% DO saturation were gently aerated by stirring the sample on a magnetic stir plate. Following water quality measurements and adjustments, the samples were stored overnight at 4 °C but returned to 20 ± 1 °C before the start of the toxicity tests.

Pore water was tested for toxicity with the sea urchin (*A. punctulata*) fertilization and embryological development tests. The experiments were performed with water quality–adjusted pore water (100%) and with 50% and 25% dilutions of full strength for each sample. Samples were diluted with 30‰ filtered (0.45 μm) seawater from the toxicity testing laboratory at the University of Texas, Marine Science Institute, Port Aransas. Five replicates were tested for each sample from each site. Reference toxicity (positive control) tests with sodium dodecyl sulfate (SDS) were run with each series of tests to assess the sensitivity of the gametes.

The tests were conducted with gametes of the sea urchin *A. punctulata*, following the methods of Carr and Chapman (1992, 1995); Carr, Chapman, Howard, Biedenbach (1996); Carr, Chapman, Presley, et al. (1996); and Carr, Long et al. (1996). Pore water from a reference area in Redfish Bay, Texas, previously documented to be nontoxic, was tested with each batch and used as a negative control. Adult male and female urchins were stimulated to spawn with a mild electric shock, and the gametes were collected separately. Prior to each series of tests, a pretest was conducted to determine the optimum sperm/egg ratio for maximizing the sensitivity of the test. The test involves exposing the sperm in 5 ml of the test solution for 30 minutes, followed by the addition of approximately 2,000 eggs. After an additional 30-minute incubation period, the test was terminated by the addition of formalin. An aliquot of the egg suspension was examined under a compound microscope to determine the presence or absence of a fertilization membrane surrounding the egg, and percent fertilization was recorded for each replicate. In the embryological development test, the embryos were allowed to develop for 48 hours before the test was terminated and the percentage of normally developing embryos determined.

Statistical analyses

For both the fertilization and embryological development tests, statistical comparisons among treatments were made using analysis of variance (ANOVA) and Dunnett's one-tailed *t*-test on the arcsine square root–transformed data with the aid of SAS (SAS 1991). The trimmed Spearman-Karber method (Hamilton et al. 1977) with Abbott's correction (Morgan 1992) was used to calculate the EC50 (effective concentration to 50% of the test population) values for dilution series tests. Prior to statistical analyses, the transformed datasets were screened for outliers (SAS 1992). Outliers were detected by comparing the studentized residuals to a critical value from a *t*-distribution chosen using a Bonferroni-type adjustment. The adjustment is based on the number of observations, *n*, so that the overall probability of a type I error is at most 5%. The critical value, *cv*, is given by the following equation:

$$cv = t(\text{df}_{\text{Error}}, 0.05/(2 \times n)).$$

After omitting outliers but before further analyses, the transformed datasets were tested for normality and for homogeneity of variance using SAS/LAB software (SAS 1992).

A second criterion was also used to compare test means to reference means for the 2 sea urchin tests. The detectable significance criteria were developed to determine the 95% confidence value based on power analysis of all similar tests performed by our laboratory (Carr and Biedenbach 1999). This value is the minimum significant difference that is necessary to accurately detect a difference from the reference (β = 0.05). The minimum significant difference value for the sea urchin fertilization assay at α = 0.05 is 15.5% of the control. At α = 0.01, the minimum significant difference value is 19%. For the sea urchin embryological development assay, the minimum significant difference values are 16.4% and 20.6% for α = 0.05 and α = 0.01, respectively.

Calculation of SECs

Procedures similar to those used by Long et al. (1995), MacDonald et al. (1996), and Ingersoll et al. (1996) were used to screen the data, which included 1) at least a 10-fold difference in concentration for at least 1 chemical among the samples for a particular study and 2) twice the mean of the no-effect data as a no-concordance screen for the effects data. After the data were screened, the number of co-occurring datasets ranged from 477 to 776 samples for fertilization and from 399 to 650 for embryological development for the different contaminants. These data were then used to calculate SECs that correspond to the values calculated by Long et al. (1995) and MacDonald et al. (1996). The calculated porewater-based SECs are referred to as "porewater-based threshold effects level" (PTEL), "porewater-based effects concentration" (PEC), "porewater-based effects range low" (PERL), and "porewater-based effects range median" (PERM). The purpose of this paper is not the intended development of an independent set of SECs based on porewater data but to evaluate the feasibility of using porewater data in the Biological Effects Data Set (BEDS) which has been used by Long et al. (1995) and MacDonald (1996). For this reason, only a few examples have been calculated for comparison purposes for some of the more common metals, polycyclic aromatic hydrocarbons (PAHs), and polychlorinated biphenyls (PCBs) for which SECs have been calculated previously.

Results

The porewater-based SECs for 5 metals, 3 combinations of PAHs, and total PCBs are shown in Table 11-1. The porewater-based SECs for the sea urchin fertilization and embryological development data, as compared with existing guideline values for copper, are shown in Figure 11-1. There was excellent agreement between threshold effects level (TEL)/PTEL and effects range low (ERL)/PERL values. The probably effects level (PEL)/PEC and effects range median (ERM)/PERM were within a factor of 2 for the fertilization data and only slightly higher for the embryological development data. Because sea urchin gametes and embryos are particularly sensitive to copper (Carr, Chapman, Presley et al. 1996), this result is not unexpected.

Table 11-1 Porewater-based effect concentrations for selected contaminants

Contaminant	Fertilization				Embryological development			
	PTEL	PEC	PERL	PERM	PTEL	PEC	PERL	PERM
Cd[a]	0.43	1.47	1.6	2.63	0.34	0.92	1.24	2.1
Cu[a]	28.86	68.98	73	162	22.7	43.1	41	99
Hg[a]	0.14	0.36	0.395	0.513	0.104	0.259	0.26	0.56
Pb[a]	44.5	101	93.5	190	40.1	69.9	79.9	136.5
Zn[a]	152	281	297	508	141	294.8	263	611
Low molecular weight PAHs[b]	194.5	701.6	781.7	1080.3	128.4	453.6	537	1075
High molecular weight PAHs[b]	647.2	3611.8	2829	7714	520.7	2185.4	2569	5784
Total PAHs[b]	1285.7	8702.4	6554	23866	796	3178.3	3955	8650
Total PCBs[b]	6.48	120.01	3.16	202	24.1	100.98	127.92	266.4

[a] In mg/kg.
[b] In µg/kg.

The data for mercury show a similar trend with excellent agreement for the threshold values (Figure 11-2). Somewhat surprisingly, the PERL was considerably higher than the ERL value for the fertilization data. The PEL/PEC and ERM/PERM ratios were again within a factor of 2 for the fertilization data, while the PEC was considerably lower than the PEL for the embryological development endpoint. The PEC was lower than the PERL for the fertilization data and nearly identical for the embryological development data. The PERM for the fertilization data was lower than the PERM

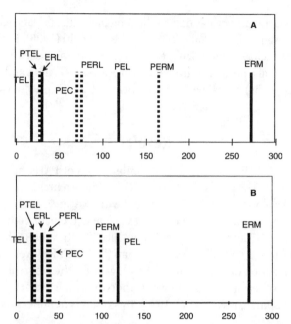

Figure 11-1 Comparison of SECs derived using sea urchin fertilization **(A)** and embryological development **(B)** test data with other guideline values for copper (mg/kg)

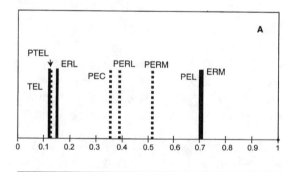

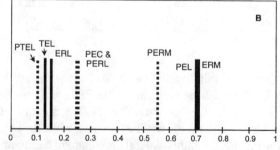

Figure 11-2 Comparison of SECs derived using sea urchin fertilization **(A)** and embryological development **(B)** test data with other guideline values for mercury (mg/kg)

for the embryological data, which is not particularly surprising in light of the fact that the fertilization assay is more sensitive than the embryological development test by more than 2 orders of magnitude, based on a comparison of EC50 values (Carr, Chapman, Presley et al. 1996).

For lead, there was again excellent agreement between the porewater-based threshold values and the existing guidelines (Figure 11-3). The PEL/PEC and ERM/PERM values were very similar for the fertilization data and were slightly lower for the embryological development data, but still the difference between corresponding values was less than a factor of 2. For zinc, there was excellent agreement between all corresponding values (Figure 11-4). For both the fertilization and embryological development data, the PECs and PERMs were higher than their corresponding guideline values, and the fertilization PERM was even lower than the embryological development PERM. The concordance between the threshold porewater-based SECs and their respective guidelines for cadmium was excellent (Figure 11-5). The PECs and PERMs, however, did not agree with their corresponding guidelines as well as did the other metals, with the ratios between ERM/PERM and PEC/PEL values ranging as high as 3 to 4. This is somewhat surprising in light of the fact that the sea urchin assays appear to be relatively insensitive to cadmium, compared with the other metals (Carr, Chapman, Presley et al. 1996).

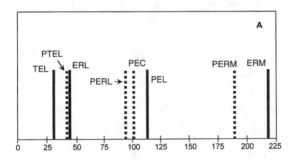

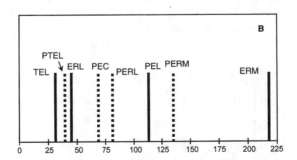

Figure 11-3 Comparison of SECs derived using sea urchin fertilization **(A)** and embryological development **(B)** test data with other guideline values for lead (mg/kg)

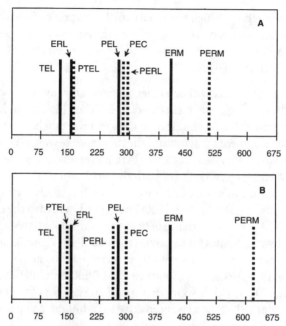

Figure 11-4 Comparison of SECs derived using sea urchin fertilization **(A)** and embryological development **(B)** test data with other guideline values for zinc (mg/kg)

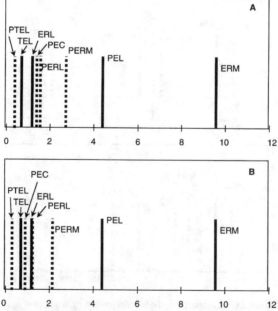

Figure 11-5 Comparison of SECs derived using sea urchin fertilization **(A)** and embryological development **(B)** test data with other guideline values for cadmium (mg/kg)

Figures 11-6 to 11-9 show comparisons between the porewater-based SECs and the existing guidelines for high molecular weight, low molecular weight, and total PAHs and total PCBs. Again there was excellent concordance on the low end for all 3 groups of PAHs. For the high molecular weight PAHs, the PEL/PEC for the embryological development was the only corresponding value that was not within a factor of 2 (Figure 11-6B). For the low molecular weight PAHs, the fertilization PERM was considerably lower than the embryological development PERM and lower than the ERM by approximately a factor of 3 (Figure 11-7). In contrast, for the total PAHs, the PERMs were nearly identical for the 2 porewater assays (Figure 11-8). The only corresponding values that were not within a factor of 2 were the embryological development PEL/PEC values. For total PCBs, fertilization PERL and PTEL were considerably lower than their corresponding guideline values by as much as a factor >7 (Figure 11-9A). In contrast, the embryo PERL was greater than the fertilization PERL by a factor of 40 and was even higher than the embryo PEC (Figure 11-9B). This apparent anomaly is related to an artifact resulting from the data screening process. The effects database was reduced to 53 paired observations after the data screening, and there was a large difference between the 8th and 10th percentile. The PEL/PEC and ERM/PERM were all within a factor of 2, and the PERMs were higher than the ERMs for both the fertilization and embryo SECs for total PCBs.

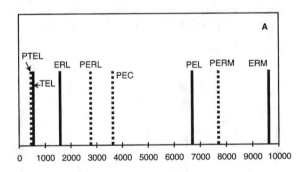

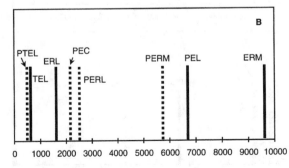

Figure 11-6 Comparison of SECs derived using sea urchin fertilization **(A)** and embryological development **(B)** test data with other guideline values for high molecular weight PAHs (µg/kg)

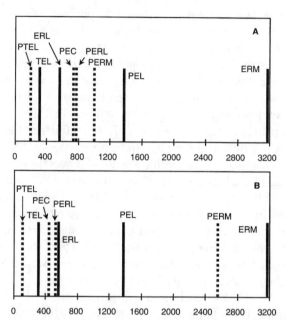

Figure 11-7 Comparison of SECs derived using sea urchin fertilization **(A)** and embryological development **(B)** test data with other guideline values for low molecular weight PAHs (µg/kg)

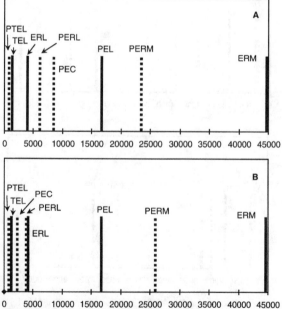

Figure 11-8 Comparison of SECs derived using sea urchin fertilization **(A)** and embryological development **(B)** test data with other guideline values for total PAHs (µg/kg)

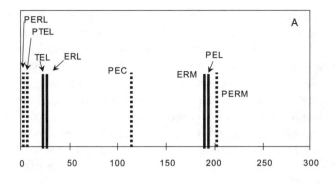

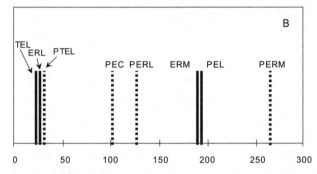

Figure 11-9 Comparison of SECs derived using sea urchin fertilization **(A)** and embryological development **(B)** test data with other guideline values for total PCBs (μg/kg)

Discussion

Using this group of contaminants as an example, it appears that there is excellent concordance between the porewater-based SECs and the existing sediment guidelines, which are based on a completely independent dataset. Before this exercise was conducted, we surmised that the porewater-based guidelines would be lower than the existing guidelines by an order of magnitude or more because the sea urchin assays are generally thought to be an order of magnitude or more sensitive than the amphipod test (which comprises the majority of the toxicity data in the BEDS), based on numerous surveys in which these tests have been conducted concurrently. The results of this exercise have demonstrated that the porewater-based threshold values (TEL and ERL) exhibited excellent concordance with the existing guidelines. The inclusion of porewater-based toxicity data in the BEDS, therefore, would appear to have little effect on the threshold guidelines for most contaminants.

Porewater-based effects guidelines values (PECs and PERMs) were generally within a factor of 2 of their respective sediment guideline with some exceptions (e.g., total PAHs). For certain contaminants (e.g., zinc) the PECs and PERMs were higher than their corresponding sediment guideline values. In general, the embryological development–derived SECs were lower than the corresponding SECs, based on the fertilization data, but there were notable exceptions to this generalization as well (e.g., mercury). The inclusion of porewater-based data would tend to lower the effects-level guidelines if they were included in the BEDS, but not significantly, based on the relative proportion of amphipod to porewater data. The resulting effects guidelines might provide better predictions of impacts on benthic communities because porewater tests provide toxicity estimates based on sensitive endpoints of sensitive life stages, rather than the acute mortality data with an adult organism obtained from the amphipod test. The close correspondence among the comparable values for these 2 distinctly different datasets, however, supports the validity of the co-occurrence approach for developing SQG values.

References

Carr RS. 1993. Sediment quality assessment survey of the Galveston Bay System. Clear Lake TX, USA: Galveston Bay National Estuary Program report. GBNEP-30. 101 p.

Carr RS. 1999. Toxicity assessment survey and toxicity identification evaluation (TIE) studies in Lavaca Bay, Texas. Corpus Christi TX, USA: Final report prepared for the U.S. Fish and Wildlife Service, Ecological Services Office. 31 p, 7 tables, 3 figures, 8 appendices, 4 attachments.

Carr RS, Biedenbach JM. 1999. Use of power analysis to develop detectable significance criteria for sea urchin porewater toxicity tests. *Aquat Ecosyst Health Manag* 2:413–418.

Carr RS, Biedenbach JM, Hooten R. 2001. Sediment porewater toxicity test survey and phase I sediment toxicity identification evaluation studies in Lavaca Bay, Texas: An estuarine Superfund site. *Environ Toxicol* 16:20–30.

Carr RS, Chapman DC. 1992. Comparison of whole sediment and pore-water toxicity tests for assessing the quality of estuarine sediments. *Chem Ecol* 7:19–30.

Carr RS, Chapman DC. 1995. Comparison of methods for conducting marine and estuarine sediment porewater toxicity tests: Extraction, storage, and handling techniques. *Arch Environ Contam Toxicol* 28:69–77.

Carr RS, Chapman DC, Howard CL, Biedenbach J. 1996. Sediment Quality Triad assessment survey in the Galveston Bay Texas system. *Ecotoxicology* 5:341–361.

Carr RS, Chapman DC, Presley BJ, Biedenbach JM, Robertson L, Boothe P, Kilada R, Wade T, Montagna P. 1996. Sediment porewater toxicity assessment studies in the vicinity of offshore oil and gas production platforms in the Gulf of Mexico. *Can J Fish Aquat Sci* 53:2618–2628.

Carr RS, Long ER, Chapman DC, Thursby G, Biedenbach JM, Windom H, Sloane G, Wolfe DA. 1996. Toxicity assessment studies of contaminated sediments in Tampa Bay, Florida. *Environ Toxicol Chem* 15:1218–1231.

Carr RS, Montagna PM, Kennicutt MC. 1998. Sediment quality assessment of storm-water outfalls and other sites of concern in the Corpus Christi Bay National Estuary Program study area. Corpus Christi TX, USA: Report prepared for the Corpus Christi Bay National Estuary Program, CCBNEP-32. 104 p, 23 appendices.

Ingersoll CG, Haverland PS, Brunson EL, Canfield TJ, Dwyer FJ, Henke CE, Kimble NE, Mount DR, Fox RG. 1996. Calculation and evaluation of sediment effect concentrations for the amphipod *Hyalella azteca* and the midge *Chironomus riparius. J Great Lakes Res* 22:602–623.

Kennicutt MC, editor. 1995. Gulf of Mexico offshore operations monitoring experiment. Phase I: Sublethal responses to contaminant exposure. New Orleans LA, USA: U.S. Department of the Interior, Minerals Management Service, Gulf of Mexico OCS Regional Office. OCS Study MMS 94-0045. 709 p.

Long ER, Scott GI, Kucklick J, Fulton M, Thompson B, Carr RS, Biedenbach J, Scott KJ, Thursby GB, Chandler GT, Anderson JW, Sloane GM. 1998. Magnitude and extent of sediment toxicity in selected estuaries of South Carolina and Georgia. Silver Spring MD, USA: Coastal Monitoring Bioeffects Assessment Division. NOAA Technical Memorandum NOS ORCA 128.

Long ER, Sloane GM, Carr RS, Johnson T, Biedenbach J, Scott KJ, Thursby GB, Crecelius E, Peven C, Windom HL, Smith RD, Loganathon R. 1997. Magnitude and extent of sediment toxicity in four bays of the Florida Panhandle: Pensacola, Choctawhatchee, St. Andrew and Apalachicola. Silver Spring MD, USA: Coastal Monitoring Bioeffects Assessment Division. NOAA Technical Memorandum NOS ORCA 117.

Long ER, Sloane GM, Carr RS, Scott KJ, Thursby GB, Wade T. 1995. Sediment toxicity in Boston Harbor: Magnitude, extent, and relationships with chemical toxicants. Silver Spring MD, USA: Coastal Monitoring and Bioeffects Assessment Division. NOAA Technical Memorandum NOS ORCA. 85 p, 31 figures, 4 appendices.

Long ER, Sloane GM, Scott GI, Thompson B, Carr RS, Biedenbach J, Wade TL, Presley RJ, Scott KJ, Mueller C, Brecken-Fols G, Albrecht B, Anderson JW, Chandler GT. 1998. Magnitude and extent of chemical contamination and toxicity in sediment of Biscayne Bay and vicinity. Silver Spring MD, USA: Coastal Monitoring and Bioeffects Assessment Division. NOAA Technical Memorandum NOS ORCA 117.

Long ER, Wolfe DA, Carr RS, Scott KJ, Thursby GB, Windom HL, Lee R, Calder FD, Sloane GM, Seal T. 1995. Magnitude and extent of sediment toxicity in Tampa Bay, Florida. Silver Spring MD, USA: Coastal Monitoring and Bioeffects Assessment Division. NOAA Technical Memorandum NOS ORCA 78. 84 p, 2 appendices.

[NBS] National Biological Survey. 1993. Toxicity testing of sediments from Charleston Harbor, South Carolina and vicinity. Corpus Christi TX, USA: NBS. Final report submitted to National Oceanic and Atmospheric Administration. 7 p, 16 tables, 4 attachments.

[NBS] National Biological Service. 1994a. Toxicity testing of sediments from Boston Harbor, Massachusetts. Corpus Christi TX, USA: NBS. Final report submitted to National Oceanic and Atmospheric Administration. 6 p, 10 tables, 4 attachments.

[NBS] National Biological Service. 1994b. Survey of sediment toxicity in Pensacola Bay and St. Andrew Bay, Florida. Corpus Christi TX, USA: NBS. Final report submitted to National Oceanic and Atmospheric Administration. 12 p, 24 tables, 5 attachments.

[NBS] National Biological Service. 1995a. Toxicity testing of sediment from western Florida and coastal South Carolina and Georgia. Corpus Christi TX, USA: NBS. Final report submitted to National Oceanic and Atmospheric Administration. 14 p, 35 tables, 4 attachments.

[NBS] National Biological Service. 1995b. Toxicity testing of sediment from Biscayne Bay, Florida and surrounding areas. Corpus Christi TX, USA: NBS. Final report submitted to National Oceanic and Atmospheric Administration. 11 p, 17 tables, 11 figures, 4 attachments.

[NBS] National Biological Service. 1996. Toxicity testing of sediment from Sabine Lake, Texas. Corpus Christi TX, USA: NBS. Final report submitted to National Oceanic and Atmospheric Administration. 11 p, 10 tables, 4 figures, 4 attachments.

[USFWS] U.S. Fish and Wildlife Service. 1992. Amphipod solid-phase and sea urchin porewater
    toxicity tests with Tampa Bay, Florida sediments. Corpus Christi TX, USA: USFWS. Final report
    submitted to National Oceanic and Atmospheric Administration. 9 p, 16 tables, 3 attachments.
[USGS] U.S. Geological Survey. 1997a. Toxicity testing of sediment from Biscayne Bay, Florida and
    surrounding areas, phase II. Corpus Christi TX, USA: USGS. Final report submitted to National
    Oceanic and Atmospheric Administration. 10 p, 8 tables, 10 figures, 4 attachments.
[USGS] U.S. Geological Survey. 1997b. Toxicity testing of sediments from Galveston Bay, Texas and
    surrounding areas. Corpus Christi TX, USA: USGS. Final report submitted to National Oceanic
    and Atmospheric Administration. 11 p, 10 tables, 10 figures, 4 attachments.

CHAPTER 12

Regulatory Applications of Porewater Toxicity Testing

Richard Scroggins (Co-Workgroup Leader), Walter J Berry (Co-Workgroup Leader), Robert A Hoke, Kristen Milligan, Donald J Morrisey, Linda Porebski

The purpose of this chapter is to evaluate the use of porewater toxicity tests in regulatory applications, including their potential use in the development of sediment quality guideline (SQG) values. Specifically, the following discussion focuses on the appropriateness and readiness of porewater tests for use in regulatory applications as well as on important factors that must be considered before a porewater toxicity test is used in these applications.

In the regulatory arena, decisions are often based as much on legal and programmatic considerations as on scientific considerations. It is not our purpose here to recommend how specific tests can be used in particular programs. However, it may be useful to regulators to consider the scientific criteria required for the selection of, and potential uses for, porewater toxicity tests in regulatory situations. It may also be useful for scientists to consider some of the unique regulatory requirements associated with use of these tests. Regulatory use of these tests often does not allow for professional judgment in the interpretation of the results. This is because standards for test acceptability and "pass or fail" criteria must be established a priori, to guarantee fair and consistent application of a guideline, standard, or regulation.

We cannot stress enough the fact that the choice of a test must be driven by the regulatory context in which the results are to be used or by the question that is being asked. This admonition also applies to choice of an appropriate sampling design, reference sediment, and test endpoints. Factors to consider include whether dietary uptake is important, whether the test will be used as a stand-alone test or as part of a screening tier, and whether particular porewater constituents are considered confounding factors or contaminants of concern.

Porewater Toxicity Testing: Biological, Chemical, and Ecological Considerations. R. Scott Carr and Marion Nipper, editors.
© 2003 Society of Environmental Toxicology and Chemistry (SETAC). ISBN 1-880611-65-1

Are Porewater Toxicity Tests Useful for Regulatory Purposes?

One of the principal reasons for using porewater toxicity testing in a regulatory framework is that these tests provide additional information not currently provided by solid-phase, elutriate, or sediment extract tests. At the present time, there are also more aqueous-phase tests and test species available for sublethal or chronic effects measurement in pore water than there are epifaunal or infaunal test species for equivalent solid-phase tests. The use of additional test species also increases the number of potential endpoints for evaluation (e.g., growth, production of young, development, fertilization success).

Porewater toxicity tests provide a direct assessment of the water phase of sediments and may be useful in determining potential exposure routes for contaminants in sediments. In contrast, elutriate tests, which are not designed to assess sediments but to mimic conditions during open-water disposal of dredged materials, dilute the porewater and may introduce artifacts because of high concentrations of solids. The porewater phase is more closely related to in-place sediment conditions in the field than is an elutriate. Tests with pore water have been described as presenting a "worst case," relative to elutriates, because elutriates by design represent a 4-fold dilution of the aqueous phase of sediments (Burton 1998). Pore water, as a phase for toxicity testing, is also less altered than organic solvent extracts of sediment and has no potential concerns related to residual concentrations of solvent.

Determination of causal effects through toxicity identification evaluations (TIEs) is logistically more straightforward using porewater tests than solid-phase tests. Because pore water is also an aqueous phase, the use of aqueous-phase (i.e., effluent) TIE procedures is more readily transferred to porewater testing than to solid-phase testing. Similarly, dilution series can be created more easily using the porewater phase than the solid phase of sediments.

Another advantage of porewater tests is that they are generally cheaper and more rapid to conduct than solid-phase tests. This is particularly important in a regulatory context because cost and ease of test performance are important considerations for both performing laboratories and regulated entities. Short-term ex situ porewater tests may allow several tests to be conducted (either with a single test or a battery of tests) in a short time. This can facilitate a tiered assessment within the recommended holding times for solid-phase sediments (i.e., 6 weeks) using various chemical analyses and porewater toxicity tests (U.S. Environmental Protection Agency/U.S. Army Corps of Engineers [USEPA/USACE] 1991; Environment Canada 1994). Tests that can be conducted quickly and cheaply are also more attractive for screening purposes because they facilitate efficient screening of large numbers of samples.

Regulatory Objectives and Practical Constraints

Potential regulatory uses for porewater toxicity tests include screening, compliance testing, TIEs, and environmental monitoring and assessment programs. Before any sediment assessment tool is used in a regulatory context, it is imperative that the purpose and specific objectives of the investigation be clearly defined. A clear statement of objectives is necessary because these will determine the science-based approach to be followed. For example, in a study in which ammonia or volatile compounds such as hydrogen sulfide are not contaminants of concern, it may be appropriate to use aeration to eliminate them from the sample before testing or to use a test species tolerant of the observed concentrations of ammonia or hydrogen sulfide in the sample. Alternatively, if ammonia is a contaminant of concern, the study design should use a test species sensitive to ammonia.

The principal constraints to the use of pore water for routine toxicity testing are the small volume of pore water typically available for testing, possible confounding factors (e.g., H_2S, NH_3), and potential artifacts introduced by sampling, handling, and storage of pore water. Depending on the purpose of the test, substances that are not the primary contaminants of concern in an investigation may be viewed as confounding factors, which must be accounted for in the test design and data interpretation.

In general, the regulatory use of porewater toxicity testing appears to be well suited to rapid screening applications and to providing additional support to existing regulatory frameworks. At this time, we do not see porewater tests replacing solid-phase tests or being used alone in pass/fail decisions.

Chapter Outline

The next section of this chapter discusses potential regulatory uses of porewater toxicity tests, how they might be incorporated into regulatory programs, and the types of situation in which their use would not be advisable. "Sediment Porewater Usage and Considerations for Regulatory Application" (p 268) takes a regulatory perspective on selected methodological aspects of porewater testing that are of special significance in a regulatory context (broader and more detailed discussions of methods are presented in other chapters). It is implicit in this chapter's discussions that individual tests must have reached a suitable stage of development and quality assurance before they are suitable for consideration in regulatory programs. The subsequent section, "Readiness of Porewater Toxicity Test Methods for Regulatory Use" (p 272), reviews the state of development of the various tests currently available, with particular reference to whether adequately standardized and prescriptive methods have been developed, such as are needed for regulatory use. "Application of Porewater Toxicity Testing for the Development of Sediment Quality Guidelines" (p 276) discusses the pros and cons of using porewater toxicity data to refine existing SQGs or develop new ones. The final 2 sections, "Information Gaps" (p 278) and "Summary Points for the Use of Porewater Tests in a Regulatory Context (p 279), identify current gaps in the available information on porewater testing that need to be filled before regulatory use is feasible and summarize the discussions in the preceding sections.

Are Porewater Toxicity Tests Appropriate for Regulatory Purposes?

In which regulatory programs could porewater toxicity testing be used?

Table 12-1 summarizes regulatory programs that could use porewater toxicity testing data to develop action limits or in monitoring programs. We define action limits as specific pass/fail levels that may trigger particular actions, such as additional testing or studies, or contribute to decisions concerning issuing of permits. Monitoring programs could use porewater toxicity tests in many different ways, such as measuring changes in sediment contamination or identifying areas of concern for additional investigation.

Table 12-1 Potential regulatory applications of porewater toxicity tests

	Potential regulatory application	
Types of regulatory programs	Action limits	Monitoring
Aquatic sediment removal (e.g., dredging and disposal, subaqueous sand or gravel mining)	√	
Mine decommissioning (e.g., closure plan and sediment reclamation)		√
Sediment remediation programs	√	√
New chemical registration programs (sediment spiking)	√	
Effluent point source controls or cumulative impact assessments (e.g., watershed management, non-point source monitoring)	√	√
Sediment quality criteria		√
Disposal site monitoring		√
Environmental quality monitoring		√
Impact assessment of existing operations (e.g., sediment monitoring of aquaculture activities)		√
Environmental assessment programs (e.g., environmental assessments, environmental impact statements)		√
Fish habitat protection programs		√

How could porewater toxicity testing be incorporated into a regulatory framework?

There are many potential regulatory applications for porewater toxicity tests, once they have been shown to be ecologically relevant and standard methods are available. The goals and purposes of each regulatory program must be taken into account before porewater toxicity testing is incorporated. In general, these tests may be useful in an action or decision-making mode, such as supporting permit decisions, or in a more passive information-gathering mode, such as monitoring marine environmental quality. In the action mode, these tests could be used with various combinations of solid-phase tests and chemistry and/or benthic information to provide weight-of-evidence decisions at the lower levels of tiered testing schemes. For example, the addition of porewater toxicity information to chemical or other screening information can help to integrate potential effects of both measured and unmeasured chemicals.

One regulatory use of porewater toxicity testing is in Canada, as part of a pass/fail battery for granting or denying permits for disposal of dredged material at sea. A discussion of this application appears in the plenary chapter on regulatory uses of porewater toxicity tests (Chapter 10). This test battery could fail a sediment sample that is above SQGs and that is toxic in any 2 tests in the test battery. The test battery includes solid-phase amphipod, bioaccumulation, and bacterial luminescence tests, as well as a porewater echinoid fertilization test (Porebski and Osborne 1998).

In the environmental monitoring mode, these tests can provide rapid and cost-effective screening information that allows regulators to prioritize areas for further assessment or provide time series information for trends in toxicity. In this type of framework, where the goal is broad, "ball-park" assessment, it may be acceptable to use porewater tests alone.

Are there situations in which porewater toxicity testing should not be used?

The use of porewater toxicity testing alone is likely appropriate only for certain screening applications or TIE evaluations. In all other applications, the purpose of the investigation or nature of the questions of interest will require the use of additional toxicity testing approaches or higher-tier biological analyses (e.g., solid-phase tests, community assessments).

There will also be specific situations in which the results of porewater toxicity tests should not be used to extrapolate to potential solid-phase effects, for example, when the porewater samples originate in dynamic erosional habitats. Porewater tests are also not appropriate when the issue of concern involves dietary uptake of contaminants of concern via the particulate phase or when a species of interest is unsuitable for porewater testing. Longer-term studies (>48 hours) may also be problematic unless provisions are made for renewal of porewater test solutions and feeding of test organisms. The majority of porewater toxicity test designs are less than 2 days in length (although 4- and 7-day designs are available) and involve small test species to minimize the volume of porewater necessary for testing.

Sediment Porewater Usage and Considerations for Regulatory Application

When the potential utility for porewater toxicity tests in regulatory applications is assessed, it is important to consider and carefully develop a study design to meet specific regulatory needs. In addition to the study design considerations for traditional toxicity tests, there are study design issues specific to toxicity testing of pore water. The availability of appropriate, standardized methods and the relevance of the test to the specific application must be assessed. Porewater tests can be considered for use in a variety of environments, including depositional and dynamic or erosional environments. However, the type of environment may affect the study design (see "Porewater collection, handling, and storage," p 270). Guidance on study design for sediment collection and testing (including porewater testing) for the purposes of environmental monitoring and evaluation of the suitability of dredged sediments for open water disposal is outlined by Environment Canada (1994).

This section discusses the various issues that must be considered when developing study design for the regulatory application of porewater toxicity tests. These issues include contaminants of concern, confounding factors, reference and control sediment comparisons, porewater collection, handling and storage, test methodology, data analysis and reporting, and data interpretation.

Contaminants of concern versus confounding factors

Sometimes a particular chemical in the porewater (e.g., ammonia) may not be a contaminant of concern for regulatory purposes but can influence the outcome of a porewater toxicity test. If ammonia is a confounding factor and not a contaminant of concern, it may be appropriate to either aerate the sample to volatilize the ammonia or use a porewater test species whose tolerance to ammonia is within the range found in the samples. In other situations, however, ammonia may be a contaminant of concern, for example, in testing the sediments under floating fish pens. Another factor is pH, which probably would be considered a confounding factor in most situations but which may be a major contaminant of concern in studying an acid mine discharge. Additional examples of confounding factors are given in Table 12-2. Similar issues involving confounding factors must also be considered when appropriate reference samples for porewater toxicity tests are chosen.

Table 12-2 Potential applications where confounding factors may be of concern

Confounding factor	Potential application
NH$_3$	Aquaculture, wastewater treatment plant (WWTP)
H$_2$S	WWTP
Low dissolved oxygen (DO)	Aquaculture, WWTP
pH	Acid mine drainage
Ion imbalance	Oil and gas productions

Reference versus control sediment comparisons

Two types of negative controls (i.e., controls in which no effect on the test species is expected) can be used in sediment porewater toxicity testing. The first type is a "reference control" and the second type is a "performance control."

A porewater reference control is prepared by extracting the pore water from clean sediment collected in the vicinity of the test sediment collection location (spatial control). Ideally, the physicochemical characteristics of the reference sediment should be similar to those of the test sediment but without the contaminants of concern (USEPA/USACE 1991, 1998; Environment Canada 1994).

Often, it is necessary to ensure that the reference control also accounts for confounding factors (i.e., not contaminants of concern) that may be important in the sample pore waters. A separate reference control site for this purpose may be needed if confounding factors are not adequately represented in the spatial control described above or if the spatial control is not sufficiently free of contaminants.

The second type of negative control is a performance control, which often may be the pore water from the sediment used for holding or culturing the test species. Alternatively, in some situations, it also may be appropriate to use laboratory dilution water as a performance control, for example, if the test species is cultured in formulated sediment prepared using laboratory dilution water or if a nonbenthic species or life stage is used in the porewater test. The principal purpose of the performance control is to provide a direct measure of test species health. The number and type of control samples for the sampling area should be chosen on the basis of the purposes of the regulatory program.

Sometimes, for a particular species, the reference sediment may not be suitable for generating a reference porewater sample. In these situations, it may be necessary to use the performance control of the porewater test for comparison with porewater test results for environmental samples. However, if only a performance control is used, there is no control for confounding factors, and interpretation of test results will require an understanding of the tolerance limits of the test species for potential confounding factors in the test samples.

Typically, both negative performance and reference controls are used in solid-phase tests of contaminated sediments. It seems intuitively obvious that the 2 types of controls also would be used in toxicity tests of pore waters from contaminated sediments.

Porewater collection, handling, and storage

Subject to collection, handling, and storage concerns, porewater toxicity tests should be applicable to sediment samples from most areas. Porewater toxicity testing usually is conducted by collecting a solid-phase sediment sample and removing the pore water for subsequent testing. A more limited set of situations may involve the use of in situ porewater toxicity testing (Sarda and Burton 1995). In situ testing may be particularly relevant where the quality of pore water may fluctuate drastically in a short period of time and an integrative measure of exposure is of interest (e.g., groundwater recharge from mine wastes after storm events, intertidal sediment assessments).

The amount of consideration given to methodological issues, such as oxidation changes in porewater samples, the volume of pore water reasonably obtainable from a sediment sample, or porewater shelf life, may be influenced by the objective of the regulatory program.

The collection, handling, and storage of sediment and the generation of porewater samples from this sediment should be defined for the regulatory program in question. Previous chapters discussed the advantages and disadvantages of various collection, handling, and storage methods. The potentially different artifacts introduced by different methods of collection, handling, and storage emphasize the need for standardized procedures in regulatory programs. The results of any biological testing of sediment or its pore water would be of little value if the sample collection and/or sample preparation did not follow a standardized procedure. For example, Environment Canada (1994) has stipulated a set of procedures for sediment collection, handling, and toxicity testing of whole sediment and pore water, which dredging permittees should follow under the Ocean Disposal Program.

Test methodology

There are several categories of criteria against which a porewater toxicity test must be evaluated before a method is ready for regulatory use. Key criteria for evaluation of porewater toxicity tests include sensitivity, degree of standardization, usefulness, ecological relevance, and logistical considerations.

Sensitivity

- Is the test species sensitivity known for typical confounding factors (NH_3, H_2S, pH, dissolved oxygen (DO), ion imbalance, etc.)?

- Is the test species sensitivity known for the contaminants of concern (polycyclic aromatic hydrocarbons (PAHs), polychlorinated biphenyls (PCBs), metals, etc.)?
- Does the test method measure relevant endpoints (e.g., survival, growth, production of young, development)?

Degree of standardization

- Is a written test method available (e.g., research method or known nonstandardized method)?
- Has the test method been published by a government organization or endorsed by a standardization group (Environment Canada, USEPA, American Society for Testing and Materials [ASTM], etc.)?
- Does the test method require reference toxicant testing (i.e., positive controls)?
- Does the test method have clear test validity criteria?
- Was an interlaboratory validation (i.e., round-robin or ring) study completed as part of method development?

Usefulness and relevance

- Has the porewater toxicity test method of interest been demonstrated to be useful in an integrated assessment program?
- Could the test method be adapted for use in porewater TIEs?

Logistical considerations

- Is the cost of conducting the test reasonable?
- Is the test method rapid?
- Is the test easy to conduct?
- Is the test conducted with field-collected or laboratory-cultured test species?

Data analysis and reporting

As with all regulatory methods, it is desirable that data analysis and reporting be standardized, including test performance relative to negative and positive control performance criteria. Because of the importance of confounding factors (e.g., H_2S, pH, DO, and NH_3) in porewater testing, reporting of these measurements is especially important. Reporting should also include sample and test organism identification, description of test facilities and apparatus, type of control and dilution water, test method, test conditions, and test results.

In a standardized regulatory procedure, it is critical that a minimum level of data be reported to ensure the regulatory agency of the integrity and validity of the test result. Other supporting data associated with the test may not need to be reported but should be kept on file by the performing laboratory. An example of the type of test data that must be reported is outlined by Environment Canada (1998).

Interpretation

Some degree of standardization of interpretation is usually required in a regulatory context to ensure consistency and fairness. Where a regulatory framework sets fixed criteria for the evaluation of individual tests or for the way that the whole framework should be used or judged, opportunities to use best professional judgment and weight of evidence in interpreting the results of a particular test are reduced or eliminated. Therefore, it is critical that this judgment is incorporated into the test development process and that any verification, validation, and standardization be done before the setting of the test evaluation criteria.

The greater the information gaps for a regulatory test, the greater the risk that the test will not consistently achieve its regulatory objectives (low false positives and false negatives). The development and understanding of porewater toxicity tests and their integration into field assessments is still in relatively early stages, compared to more commonly used solid-phase tests. A better understanding of medium-related issues, such as effects of extraction and storage on test results, would be useful. A basic list of criteria for assessing the appropriateness and readiness of a porewater toxicity test for regulatory use is provided in Table 12-3.

The preceding discussion is especially important in programs if the test may be used as a stand-alone decision criterion. It is therefore recommended that any planned use of porewater toxicity tests for pass/fail regulatory programs be predicated on adequate assessment of the suitability of the test organisms, test method, and test endpoints. Decisions must be made ahead of time about test acceptability, and criteria established for "pass" or "fail." These criteria are requirements for the use of these tests in many programs (Porebski and Osborne 1998; *Puget Sound Dredged Disposal Analysis-User's Manual* [PSDDA-UM] 2000).

Readiness of Porewater Toxicity Test Methods for Regulatory Use

Porewater toxicity tests are not commonly employed for regulatory purposes. Some of the reasons for this include the following:

1) A distinction needs to be maintained between regulatory methods and standard guides or research methods. Regulatory methods must be definitive and standardized so they can be used to generate data which support

Table 12-3 Criteria for assessing readiness of toxicity testing procedures for porewater toxicity tests[a]

Test	Known tolerance to confounding factors		Relevant endpoints			Cost[b,d]	Short test duration[c,d]	Available or published[d]	Test method Published standard method[d]	Reference toxicant requirement[e]	Validity criteria	Interlab validation	Lab accredit	Use for TIEs[f]	Demonstrated use in integrated programs[f]
	NH₃	H₂S	Growth	Surv	Repr										
Saltwater															
Sea urchin fertilization: *Strongylocentrotus purpuratus, Arbacia punctulata, Lytechinus pictus*	Y	Y	Y	N	N	1-2	Y	Y (1,3,7,8,11)	Y (11)	Y	Y	Y	Y	1	1-2
Sea urchin development: *A. punctulata, S. purpuratus*	Y	Y	Y	N	N	1-2	Y	Y (1,3,7,8)	N	Y	Y	Y	Y	1	1-2
Sand dollar fertilization: *Dendraster excentricus*	Y	U	Y	N	N	1-2	Y	Y (1,11)	Y (11)	Y	Y	Y	Y	1	1-2
Mollusk embryo: *Haliotis rufescens, Mytilus galloprovincialis*	Y	Y	Y	Y	N	1-2	Y	Y (4,15)	N	Y	Y	Y	Y	1	1-2
Polychaete survival and reproduction: *Dinophilus gyrociliatus*	Y	U	N	Y	Y	2-3	N	Y (6,7,9)	N	Y	U	U	U	2	2
Harpacticoid nauplii survival	U	U	N	Y	N	2	Y	Y (9)	N	U	U	U	U	1	2

Table 12-3 *(cont'd.)*

Test	Known tolerance to confounding factors		Relevant endpoints			Cost[b,d]	Short test duration[c,d]	Test method						Use for TIEs[f]	Demonstrated use in integrated programs[f]
	NH$_3$	H$_2$S	Growth	Surv	Repr			Available or published[d]	Published standard method[d]	Reference toxicant requirement[e]	Validity criteria	Interlab validation	Lab accredit		
Saltwater *(cont'd.)*															
Algal zoospore germination and growth: *Ulva* spp.	Y	U	Y	Y	N	2–3	N	Y (14)	N	Y	U	U	U	1–2	2
Crustacean larvae survival	U	U	N	Y	N	2	U	Y (9)	N	U	U	U	U	1	3
Microtox bioluminescence inhibition	Y	Y	Metabolic activity			1–2	Y	Y (10)	Y (10)	Y	Y	Y	Y	1–2	1–4
Fish egg development	Y	U	Y	Y	N	1	Y	Y (7)	N	U	U	U	U	1	3
Dinoflagellate-bioluminescence activity: *Gonyaulax polyedra*	Y	U	Metabolic activity			1–2	Y	Y (2)	N	U	Y	U	U	2	3–4
Mysid shrimp survival: *Americamysis bahia*	Y	Y	N	Y	N	2	Y	Y (5,12)	N	Y	Y	N	N	1	4
Amphipod survival: *Ampelisca abdita*	Y	Y	N	Y	N	2	Y	Y (5,12)	N	Y	Y	N	N	1	4
Freshwater															
Daphnid reproduction: *Daphnia, Ceriodaphnia*	Y	Y	N	Y	Y	1–4	N	N	N	Y	Y	Y	Y	1–3	1–4

Table 12-3 *(cont'd.)*

Test	Known tolerance to confounding factors — NH₃	H₂S	Relevant endpoints — Growth	Surv	Repr	Cost[b,d]	Short test duration[c,d]	Test method — Available or published[d]	Published standard method[d]	Reference toxicant requirement[e]	Validity criteria	Interlab validation	Lab accredit	Use for TIEs[f]	Demonstrated use in integrated programs[f]
Freshwater *(cont'd.)*															
Daphnid survival: *Daphnia, Ceriodaphnia*	Y	U	N	Y	N	1–3	Y	Y (16,17)	N	Y	Y	Y	Y	1–2	1–4
Microtox bioluminescence inhibition	Y	Y	Metabolic activity			1	Y	Y (13,17)	N	U	Y	Y	Y	2–4	1–4
Amphipod survival and growth: *Hyalella azteca*	Y	U	Y	Y	N	1–4	N	Y (16)	N	Y	Y	Y	Y	2–3	1–4
Algal cell growth: *Selenastrum capricornutum*	Y	U	Y	N	Y	1–2	U	Y (18)	N	Y	Y	Y	Y	2–3	2–4
Fish survival and growth: *Pimephales promelas*	Y	Y	Y	Y	N	2–3	N	Y (16)	N	Y	Y	Y	Y	1	4
Chironomid larval survival and growth: *Chironomus tentans, C. riparius*	Y	U	Y	Y	U	2–4	N	Y (17)	N	Y	Y	Y	Y	3	2–4

a Y = Yes, U = Unknown, N = No.
b 1 = low; 4 = high.
c ≤48 h.
d References in parentheses: 1) APHA 1998; 2) ASTM 1998; 3) Anderson et al. 1997; 4) Anderson et al. 2001; 5) Burgess et al. 1996; 6) Carr et al. 1989; 7) Carr and Chapman 1992; 8) Carr, Chapman, Howard, Biedenbach 1996; 9) Carr, Chapman, Presley et al. 1996; 10) Environment Canada 1992a; 11) Environment Canada 1992b; 12) Ho et al. 1997; 13) Hoke et al. 1990; 14) Hooten and Carr 1998; 15) Hunt et al. 2001; 16) Ankley et al. 1991; 17) Giesy et al. 1990; 18) Ankley et al. 1990.
e Published literature available on use of reference toxicant.
f 1 = good; 4 = poor.

regulatory actions that may be challenged in the courts. Standard guides for porewater toxicity, while useful for general monitoring purposes, may not be prescriptive enough or implemented uniformly enough to survive legal challenge.

2) Table 12-3 presents criteria for assessing the readiness of porewater toxicity procedures for use in regulatory programs. Workshop participants with experience using the listed porewater toxicity tests completed the table. Based on the responses of the workshop experts, it was clear that there are several standard test methodologies currently available for use. Published standard test methods currently available for porewater toxicity testing in regulatory applications include the Microtox test, the sea urchin fertilization test, and the sand dollar fertilization test (Table 12-3). Some methodological deficiencies are also obvious from Table 12-3, such as the lack of information on the sensitivity of test species to some confounding factors (e.g., H_2S). It is also apparent that although research or lab-specific test methods are available, standardized published methods for porewater toxicity testing are currently not available. Laboratory accreditation and certification programs that aid in judging the performance and quality of regulatory testing also are not currently focused on assessing the ability of laboratories to conduct porewater toxicity tests. Laboratory accreditation and certification programs for toxicity testing typically are directed toward the performance of effluent toxicity testing. However, there was reasonably good agreement among the workshop experts for most criteria used to rank the porewater toxicity tests.

Application of Porewater Toxicity Testing for the Development of Sediment Quality Guidelines

In addition to being used directly to assess sediments in a regulatory context, porewater tests can contribute to the development and verification of other assessment tools, such as effects-based chemical guidelines. This section will review the types of SQGs that currently exist and briefly discuss how porewater toxicity tests potentially could be used in relation to these guidelines.

Uses related to empirical guidelines

Sediment quality guidelines are currently used in a variety of regulatory programs such as those outlined in Table 12-1. SQGs are generally divided into 2 groups: empirical guidelines such as effects range low/effects range median (ERL/ERM), threshold effects level/probable effects level (TEL/PEL), and apparent effects threshold (AET) values (Neff et al. 1986; USEPA 1989; Long and Morgan 1991; Jaagumagi 1992; MacDonald et al. 1992; Reynoldson and Zarull 1993; Long et al.

1995) and theoretical, equilibrium partitioning (EqP) guidelines such as acid volatile sulfide/simultaneously extractable metals (AVS/SEM)–based guidelines for metals and equilibrium partitioning sediment guidelines (ESGs) for nonionic compounds (Di Toro et al. 1991). The empirical methods are based on large databases of synoptic chemistry and biological effects data. These approaches have not included porewater toxicity data. A comprehensive dataset using matched porewater toxicity and bulk sediment chemistry was used to derive some select SQGs presented in Chapter 11. Using the most popular empirical models, we employed porewater data only to explore the effects on guidelines values. The exercise of generating "porewater SQGs" highlighted the interesting questions of what these guidelines would actually be representing and what is the best use of the data. Would this information be best integrated into the current SQGs, or kept separate to provide different or complementary information on sediment status, or omitted from SQG development altogether?

A decision to systematically incorporate porewater toxicity test data into databases used to derive empirical solid-phase SQGs must be based on an understanding of what the numbers will represent and more important, in a regulatory context, how they are likely to be used to make decisions. It is possible that porewater test results may provide information not provided by traditional solid-phase toxicity tests. In this situation, inclusion of porewater toxicity test data could result in the development of better aggregate empirical guidelines (in combination with sediment test data) that are more protective of benthic communities. Such guidelines might also be more useful in explaining results of porewater toxicity tests.

Individuals currently involved in guideline development must be aware that the questions of "Should these numbers be developed and why?" need to be answered before the more practical questions of "Can this be done?", "Does the additional information make the effort and cost of research and development worthwhile?", or "Will adding these data increase the sensitivity?" (make the numbers go down).

The available data on the effect of including porewater data in existing guidelines is essentially limited to the comparison we have seen in Chapter 11. The ecological and environmental relevance of either incorporating porewater data into current SQGs or creating separate SQGs using only porewater data is unclear at present (e.g., do porewater tests indicate the presence of unique contaminants in the sediments that solid-phase tests do not?).

Uses related to theoretical guidelines

Equilibrium partitioning guidelines are based on EqP theory and attempt to predict biological effects on the basis of knowledge of the water-only toxicity of compounds or classes of compounds and their chemical partitioning in sediments. Most of the supporting evidence for the EqP approach has come from solid-phase toxicity tests (as opposed to porewater toxicity tests), but an important part of the EqP theory relates to the correlation of porewater chemical concentration and biological effects (Di Toro et al. 1991).

The potential exists to use porewater toxicity test results to help verify theoretical solid-phase SQGs. For example, results from tests using pore water from spiked sediments might be useful for verifying theoretical solid-phase guidelines because the use of porewater data facilitates the inclusion of additional sublethal and chronic studies in the evaluation of the EqP theory. These tests would be used to gain insight into an approach used for developing SQGs and thus provide additional value. Likewise, porewater toxicity test results from field-collected sediments could be used to corroborate predictions from empirical SQGs.

Empirical porewater guidelines: Future directions

Measurement of porewater chemistry can be very helpful in understanding porewater toxicity testing data, as it is for solid-phase tests (USEPA, in review), and these measurements could be used for developing aqueous-phase (porewater) SQGs. These guidelines could potentially be derived using porewater data from either field-collected or spiked sediments.

As with the solid-phase guidelines, discussion of the potential uses of porewater toxicity tests for development of porewater quality guidelines must be dominated by considerations of the specific application of the guidelines, added value provided by the tests, ecological relevance, and practical constraints. For example, the decision to pursue aqueous-phase (porewater) SQGs for regulatory uses must weigh the practical constraints of sampling and collection requirements necessary for accurate porewater chemistry measurements with the added regulatory value that a porewater guideline would provide. The decision to use porewater toxicity tests in SQG development will ultimately depend on the intended applications for that guideline and for the porewater tests in the development process.

The considerations discussed in this section must be reviewed and critically analyzed before deciding if, when, and how to use porewater toxicity tests in SQG development or refinement. Various guidelines could potentially benefit from the use of porewater toxicity test results. The utility of porewater toxicity tests in guideline development and the appropriateness of these approaches for regulatory applications will become more evident as more research is conducted on both SQGs and porewater toxicity tests.

Information Gaps

Porewater toxicity testing is an area of ecotoxicology that is at an early and active stage of development and refinement. From the regulatory perspective, several current information gaps relating to methodology must be filled before porewater tests can be considered for use. These are identifiable from Table 12-3 and are as follows:

1) the lack of information on the sensitivity of test species to some confounding factors (e.g., H_2S);

2) the lack of standard, published methods for porewater testing; and

3) the lack of laboratory accreditation and certification for the laboratories conducting porewater tests.

Summary Points for the Use of Porewater Tests in a Regulatory Context

1) There is a need to determine, a priori, the purposes for using porewater toxicity tests in a specific regulatory program. Regulatory authorities must determine what questions they are trying to answer and ensure that the questions are appropriate for the specific regulatory application.

2) Porewater testing can often provide "value added" information that complements whole sediment, elutriate, and sediment extract tests. We suggest that it is worthwhile to pursue the development and standardization of appropriate porewater toxicity tests for use in regulatory programs.

Advantages for the use of pore water as a test phase in a regulatory context include these:

1) direct contact with a sediment fraction (versus elutriate) ability to use or adapt small water-based tests with developed methods;

2) accessibility to a wider range of sublethal endpoints;

3) rapid, economical, simple screening tool that could be used commercially;

4) ability to use TIE methods more easily and to use dilution series test design;

5) less manipulative than solvent extracts (no residual solvent concerns); and

6) assessment of dissolved phase of sediment and route of exposure information.

Some of the constraints for using pore water as a test phase in a regulatory context include

1) small volume;

2) artifacts resulting from sampling, extraction, and storage (oxidation changes, etc.);

3) shelf life;

4) short-term tests;

5) higher frequency of reported sensitivity to contaminants such as ammonia; and

6) less developed and standardized methods.

This group considers porewater toxicity tests to be suitable for the following types of frameworks:

1) Use alone in rapid screening: Examples would be spatial identification of hot spots, prioritization of areas and triggering of more detailed investigations, and initial benchmarking of trends over time.

2) Use as a lower tier: In a tiered testing scheme that includes chemical and/or biological screening levels, the addition of rapid and cost-effective porewater tests to lower tiers can provide additional sensitive endpoints for the estimation of potential biological effects. It also provides an integrative measure of contaminants, including unmeasured factors. Responses in porewater tests could contribute to triggering additional toxicity testing or other investigations of the sediment at a higher tier.

3) Use as an integrative tool in a battery of tests, which includes whole sediment toxicity tests: Porewater toxicity testing can provide rapid, cost-effective estimates of acute and sublethal responses, genotoxicity potential, and possibly endocrine disruption responses in a tiered approach, which also includes other testing or monitoring such as solid-phase toxicity tests, benthic community surveys, and solid-phase or porewater chemistry.

In a regulatory context, sometimes parameters generally considered noncontaminants might be regarded as contaminants of concern. This has implications for study design and use of reference sediment pore waters (e.g., ammonia contamination under aquaculture pens). Porewater toxicity tests would not be suitable in the following regulatory contexts:

1) as a stand-alone pass/fail test without corroborating weight of evidence from other support criteria, including chemistry, solid-phase toxicity tests, and/or evidence of effects on benthic communities; and

2) as a substitute for a solid-phase test.

Other factors to consider include these:

1) Use of 2 negative porewater controls (i.e., reference and performance controls) is recommended, as in solid-phase sediment toxicity testing.

2) Where a suitable reference sediment for porewater extraction and testing cannot be found, the performance control can be used for the statistical comparison, which leads to calculated estimates of the endpoint. However, knowledge of the tolerance levels of the test species to major confounding factors is advisable in this situation.

3) Table 12-3 lists basic criteria for assessing the readiness of porewater toxicity tests for regulatory use. This table suggests that there are some marine and freshwater porewater toxicity tests available for application. It is important that any test slated for use in a pass/fail mode be standardized as a formal testing procedure.

4) It is theoretically possible to incorporate porewater toxicity test results into the current SQGs in several different ways. Incorporation could proceed by creating new solid-phase SQGs that incorporate porewater toxicity testing data, by combining porewater toxicity test data into the existing databases used to derive ESGs, or by developing guidelines specifically for pore water. The latter type of SQGs might be more predictive of porewater effects and better explain the results of porewater tests. They might also be more protective if, for example, the porewater toxicity tests were to pick up a biological response to a chemical or class of chemicals that was missed by the solid-phase testing.

5) EqP-derived SQGs might benefit from further evaluation provided by additional synoptic porewater chemistry and toxicity data.

6) At the present time, it is not clear whether the addition of porewater toxicity testing data to the databases used to develop solid-phase SQGs would be valuable. This effort would depend on conducting additional studies that compare porewater chemistry and toxicity data.

References

Ankley GT, Katko A, Arthur JW. 1990. Identification of ammonia as an important sediment-associated toxicant in the lower Fox River and Green Bay, Wisconsin. *Environ Toxicol Chem* 9:313–322.

Ankley GT, Schubauer-Berigan MK, Dierkes JR. 1991. Predicting the toxicity of bulk sediments to aquatic organisms with aqueous test fractions: Pore water vs. elutriate. *Environ Toxicol Chem* 10:1359–1366.

Anderson BS, Hunt JW, Phillips BM, Fairey RJ, Oakden JM, Puckett HM, Stephenson M, Tjeerdema RS, Long ER, Wilson JC, Lyons M. 2001. Sediment quality in Los Angeles harbor: A triad assessment. *Environ Toxicol Chem* 20:359–370.

Anderson BS, Hunt JW, Tudor S, Newman J, Tjeerdema RS, Fairey R, Oakden J, Bretz C, Wilson CJ, LaCaro F, Stephenson MD, Puckett HM, Long ER, Fleming T, Summers K. 1997. Chemistry, toxicity and benthic community conditions in selected sediments of the southern California bays and estuaries. Sacramento CA, USA: State Water Resources Control Board. Final Report. 140 p.

[APHA] American Public Health Association. 1998. Standard methods for the examination of water and wastewater. 20th ed. Echinoderm fertilization and development (Part 8810A-C). Washington DC, USA: APHA.

[ASTM] American Society for Testing and Materials. 1998. Standard guide for conducting toxicity tests with bioluminescent dinoflagellates. Designation E 1924-97. Volume 11.05, ASTM annual book of standards, Biological effects and environmental fate; Biotechnology; Pesticides. West Conshohocken PA, USA: ASTM. 1556 p.

Burgess R, Ho K, Morrison G, Chapman G, Denton D. 1996. Marine toxicity identification evaluation (TIE) guidance document: Phase I. Washington DC, USA: U.S. Environmental Protection Agency, Office of Research and Development. EPA/600/R-96/054.

Burton Jr GA. 1998. Assessing aquatic ecosystems using pore waters and sediment chemistry. Ottawa ON, Canada: Natural Resources Canada, Aquatic Effects Technology Evaluation Program. Final Report, Contract No. NRCan 97-0083.

Carr RS, Chapman DC. 1992. Comparison of whole sediment and pore-water toxicity tests for assessing the quality of estuarine sediments. *Chem Ecol* 7:19–30.

Carr RS, Chapman DC, Howard CL, Biedenbach J. 1996. Sediment Quality Triad assessment survey in the Galveston Bay Texas system. *Ecotoxicology* 5:341–361.

Carr RS, Chapman DC, Presley BJ, Biedenbach JM, Robertson L, Boothe P, Kilada R, Wade T, Montagna P. 1996. Sediment porewater toxicity assessment studies in the vicinity of offshore oil and gas production platforms in the Gulf of Mexico. *Can J Fish Aquat Sci* 53:2618–2628.

Carr RS, Williams JW, Fragata CTB. 1989. Development and evaluation of a novel marine sediment pore water toxicity test with the polychaete *Dinophilus gyrociliatus*. *Environ Toxicol Chem* 8:533–543.

Di Toro DM, Zarba CS, Hansen DJ, Berry WJ, Swartz RC, Cowan CE, Pavlou SP, Allen HE, Thomas NA, Paquin PR. 1991. Technical basis for establishing sediment quality criteria for nonionic organic chemicals using equilibrium partitioning. *Environ Toxicol Chem* 10:1541–1583.

Environment Canada. 1992a. Biological test method: Toxicity test using luminescent bacteria (*Photobacterium phosphoreum*). Ottawa ON, Canada: Environment Canada. Report EPS 1/RM/24. 61 p.

Environment Canada. 1992b. Biological test method: Fertilization assay using echinoids (sea urchins and sand dollars). Ottawa ON, Canada: Environment Canada. Report EPS 1/RM/27. 97 p.

Environment Canada. 1994. Guidance document on collection and preparation of sediments for physicochemical characterization and biological assessment. Ottawa ON, Canada: Environment Canada. EPS 1/RM/29.

Environment Canada. 1998. Reference method for determining the acute lethality of sediment to estuarine or marine amphipods. Ottawa ON, Canada: Environment Canada. EPS 1/RM/35.

Giesy JP, Rosiu CJ, Graney RL, Henry MG. 1990. Benthic invertebrate bioassays with toxic sediment and pore water. *Environ Toxicol Chem* 9:233–248.

Ho KT, McKinney R, Kuhn A, Pelletier M, Burgess R. 1997. Identification of acute toxicants in New Bedford Harbor sediments. *Environ Toxicol Chem* 16:551–558.

Hoke RA, Giesy JP, Ankley GT, Newsted JL, Adams JR. 1990. Toxicity of sediments from western Lake Erie and the Maumee River at Toledo, Ohio, 1987: Implications for current dredged material disposal practices. *J Great Lakes Res* 16:457–470.

Hooten RL, Carr RS. 1998. Development and application of a marine sediment pore-water toxicity test using *Ulva fasciata* zoospores. *Environ Toxicol Chem* 17:932–940.

Hunt JW, Anderson BS, Phillips BM, Newman J, Tjeerdema R, Fairey R, Puckett M, Stephenson M, Smith RW, Wilson CJ, Taberski KM. 2001. Evaluation and use of reference sites in determining statistically significant sediment toxicity. *Environ Toxicol Chem* 20:1266–1275.

Jaagumagi R, Persaud D, Hayton A. 1992. Guidelines for the protection and management of aquatic sediment quality in Ontario. Toronto ON, Canada: Ontario Ministry of the Environment. ISBN 0-7729-9248-7. 23 p.

Long ER, MacDonald DD, Smith SL, Calder FD. 1995. Incidence of adverse biological effects within ranges of chemical concentrations in marine and estuarine sediments. *Environ Manag* 19:81–97.

Long ER, Morgan LG. 1991. The potential for biological effects of sediment-sorbed contaminants tested in the National Status and Trends Program. Seattle WA, USA: National Oceanic and Atmospheric Administration. NOS OMA 52. 175 p.

MacDonald DD, Smith SL, Wong MP, Mudroch P. 1992. The development of Canadian marine environmental quality guidelines. Ottawa ON, Canada: Environment Canada. Marine Environmental Quality Series No. 1. 121 p.

Neff JM, Bean DJ, Cornaby BW, Vaga RM, Gulbransen TC, Scanlon JA. 1986. Sediment quality criteria methodology validation: Calculation of screening level concentrations from field data. Work Assignment 56, Task IV. Duxbury MA, USA: Battelle. Report to U.S. Environmental Protection Agency, Office of Water, Washington DC. 225 p.

Porebski LM, Osborne JM. 1998. The application of a tiered testing approach to the management of dredged sediments for disposal at sea in Canada. *Chem Ecol* 14:197–214.

[PSDDA-UM] Puget Sound Dredged Disposal Analysis-Users Manual. 2000. Dredged material evaluation and disposal procedures, a users manual for the Puget Sound dredged disposal analysis (PSDDA) program. Seattle WA, USA: Prepared by the U.S. Army Corps of Engineers in cooperation with the U.S. Environmental Protection Agency, Region 10, and the Washington State Departments of Ecology and Natural Resources.

Reynoldson TB, Zarull MA. 1993. An approach to the development of biological sediment guidelines. In: Francis G, Kay J, Woodley S, editors. Ecological integrity and the management of ecosystems. Boca Raton FL, USA: St. Lucie. p 177–199.

Sarda N, Burton Jr GA. 1995. Ammonia variation in sediments: Spatial, temporal and method-related effects. *Environ Toxicol Chem* 9:1499–1506.

[USEPA] U.S. Environmental Protection Agency. 1989. Evaluation of the apparent effects threshold (AET) approach for assessing sediment quality. Washington DC, USA: USEPA Office of the Administrator, Science Advisory Board. Report of the Sediment Criteria Subcommittee. 18 p.

[USEPA] U.S. Environmental Protection Agency. Technical basis for the derivation of equilibrium-partitioning sediment guidelines (ESGs) for the protection of benthic organisms: Nonionic organics. Washington DC, USA: USEPA (in review).

[USEPA/USACE] U.S. Environmental Protection Agency and U.S. Army Corps of Engineers. 1991. Evaluation of dredged material proposed for ocean disposal: Testing manual. Washington DC, USA: USEPA Office of Water, USACE, Dept of the Army. EPA-503/8-91/001.

[USEPA/USACE] U.S. Environmental Protection Agency and U.S. Army Corps of Engineers. 1998. Evaluation of dredged material proposed for discharge in waters of the U.S.: Testing manual, inland testing manual. Washington DC, USA: USEPA, Office of Water and USACE, Dept of the Army. EPA 823-F-98-002.

Recommendations for Research Related to Biological, Chemical, and Ecological Aspects of Sediment Pore Water: The Way Forward

Marion Nipper, R Scott Carr, William J Adams, Walter J Berry, G Allen Burton Jr, Kay T Ho, Donald D MacDonald, Richard Scroggins, Parley V Winger

Toxicity tests are useful and reliable tools for evaluating the adverse effects of chemicals discharged into aquatic ecosystems. The science of sediment toxicology evolved rapidly following the realization that sediments are a sink and a source for contaminants and that they can be used to assess past and present inputs of xenobiotics in aquatic environments. As part of the development and application of sediment toxicity tests for environmental assessments and monitoring of aquatic ecosystems, porewater tests quickly evolved into important tools in environmental assessments, as verified throughout the chapters of this book. They have become integral to numerous sediment assessment programs in the U.S. (and possibly elsewhere), as described in Chapter 8.

An advantage of porewater and whole sediment toxicity testing, as opposed to surfacewater testing, is that sediments time-integrate fluctuations in contaminant concentrations. Sediments equilibrate with the interstitial water, and although this equilibrium is dynamic, the concentrations of contaminants in the pore water generally reflect sediment quality and change much more slowly than in overlying water. Sediment pore water is a major route of exposure of infaunal and epibenthic organisms to sediment contaminants (Adams et al. 1985; Di Toro et al. 1991; U.S. Environmental Protection Agency [USEPA] 2000), and there are several methods available for the extraction (Presley et al. 1967; Edmunds and Bath 1976; Hesslin 1976; Jahnke 1988; Winger and Lasier 1991; Carr and Chapman 1995) and toxicity testing of pore water with aquatic organisms (Carr et al. 1989; Ankley et al. 1992; Carr 1998; Hooten and Carr 1998). In addition, the sensitivity of porewater toxicity test methods has been compared with that of solid-phase tests (Carr and Chapman 1992; Sarda and Burton 1995; Carr, Chapman et al. 1996; Carr, Long et al. 1996; Nipper et al. 1998; Carr et al. 2000), showing that in most cases porewater tests have elevated sensitivity when compared to the most common whole sediment tests. The elevated sensitivity is likely due to the direct exposure to contaminants in the pore water, use of early life stages of aquatic test organisms, and incorporation of sublethal test endpoints. Therefore, porewater tests can serve as early warning signals to predict potential adverse effects to benthic organisms.

Information gaps still exist regarding the most efficient and realistic methodology and applications of porewater toxicity tests. Numerous factors related to porewater sampling, toxicity testing, and application of results are strongly dependent on the environmental conditions at each study site. Pore waters should not be looked at in isolation of whole sediments, but as an additional tool in the assessment of sediment quality, because the ultimate objectives of porewater testing are often to assess benthic conditions.

Particle size, currents, upwelling, and benthic organisms influence the equilibrium between sediment and pore water. Important factors vary from small streams to high-energy estuarine and oceanic areas. In some cases, the concentrations of sediment-associated contaminants may never reach equilibrium. Sampling pore water from these environments does not retain the same time-integration of contaminant concentrations observed in less dynamic environments, but this holds true for the whole sediment as well.

The important aspects of porewater toxicity testing were identified and discussed by the 5 workgroups that composed the Society of Environmental Toxicology and Chemistry (SETAC) Technical Workshop on Porewater Toxicity Testing. Although the workgroups were assigned distinct discussion themes, several of the discussions overlapped (Figure 13-1), but the topics were approached from different perspectives, resulting in interdependency and multidisciplinarity rather than redundancy. The most significant and relevant recommendations made by the discussion groups and the information gaps identified by them are summarized in this chapter.

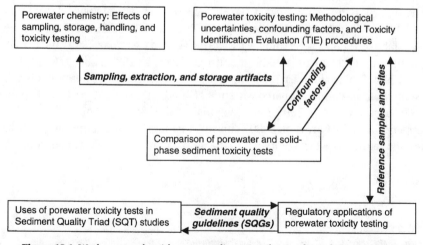

Figure 13-1 Workgroup tasks with common discussion themes shown between arrows

The Status Quo: Current Applications of Porewater Toxicity Tests

The reliability of porewater toxicity tests as indicators of sediment quality and their correlation with sediment effects to benthic communities has been the subject of much discussion. The results of porewater toxicity tests often differ significantly from those of solid-phase toxicity tests. However, concordance between the results of solid-phase and porewater toxicity tests should not always be expected because they represent different routes of exposure (Chapter 3). Discordance can also be indicative of different species or method sensitivity, rather than inaccuracy in one of the tests. Increased sensitivity of solid-phase tests and stronger concordance with porewater tests occurred when chronic or sublethal solid-phase test endpoints were included (Chapter 2), suggesting that chronic endpoints may be more indicative of potential impacts to the benthic biota than acute tests. The relationship between porewater toxicity test results and benthic community is highlighted in the chapters that discuss the use of porewater testing in the Sediment Quality Triad (SQT) approach (Chapters 8 and 9). Among the desirable attributes of porewater toxicity tests for use in the SQT approach is the ability to identify causality. The use of both porewater and solid-phase tests whenever possible enhances the ability to discriminate sediment quality, contributes to the weight of evidence, and supports concordance among the triad components (see Chapters 3 and 9).

Although porewater toxicity tests are not commonly used for regulatory purposes (Chapter 10), several situations in which they would be useful were identified (Chapter 12). In order to apply porewater toxicity tests in regulatory programs, the purpose of the testing and the questions to be answered need to be established a priori, ensuring that the questions are appropriate for each specific regulatory application. Porewater tests were considered suitable for several types of regulatory frameworks because they often provide additional information that complements whole sediment, elutriate, and sediment extract tests. However, porewater tests should not be used as stand-alone, pass/fail methods or as substitutes for solid-phase tests (Chapter 12). This is corroborated by the fact that the 2 tests represent different routes of exposure and that the feeding mode of a test species can be of critical importance in the exposure to certain chemicals. In addition to their use in regulatory programs, the incorporation of porewater toxicity test results into the empirical derivation of sediment quality guidelines (SQGs) might generate values that are more predictive of porewater exposures and protective of biota to chemicals missed by the solid-phase exposures (see Chapters 11 and 12).

Concerns regarding the reliability of porewater toxicity tests were extensively discussed, particularly issues related to methodological uncertainties and artifacts associated with sediment and porewater sampling, extraction, and storage. These are critically important for obtaining the most field-representative samples of pore water (see Chapters 4 through 7). Artifacts and chemical changes are difficult to avoid when sediment is collected and pore water isolated for toxicity testing. To aid in the assessment of the artifacts, chemical concentrations in the pore water should be determined along with regular measurements of contaminants in the whole sediment. This information highlights the routes and level of exposure and facilitates interpretation of test results. Several porewater sampling methods were suggested in Chapters 5 and 7, but method selection should be based on the objective of the study. Several porewater features, a number of which can act as confounding factors (e.g., salinity, alkalinity, pH, conductivity, dissolved oxygen [DO], NH_3, H_2S, Eh), should be routinely measured after porewater collection and after storage. These measurements aid in interpreting test results and understanding the contribution of these factors to the concordance or discordance between solid-phase and porewater test methods. Statistically, confounding factors are best accounted for by including them as variables in a multivariate analysis (Chapters 8 and 9). However, some potential confounding factors (e.g., salinity) should be adjusted before testing to ensure that test conditions are compatible with the requirements of the test species (Chapter 7).

Several aqueous-phase toxicity test methods can be used with pore waters from freshwater, estuarine, and marine environments (see Table 12-3, Chapter 12). Pore waters can also be used to identify the types and/or the sources of contaminants in a particular area. Toxicity identification evaluation (TIE) procedures for whole sediments are under development but are not as advanced as those for pore waters (Chapters 6 and 7). A battery of tests that includes multispecies and different life stages is recommended for porewater assessments. Information from these tests will enrich the database and help account for different modes of action and species sensitivity. Although the use of indigenous species may be desirable under certain circumstances (e.g., for in situ tests or with a particular species of concern), their use is not recommended or considered important in understanding potential biological impacts. The use of water column organisms for porewater toxicity tests was considered scientifically appropriate (Chapter 7).

Study design is critical in understanding the biological effects of contaminants as identified by porewater toxicity tests. For example, the depth of the sediment sampled should match the depth of interest for each particular survey in order to reflect exposure conditions of organisms in a real-world situation. Such factors are dependent upon the local dynamics of the aquatic system (currents, grain size, severity of storm events, etc.) and the objectives of the sediment assessment program (e.g., dredging, assessment of exposure of benthic biota) (Chapter 7).

Test controls and reference sites are critical components of environmental assessments. Adverse effects at a test site cannot be determined if good performance of toxicity test organisms is not achieved in the controls and references (Chapter 7). Reference sites should have similar sediment (and therefore, porewater) characteristics and should be selected from a location near the study site or at least in the same ecoregion. However, it was suggested that if a suitable reference sediment cannot be found, the performance control (e.g., filtered fresh water or sea water, reconstituted water) could be used for statistical comparisons for regulatory purposes (Chapters 10 and 12). Tolerance levels of the test species to major confounding factors were emphasized, and they clearly are important when the performance control is used for comparison in the analysis of porewater tests.

In spite of the intense research that has been conducted in the field of porewater chemistry and toxicity testing in recent years, several information gaps and research needs were identified during the workshop. Based on the variety of applications of porewater toxicity tests and on the methodological uncertainties still associated with these tests, a number of areas of research were identified for the improvement of accuracy and enhanced understanding of porewater toxicity test results. Some of the fundamental needs identified are described below.

The Way Forward: Areas of Research Need

Porewater chemistry and toxicity testing

The toxicological effects of artifacts and chemical changes introduced when pore water is extracted require some additional research. Studies are necessary to improve the understanding of the effects of storage methods and time on the fundamental constituents in both pore water and whole sediments (e.g., DO, Eh, CO_2, dissolved organic carbon [DOC], H_2S, NH_3, Fe, and Mn). These may act as confounding factors and influence toxicity test results. Freezing was considered appropriate for the storage of marine pore waters by Carr and Chapman (1995), but the effect of this and other storage methods on chemistry and toxicity of pore waters extracted from a broader variety of sediments, containing a wide array of contaminants, was identified as a critical need (Chapter 7).

Identification of changes in porewater chemistry and toxicity caused by precipitation of iron and manganese oxides during and after extraction is critical to the interpretation of the results of porewater toxicity tests. Pore waters are frequently anoxic, and oxidation and precipitation reactions are inevitable in many porewater toxicity tests. Allowing these reactions to occur in a prescribed manner during the test procedure may be an effective way to account for these processes, but this should be verified. A better understanding of the concentrations and distributions of prominent colloid species in pore water, including DOC, and iron and manganese oxides, is needed. More specifically, a quantitative understanding of the interactions of colloids with surfaces and with dissolved contaminants is desirable. Sorption of contaminants to test and storage vessels and the importance of surface area to water volume on adsorption of contaminants also need to be assessed. Appropriate volumes of pore water needed for testing should be determined based on test species and test exposure conditions (e.g., size of test animal, types of suspected contaminants, oxygen needs, metabolic waste). Numerous aqueous tests can be readily applied to pore waters, but the development of more short-term chronic tests is also needed.

In addition to identifying the effects of sediment storage on porewater chemistry, the significance of degassing and volatilization of CO_2, H_2S, and NH_3 associated with sampling, extraction, and storage procedures also must be determined. The importance of microbial processes to porewater chemistry, contaminant degradation, and ammonia production must be evaluated. This information would provide guidance for interpretation of the results from toxicity tests and for method standardization. Spiking experiments may be useful in assessing the effects of the various extraction methods on volatilization, sorption of hydrophobic organic compounds (HOCs), metal availability, and NH_3 concentrations.

In situ porewater extraction and/or toxicity testing was recommended, but the need for the development of large-volume in situ porewater collection devices was identified. Because of the difficulty in obtaining large volumes of pore water, there is a need to develop methods for chemical analyses that can provide low detection limits with small sample volumes.

Porewater versus solid-phase tests

Porewater toxicity tests were identified as useful tools for sediment quality assessments, but the role of porewater toxicity testing needs to be defined for situations in which the aqueous phase is not the primary route of exposure, and the comparative sensitivities of porewater and solid-phase test species to contaminants also need to be evaluated. The equilibrium partitioning (EqP) approach was adopted by USEPA for the establishment of sediment guidelines for nonionic organics (USEPA 2000), but its validity under a variety of conditions (e.g., dynamic sediments) is not fully understood. Combining the use of porewater testing with chemical analyses could help validate this approach and assess the circumstances at which its application is (or is not) appropriate.

Synoptic measurements of porewater chemistry and toxicity data are needed to provide a better understanding of the relationships between concentrations of chemical mixtures in sediments and pore water and porewater toxicity. These efforts would facilitate the validation and standardization of porewater toxicity test methods, which was also identified as necessary for the SQT and for use in regulatory programs.

Sediment Quality Triad studies

Identification of benthic indicator species and toxicity endpoints that reflect potential impacts on the benthic populations are needed. Porewater toxicity tests can be useful components of the SQT approach, as long as the relationships between concentrations of chemical mixtures in sediments and pore water and porewater toxicity are better understood. Development of chemical-specific porewater toxicity tests and identification of toxicological mechanisms at the cellular level were also identified as research needs. Validation studies with sediment toxicity tests should be performed for the enhancement of the use of porewater tests in SQT assessments. This would increase our understanding of relationships with porewater exposure and verify whether these are indeed accurate metrics, representative of conditions where benthic degradation would be expected. From the statistical point of view, generalized multiparameter patterns in ecological structure and function relative to porewater toxicity, as opposed to species differences with location, need to be determined.

Regulatory issues

In addition to the research needs, the development of accreditation and certification programs for the laboratories that conduct porewater testing is a critical step in the establishment of porewater toxicity testing as a standard regulatory tool. Although porewater toxicity tests are already applied by a number of agencies for the assessment of sediment quality, the authors of this book hope that the recommendations presented herein will reinforce the use of porewater toxicity testing. The authors also hope that the identified research needs will provide direction for advances in this field, increase the utility of porewater toxicity testing in the regulatory arena, facilitate the assessment of sediment quality and the development of SQGs, and lead to a better understanding of the relationships between sediment contaminants and effects in aquatic environments.

References

Adams WJ, Kimerle RA, Mosher RG. 1985. Aquatic safety assessment of chemicals sorbed to sediments. In: Cardwell RD, Purdy R, Bahner RC, editors. Aquatic toxicology and hazard assessment: Seventh Symposium. Philadelphia PA, USA: American Society for Testing and Materials. ASTM STP 854. p 429–453.

Ankley GT, Lodge K, Call DJ, Balcer MD, Brooke LT, Cook PM, Kreis Jr RJ, Carlson AR, Johnson RD, Niemi GJ, Hoke RA, West CW, Giesy JP, Jones PD, Fuying ZC. 1992. Integrated assessment of contaminated sediments in the lower Fox River and Green Bay, Wisconsin. *Ecotoxicol Environ Saf* 23:46–63.

Carr RS. 1998. Marine and estuarine porewater toxicity testing. In: Wells PG, Lee K, Blaise C, editors. Microscale testing in aquatic toxicology: Advances, techniques, and practice. Boca Raton FL, USA: CRC. p 523–538.

Carr RS, Chapman DC. 1992. Comparison of solid-phase and pore-water approaches for assessing the quality of marine and estuarine sediments. *Chem Ecol* 7:19–30.

Carr RS, Chapman DC. 1995. Comparison of methods for conducting marine and estuarine sediment porewater toxicity tests: Extraction, storage, and handling techniques. *Arch Environ Contam Toxicol* 28:69–77.

Carr RS, Chapman DC, Howard CL, Biedenbach JM. 1996. Sediment Quality Triad assessment survey of the Galveston Bay, Texas system. *Ecotoxicology* 5:341–364.

Carr RS, Long ER, Windom HL, Chapman DC, Thursby G, Sloane GM, Wolfe DA. 1996. Sediment quality assessment studies of Tampa Bay, Florida. *Environ Toxicol Chem* 15:1218–1231.

Carr RS, Montagna PA, Biedenbach JM, Kalke R, Kennicutt MC, Hooten R, Cripe G. 2000. Impact of storm-water outfalls on sediment quality in Corpus Christi Bay, Texas. *Environ Toxicol Chem* 19:561–574.

Carr RS, Williams JW, Fragata CTB. 1989. Development and evaluation of a novel marine sediment porewater toxicity test with the polychaete *Dinophilus gyrociliatus*. *Environ Toxicol Chem* 8:533–543.

Di Toro DM, Zarba CS, Hansen DJ, Berry WJ, Swartz RC, Cowan CE, Pavlou SP, Allen HE, Thomas NA, Paquin PR. 1991. Technical basis for establishing sediment quality criteria for nonionic organic chemicals using equilibrium partitioning. *Environ Toxicol Chem* 10:1541–1583.

Edmunds WM, Bath AH. 1976. Centrifuge extraction and chemical analysis of interstitial waters. *Environ Sci Technol* 10:467–472.

Hesslin RH. 1976. An in situ sampler for close interval pore water studies. *Limnol Oceanogr* 21:912–914.

Hooten RL, Carr RS. 1998. Development and application of a marine sediment pore-water toxicity test using *Ulva fasciata* zoospores. *Environ Toxicol Chem* 17:932–940.

Jahnke RA. 1988. A simple, reliable, and inexpensive pore-water sampler. *Limnol Oceanogr* 33:483–487.

Nipper MG, Roper DS, Williams EK, Martin ML, VanDam LF, Mills GN. 1998. Sediment toxicity and benthic communities in mildly contaminated mudflats. *Environ Toxicol Chem* 17:502–510.

Presley BJ, Brooks RR, Kappel HM. 1967. A simple squeezer for removal of interstitial water from ocean sediments. *J Mar Res* 25:355–362.

Sarda N, Burton Jr GA. 1995. Ammonia variation in sediments: Spatial, temporal and method-related effects. *Environ Toxicol Chem* 9:1499–506.

[USEPA] U.S. Environmental Protection Agency. 2000. Technical basis for the derivation of equilibrium partitioning sediment guidelines (ESGs) for the protection of benthic organisms: Nonionic organics (Draft). Washington DC, USA: USEPA Office of Science and Technology and Office of Research and Development. 8 sections.

Winger PV, Lasier PJ. 1991. A vacuum-operated pore-water extractor for estuarine and freshwater sediments. *Arch Environ Contam Toxicol* 21:321–324.

Abbreviations

×g	centrifugation force of gravity
2D	2-dimensional
3D	3-dimensional
AET	apparent effects threshold
AETE	Aquatic Effects Technology Evaluation
ANOVA	analysis of variance
APHA	American Public Health Association
ASTM	American Society for Testing and Materials
AVS	acid volatile sulfide
BEDS	Biological Effects Data Set
BIBI	benthic index of biotic integrity
BOD	biological oxygen demand
CCA	canonical correlation analysis
CCME	Canadian Council of Ministers of the Environment
C_D	dissolved concentration
C_{DOC}	concentration of pollutant associated with DOC in pore water
CEFAS	Centre for Environment, Fisheries and Aquaculture Science
CEPA	Canadian Environmental Protection Act
COC	colloidal organic carbon or chemical of concern
C_P	particulate concentration
C_{PW}	total porewater concentration
C_s	sediment concentration
C_T	total concentration
cv	critical value
C_w	water capacity
DDE	dichlorodiphenyldichloroethylene
DDT	dichlorodiphenyltrichloroethane
DO	dissolved oxygen
DOC	dissolved organic carbon

DOM	dissolved organic matter
DWDS	deepwater dumpsite
EC50	effective concentration to 50% of a test population
EDTA	ethylenediaminetetraacetic acid
Eh	redox potential
EMAP	Environmental Monitoring and Assessment Program
EqP	equilibrium partitioning
ERL	effects range low
ERM	effects range median
ESG	equilibrium partitioning sediment guideline
f_{oc}	sediment organic carbon content
FW	fresh water
GOOMEX	Gulf of Mexico Offshore Operations Monitoring Experiment
HCH	hexachlorocyclohexane
HDPE	high density polyethylene
HFO	hydrous ferric oxide
HOC	hydrophobic organic compound
IC25	inhibition concentration to 25% of a test population
JAMP	Joint Assessment and Monitoring Program
K_d	equilibrium sediment–water distribution coefficient
K_{DOC}	porewater partition coefficient normalized for effects of DOC
K_{dp}	polymer–water distribution coefficient
K_{OC}	organic carbon-normalized partition coefficient
K_{OW}	octanol–water partition coefficient
K_p	partition coefficient
K_{PW}	porewater partition coefficient
K_{sa}	Measurements of K_{dp} normalized to pellet surface area

LC50	lethal concentration to 50% of a test population
LDPE	low density polyethylene
MDS	multiple dimensional scaling
MFO	mixed function oxygenase
MMS	Minerals Management Service
m_p	mass of polymer
mwu	molecular weight unit
NBS	National Biological Survey
NGO	nongovernmental organization
NIWA	National Institute for Water and Atmospheric Research
NOAA	National Oceanic and Atmospheric Administration
NOEC	no-observed effects concentration
NS&T	National Status & Trends program
NSTPAH	National Status & Trends program PAH
NTA	nitrilotriacetic acid
OECD	Organization for Economic Cooperation and Development
OM	organic matter
OSPAR	Oslo and Paris Commission
PAH	polycyclic aromatic hydrocarbon
PC	principal component
PCA	principal components analysis
PCB	polychlorinated biphenyl
PCP	pentachlorophenol
PEC	porewater-based effects concentration
PEL	probable effects level
PERL	porewater-based effects range low
PERM	porewater-based effects range median
PET	polyethylene terephthalate
PP	polypropylene
PS	polystyrene
PSDDA	Puget Sound Dredged Disposal Analysis

PSDDA-UM	*Puget Sound Dredged Disposal Analysis-User's Manual*
PTEL	porewater-based threshold effects level
PVC	polyvinyl chloride
RGS	reporter gene system
SAS	Statistical Analysis System
SCCWRP	Southern California Coastal Water Research Project
SD	standard deviation
SDS	sodium dodecyl sulfate
SEC	sediment effects concentration
SEM	simultaneously extractable metal
SETAC	Society of Environmental Toxicology and Chemistry
SQG	sediment quality guideline
SQT	Sediment Quality Triad
SW	sea water
TAN	total ammonia expressed as nitrogen concentration
TBT	tributyltin
TEL	threshold effects level
TIE	toxicity identification evaluation
TOC	total organic carbon
TSS	total suspended solids
UAN	unionized ammonia expressed as nitrogen concentration
USACE	U.S. Army Corps of Engineers
USEPA	U.S. Environmental Protection Agency
USFWS	U.S. Fish and Wildlife Service
USGS	U.S. Geological Survey
V_w	peeper volume
WWTP	wastewater treatment plant

Index

A

Abalone, See *Haliotis rufescens*
Abbreviations, 293–296
Acartia tonsa, 42
Acid volatile sulfide (AVS)
  and bioavailability, 4
  effects on metal distribution, 81
  in sediment ingestion studies, 47
  simultaneously extractable metals ratio, 239, 277
  as sorption site in sediment, 65, 126
Adsorption
  ammonia, xxvi, 117
  effects on octanol-water partition coefficient, 19
  HOC, xxvi, 100–102, 104, 105–106, 108, 113
  hydrous iron oxide, 73, 74
  metals, xxvi, 73–74, 87, 108, 151
  models for, 73, 74
  PAH, 108
  PCB, 108
  of toxicants to sulfur, 76
  zinc, 74
Aeration. *See* Dissolved oxygen
Aerobic metabolism, 69–70, 108
AET. *See* Apparent effects threshold
AETE. *See* Aquatic Effects Technology Evaluation
Air stones, 102
Algae
  cell lysis and ammonia, 116
  high concordance studies, 42
  readiness of tests for regulatory uses, 274
  as test species, 4, 13, 153
Algal blooms, 116
Algal mats, 70, 76
Alien species, 158
Alkalinity, xxx, 52. *See also* pH
Aluminum, 111–112
Americamysis bahia, 32, 136, 156, 274
American Society for Testing and Materials (ASTM), 231
Ammonia
  adsorption, xxvi, 117
  as confounding factor, 52, 132, 144, 148, 206, 268–269
  daily water renewal needed during test, 54

decrease during porewater storage, 127, 130
EC50/LC50, 36, 45, 206
in Environment Canada fertilization test, 236
formation during porewater storage, xxvi, 108, 127
interaction with DOC, 85
LOEC, 45
during porewater extraction, 144, 148
in porewater test sample, xxvi, 110, 116–117
principal components analysis, 192–197
and the reference sample, 155
removal by *Ulva lactuca*, 151, 156–157
sensitivity of different endpoints to, 45
sensitivity of organism to, xxix, 17, 18, 152
solid-phase *vs.* porewater tests, 19–21
tolerance of organisms to, 45, 54, 237
as a toxicant
  anthropogenic, 149–150
  naturally occurring, 19–21, 24, 56, 69, 149, 150
toxicity, 134, 135, 136
unionized, 20, 192, 206, 238
Ampelisca abdita
  benthic abundance
    Biscayne Bay and Miami River, FL, 175
    Delaware Bay, DE, 179
  ecological relevance of porewater test, 49
  effects of ammonia, 20, 45
  low concordance studies, 45
  readiness of tests for regulatory uses, 274
  solid-phase *vs.* porewater tests, 20, 30, 31, 45, 164–170, 175, 178–180
  TIE, 136, 156
  toxicity of selected chemicals, 34, 35, 36
Ampelisca virginiana, 30, 31, 32
Amphiascus tenuiremis, 32, 44, 46
Amphipods, See also *Ampelisca abdita*; *Amphiporeia virginiana*
  benthic abundance
    Biscayne Bay and Miami River, FL, 175, 214
    Delaware Bay, DE, 179
  confounding factors in toxicity tests, 20
  effects of infaunal species, 23
  exposure-minimization strategies, 19
  in freshwater solid-phase tests, 12, 16, 163–167
  high concordance studies, 42

SETAC

A Professional Society for Environmental Scientists and Engineers and Related Disciplines Concerned with Environmental Quality

The Society of Environmental Toxicology and Chemistry (SETAC), with offices currently in North America and Europe, is a nonprofit, professional society established to provide a forum for individuals and institutions engaged in the study of environmental problems, management and regulation of natural resources, education, research and development, and manufacturing and distribution.

Specific goals of the society are:

- Promote research, education, and training in the environmental sciences.
- Promote the systematic application of all relevant scientific disciplines to the evaluation of chemical hazards.
- Participate in the scientific interpretation of issues concerned with hazard assessment and risk analysis.
- Support the development of ecologically acceptable practices and principles.
- Provide a forum (meetings and publications) for communication among professionals in government, business, academia, and other segments of society involved in the use, protection, and management of our environment.

These goals are pursued through the conduct of numerous activities, which include:

- Hold annual meetings with study and workshop sessions, platform and poster papers, and achievement and merit awards.
- Sponsor a monthly scientific journal, a newsletter, and special technical publications.
- Provide funds for education and training through the SETAC Scholarship/Fellowship Program.
- Organize and sponsor chapters to provide a forum for the presentation of scientific data and for the interchange and study of information about local concerns.
- Provide advice and counsel to technical and nontechnical persons through a number of standing and ad hoc committees.

SETAC membership currently is composed of more than 5,000 individuals from government, academia, business, and public-interest groups with technical backgrounds in chemistry, toxicology, biology, ecology, atmospheric sciences, health sciences, earth sciences, and engineering.

If you have training in these or related disciplines and are engaged in the study, use, or management of environmental resources, SETAC can fulfill your professional affiliation needs.

All members receive a newsletter highlighting environmental topics and SETAC activities, and reduced fees for the Annual Meeting and SETAC special publications.

All members except Students and Senior Active Members receive monthly issues of *Environmental Toxicology and Chemistry (ET&C)*, a peer-reviewed journal of the Society. Student and Senior Active Members may subscribe to the journal. Members may hold office and, with the Emeritus Members, constitute the voting membership.

If you desire further information, contact the appropriate SETAC office.

SETAC North America
1010 North 12th Avenue
Pensacola, Florida 32501-3367 USA
T 850 469 1500   F 850 469 9778
E setac@setac.org

SETAC Europe
Avenue de la Toison d'Or 67
B-1060 Brussels, Belgium
T 32 2 772 72 81   F 32 2 770 53 83
E setac@setaceu.org

www.setac.org

Environmental Quality Through Science®

Other SETAC Books

Life-Cycle Impact Assessment: Striving Towards the Best Practice
Udo de Haes, Finnveden, Goedkoop, Hauschild, Hertwich, Hofstetter, Jolliet,
Klöpffer, Krewitt, Lindeijer, Müller-Wenk, Olsen, Pennington, Potting, Steen, editors
2002

Silver in the Environment: Transport, Fate, and Effects
Andren and Bober, editors
2002

*Bioavailability of Metals in Terrestrial Ecosystems: Importance of Partitioning for
Bioavailability in Invertebrates*
Allen, editor
2002

Test Methods to Determine Hazards of Sparingly Soluble Metal Compounds in Soils
Fairbrother, Glazebrook, Van Straalen, Tararzona, editors
2002

Interconnections Between Human Health and Ecological Integrity
Di Giulio and Benson, editors
2002

Community-Level Aquatic System Studies-Interpretation Criteria
Giddings, Brock, Heger, Heimbach, Maund, Norman, Ratte, Sch fers, Streloke,
editors
2002

Avian Effects Assessment: A Framework for Contaminants Studies
Hart, Balluff, Barfknecht, Chapman, Hawkes, Joermann, Leopold, Luttik, editors
2001

*Impact of Low-Dose, High-Potency Herbicides on Nontarget
and Unintended Plant Species*
Ferenc, editor
2001

Risk Management: Ecological Risk-Based Decision Making
Stahl, Bachman, Barton, Clark, deFur, Ells, Pittinger, Slimak, Wentsel, editors
2001

Ecotoxicology of Amphibians and Reptiles
Sparling, Linder, Bishop, editors
2000

Ecological Risk Assessment Decision-Support System: A Conceptual Design
Reinert, Bartell, Biddinger, editors
1998

Principles and Processes for Evaluating Endocrine Disruption in Wildlife
Kendall, Dickerson, Geisy, Suk, editors
1998

Radiotelemetry Applications for Wildlife Toxicology Field Studies
Brewer and Fagerstone, editors
1998

Sustainable Environmental Management
Barnthouse, Biddinger, Cooper, Fava, Gillett, Holland, Yosie, editors
1998

Uncertainty Analysis in Ecological Risk Assessment
Warren-Hicks and Moore, editors
1998

Chemical Ranking and Scoring: Guidelines for Relative Assessments of Chemicals
Swanson and Socha, editors
1997

*Chemically Induced Alterations in Functional Development and
Reproduction of Fishes*
Rolland, Gilbertson, Peterson, editors
1997

Ecological Risk Assessment for Contaminated Sediments
Ingersoll, Dillon, Biddinger, editors
1997

Life-Cycle Impact Assessment: The State-of-the-Art, 2nd ed.
Barnthouse, Fava, Humphreys, Hunt, Laibson, Moesoen, Owens, Todd,
Vigon, Weitz, Young, editors
1997

Public Policy Application of Life-Cycle Assessment
Allen and Consoli, editors
1997

Quantitative Structure-Activity Relationships (QSAR) in Environment Sciences VII
Chen and Schüürmann, editors
1997

Whole Effluent Toxicity Testing:
An Evaluation of Methods and Prediction of Receiving System Impacts
Grothe, Dickson, Reed-Judkins, editors
1996

Procedures for Assessing the Environmental Fate and Ecotoxicity of Pesticides
Mark Lynch, editor
1995

The Multi-Media Fate Model: A Vital Tool for Predicting the Fate of Chemicals
Cowan, D. Mackay, Feijtel, Meent, Di Guardo, Davies, N. Mackay, editors
1995

Aquatic Dialogue Group: Pesticide Risk Assessment and Mitigation
Baker, Barefoot, Beasley, Burns, Caulkins, Clark, Feulner, Giesy, Graney,
Griggs, Jacoby, Laskowski, Maciorowski, Mihaich, Nelson, Parrish, Siefert,
Solomon, van der Schalie, editors
1994

A Conceptual Framework for Life-Cycle Impact Assessment
Fava, Consoli, Denison, Dickson, Mohin, Vigon, editors
1993

Guidelines for Life-Cycle Assessment: A "Code of Practice"
Consoli, Allen, Boustead, Fava, Franklin, Jensen, Oude, Parrish, Perriman,
Postlethewaite, Quay, Seguin, Vigon, editors
1993

A Technical Framework for Life-Cycle Assessment
Fava, Denison, Jones, Curran, Vigon, Selke, Barnum, editors
1991

Research Priorities in Environmental Risk Assessment
Fava, Adams, Larson, G. Dickson, K. Dickson, W. Bishop
1987